T0192630

Industrial Machine Learning

Using Artificial Intelligence as a Transformational Disruptor

Andreas François Vermeulen

Apress®

Industrial Machine Learning: Using Artificial Intelligence as a Transformational Disruptor

Andreas François Vermeulen
West Kilbride, UK

ISBN-13 (pbk): 978-1-4842-5315-1 ISBN-13 (electronic): 978-1-4842-5316-8
https://doi.org/10.1007/978-1-4842-5316-8

Managing Director, Apress Media LLC: Welmoed Spahr
Acquisitions Editor: Susan McDermott
Development Editor: Laura Berendson
Coordinating Editor: Rita Fernando

Cover designed by eStudioCalamar

Cover image designed by Freepik (www.freepik.com)

Distributed to the book trade worldwide by Springer Science+Business Media New York, 233 Spring Street, 6th Floor, New York, NY 10013. Phone 1-800-SPRINGER, fax (201) 348-4505, e-mail orders-ny@springer-sbm.com, or visit www.springeronline.com. Apress Media, LLC is a California LLC and the sole member (owner) is Springer Science + Business Media Finance Inc (SSBM Finance Inc). SSBM Finance Inc is a **Delaware** corporation.

For information on translations, please e-mail rights@apress.com, or visit http://www.apress.com/rights-permissions.

Apress titles may be purchased in bulk for academic, corporate, or promotional use. eBook versions and licenses are also available for most titles. For more information, reference our Print and eBook Bulk Sales web page at http://www.apress.com/bulk-sales.

Any source code or other supplementary material referenced by the author in this book is available to readers on GitHub via the book's product page, located at www.apress.com/9781484253151. For more detailed information, please visit http://www.apress.com/source-code.

Printed on acid-free paper

Thank you to Denise and Laurence for their support and love.

"Time is an illusion." —Albert Einstein

Table of Contents

About the Author

Andreas François Vermeulen is Chief Data Scientist and Solutions Delivery Manager at Sopra-Steria, and he serves as part-time doctoral researcher and senior research project advisor at University of St Andrews on future concepts in health-care systems, Internet-of-Things sensors, massive distributed computing, mechatronics, at-scale data lake technology, data science, business intelligence, and deep machine learning in health informatics.

Andreas maintains and incubates the "Rapid Information Factory" data processing framework.

He is active in developing next-generation data processing frameworks and mechatronics engineering with over 36+ years of global experience in complex data processing, software development, and system architecture. Andreas is an expert data scientist, doctoral trainer, corporate consultant, and speaker/author/columnist on data science, business intelligence, machine learning, decision science, data engineering, distributed computing, and at-scale data lakes.

He holds expert industrial experience in various areas (finance, telecommunication, manufacturing, government service, public safety, and health informatics).

Andreas received his bachelor degree at the North West University at Potchefstroom, his Master of Business Administration at University of Manchester, Master of Business Intelligence and Data Science degree at University of Dundee, and Doctor of Philosophy (PhD) at University of St Andrews.

About the Technical Reviewer

Chris Hillman is a Data Science Practice lead with over 25 years of experience working with analytics across many industries including retail, finance, telecoms, and manufacturing. Chris has been involved in the pre-sales and start-up activities of analytics projects helping customers to gain value from and understand advanced analytics and machine learning. He has spoken on data science and AI at Teradata events such as Universe and Partners and also industry events such as Strata, Hadoop World, Flink Forward, and IEEE Big data conferences. Chris gained a Doctor of Philosophy (PhD) researching real-time distributed feature extraction at the University of Dundee.

Acknowledgments

To all my past tutors, thank you for the wisdom you shared with me.
To my numerous associates, thanks for sharing your established wisdom!
Our deliberating and concepts on machine learning produce these ideas.
To the people at Apress, your skills transformed an idea into a book.
Well done!

> *"A man who dares to waste one hour of time has not discovered the value of life."*
>
> —Charles Darwin

So thank you, as the reader, for investing time into my knowledge distribution.

CHAPTER 1

Introduction

Industrialized Machine Learning (IML) is evolving as a disruptor in the world around us, and people are finally recognizing the true impact it has already had and will continue to have on our future. Throughout this book, I will share my knowledge and insights acquired from more than ten years of consulting, including setting up three data science teams that design and implementation digital transformations for business and research practices across the world.

The uses of IML techniques and algorithms have progressively accumulated in velocity, volume, and impact. The impact over just the last three years has evolved into an immense disruptor of many principal business processes. There is no industry that machine learning is not impacting daily, and predictions of IML growth patterns are 300% plus in volume and 250% in complexity.

In this book, I will provide a sample of wide-ranging insights and advice on the methodologies and techniques that I have found to be most useful in my current consulting ecosystem. These will assist you in capitalizing on the industrialization of your capabilities through the field of machine learning.

To apply machine learning to a selection of data sets, you need the following common process checklist:

- There must be a distinguishable pattern in the data. The determination of a pattern in the data is the primary goal of machine learning. Without the pattern, the process does not work.
- Patterns must not be solvable with a mathematical formula. It is always a good idea to investigate the likelihood that a data pattern is directly correlated to a mathematical calculation. With a proven mathematical formula, you do not need machine learning; you simply perform the mathematical calculation every time you need a result.

A. F. Vermeulen, *Industrial Machine Learning*, https://doi.org/10.1007/978-1-4842-5316-8_1

- You must have access to all aspects of the data sets to learn the insights. The data you need to perform your insights must be available as a “true” single source. Sadly, I have observed numerous projects performing significant machine learning solutions, just to discover the data they used is not for the area of interest and not allowed to be used due to restrictions. Take the time to validate the lineage and provenance of all data sources you include in the source data.

This book covers IML applications for the following industries:

- Health Informatics
- Hospitals and Other Medical Facilities
- Automotive
- Aerospace
- Communications
- Contact Centers
- Datacenters
- Finance and Banking
- Application and Software Development
- Gaming and Virtual Reality (VR)
- Augmented Reality (AR)
- IoT/Embedded Systems
- Mobility and Robotics
- Retail Planning
- E-commerce
- Wireless Carriers
- Cybersecurity

I will supply an introduction to specific machine learning techniques and explain how to industrialize them into real-world applications with comprehensive applied examples.

The general mainstream view of machine learning currently is that humans need protection against their abuse. There are more than a few groups active in opposing camps to actively control the impact of autonomics machines.

Please take note of these forces at play, as it will have a major impact on the level and nature of the IML techniques and algorithms you may use in the future.

One of the most famous concepts is Isaac Asimov's "Four Laws of Robotics" (Isaac Asimov drafted these in a 1942 short story called "Runaround"):

- First Law: A robot may not injure a human being or, through inaction, allow a human being to come to harm.
- Second Law: A robot must obey orders given it by human beings except where such orders would conflict with the First Law.
- Third Law: A robot must protect its own existence if such protection does not conflict with the First or Second Law.
- Zeroth Law: A robot may not harm humanity, or, by inaction, allow humanity to come to harm.

These four basic laws could be the basis of several hours of deep philosophical discussion. I will return to these laws and the meta-ethical issues created by the emerging field of Machine Ethics and Industrialized Machine Learning. My personal advice is this: Just because you can achieve it, doesn't mean you should implement it! Always do no harm! In simple terms: always act with responsibility to an improved overall outcome for everybody involved with your machine learning.

The basic machine learning theory will be covered first before I discuss its implementation and consequences with numerous legal restrictions during Chapter 14.

I will discuss the basic background information of the various methodologies and techniques next. Later on, I will offer examples to show how you can deploy and use IML in your daily life.

The readership of this book is set at an intermediate to advanced level. The information shared makes the following assumptions:

- You have working understanding of the Python programming ecosystem.
- You have knowledge of working with Jupyter Notebooks.
- You have experience using the Jupyter Notebooks as a data science tool.

- You can install missing Python libraries into the notebook without a great deal of assistance from this book.
- You have working knowledge of mathematical, scientific, and statistical calculations used by machine learning.
- You have accomplished basic machine learning concepts and understand the basic techniques.
- You will need to be able to get example code from a GitHub site to use the example code.

With the required user experience, I will discuss the basic knowledge of the techniques and algorithms at a non-beginner's pace. I will suggest you look at the additional material located in Appendix A as background if you are not sure about the more advanced concepts covered in each chapter.

Get Ready!!!

As we get ready to proceed, you should have a Jupyter Notebook ecosystem ready to run the examples. You can find the book's examples located here: `www.apress.com/9781484253151`.

If you need a Python environment, I personally use the Anaconda ecosystem found here: `https://www.anaconda.com/` When you are ready, I will start with the basic concepts.

Are You Ready?

I assume at this point that you have already been running a completely operational Anaconda Python 3 Jupyter Notebook environment on your computer or have access to the online version of Jupyter Notebook.

You should already have experience in the usage of the Jupyter Notebook ecosystem on your own machine or on a cloud ecosystem. At this point you can progress by two routes: personal computer or Cloud Base. This is exclusively your choice.

Personal Computer

If you are not ready with a Jupyter Notebook ecosystem, I suggest you install Anaconda's ecosystem from:

https://anaconda.org/

On installation completion, you simply run the Jupyter Notebook and you should get a page:

http://localhost:8888/tree

This would provide you the basic ecosystem that you need for this book.

Cloud Base

Want to go to cloud? You have many choices! I use three different cloud providers during my daily work and I have listed them here.

Microsoft Azure Notebooks

Develop and run code from anywhere with Jupyter notebooks on Azure. You can use Microsoft Azure if you have a Microsoft account:

https://notebooks.azure.com/

Google Cloud Platform

Collaboratory is a free Jupyter Notebook environment that requires no setup and runs entirely in the cloud. You can use Google Collaboratory, if you have a Google account:

https://colab.research.google.com

Amazon Web Services

You can use AWS Sagemaker if you have an AWS account:

https://console.aws.amazon.com/sagemaker/

Tip There are many other cloud providers you can use, and if they support the Python 3 and Jupyter Notebook ecosystem, they can be another option.

Note I used my own Jupyter Hub installation for the examples in this book. (See `http://jupyter.org/hub`)

Let's Get Started

At this point I suggest you run the following to get ready to perform the examples in this book. Open a web browser and link to: `www.apress.com/9781484253151`.

Download the examples from the Apress Source Code site.

Note For your ease of use, I have bundled each chapter's examples into a single ZIP file. This will help you to get the precise examples for each chapter.

Warning Without the code, the book will be not as effective in supporting your development of new abilities and honing of existing talents.

I will quickly help you check if you are ready to begin. Please open `Chapter_001_Test_System.ipynb` from the example directory for Chapter 1.

Run the complete Jupyter Notebook, which will assist you with getting all the Python libraries you will need throughout the examples in this book. When you have all the steps complete, you can close it and progress with your learning process.

What's Next?

The rest of this book will guide you through theoretical knowledge, supported with examples, to empower you with the knowledge to understand the background skills you need to perform IML. The next chapter will provide background knowledge before you start with the IML theory.

CHAPTER 2

Background Knowledge

The next digital evolution of the world around us is here. Companies are now using industrialized machine learning daily, driven by ever-evolving artificial intelligence capabilities as a Transformational Disruptor of traditional business models. The ability of machine learning to improve the models and methodologies that drive our world around us is enabling machine learning to adapt and evolve to these needs of the next generation of business requirements.

People are storing large amounts of their personal and company assets in massive data lakes. The customers I am consulting with are now openly admitting that without an advanced and well-designed machine learning strategy to effectively and efficiently handle these ever-expanding lakes full of critical business information, they will not survive the fourth industrial revolution.

I consult with organizations on a regular basis on how to develop their data lake, data science strategy, and machine learning to serve their evolving and ever-changing business strategies. These disruptors require agile and cost-effective, machine-driven information management to handle the priority list of senior managers worldwide.

It is a fact that many unknown insights are captured and stored in a massive pool of unprocessed data in the enterprise. These data lakes have major implications for the future of the business world. It is projected that combined data scientists worldwide will have to handle 40 zettabytes of data by 2020, an increase of 300+ times since 2005.

There are numerous data sources that still need to be converted into actionable business knowledge. The achievement will safeguard the future of the business that can achieve it.

The world's data producers are generating 2.5 quintillion bytes of new data every day. The Internet of Things will cause this volume to be substantially higher. Data scientists and engineers are falling behind on an immense responsibility. The only viable solution is an active drive to enable machine learning to adapt and evolve to these new data needs while data scientists become the trainers of the next generation of artificial intelligence capabilities.

A. F. Vermeulen, *Industrial Machine Learning*, https://doi.org/10.1007/978-1-4842-5316-8_2

The purpose of this book is to prepare you to understand how to use these incredible and powerful processing engines to act as a disruptor of your current business environments. By reading the introduction plus this background, you are already proving to be an innovative person who wants to understand, and perhaps tame, this advanced artificial intelligence. To tame your data lake with artificial intelligence, you will need practical advice on the data science, machine learning, and transformational disruptors.

I propose to teach you how to tame this beast!

I am familiar with the skills it takes to achieve this goal, and I will guide you with the sole purpose of helping you to learn and expand while understanding the practical guidance in this book.

I will get you started using machine learning theory and then advance to deployment against the data lakes from several business applications.

You will then understand the following:

- What machine learning models tame your business' data lake?
- How do you apply data science and machine learning to succeed in this undertaking?

Think of the process as comparable to a natural lake. It is vital to accomplish a sequence of proficient techniques with the lake water to obtain pure water in your glass.

By the end of this book, you will have shared in over 30+ years of working experience with data and extracting actionable business knowledge. I will share the experience I gained in working with data on an international scale with you. You will understand the processing framework that I use on a regular basis to tame data lakes and the collection of monsters that live in and around the lake.

I have included examples at the end of each chapter, along with code, which more serious data scientists can use as you progress throughout the book. But please note that it is not required for you to complete the examples in order to understand the concepts in each chapter.

So welcome to a walk-through of a characteristic machine learning of a data lake project using practical data science techniques and machine learning insights. The objective of the rest of this background chapter is to explain the fundamentals of data science and machine learning.

Data Science

In 1960, Peter Naur started using the term "data science" as a substitute for computer science. He stated to work with data, you need more than just computer science. I agree with his declaration.

Data science is an interdisciplinary science incorporating practices and methods to action knowledge and insights from data in heterogeneous schemas (structured, semi-structured, or unstructured). It amalgamates the science fields of data exploration from thought-provoking research fields like data engineering, information science, computer science, statistics, artificial intelligence, machine learning, data mining, and predictive analytics.

As I enthusiastically researched into the future usage of data science by translating multiple data lakes, I discovered several valuable insights. I will explain with end-to-end examples and share my insights on data lakes. This book explains vital elements from these sciences that you can use to process your data lake into actionable knowledge. I will guide you through a series of recognized science procedures for data lakes. These core skills are a key set of assets to perfect as you start into your encounters using data science.

Data Analytics

Data analytics is the science of fact-finding analysis of raw data with the goal of drawing conclusions from the data lake. It is driven by certified algorithms to statistically define associations between data that produce insights.

The perception of certified algorithms is exceptionally significant when you want to sway other business people about the importance of the data insights you have uncovered.

You should not be surprised if you are asked regularly to substantiate it and explain how you know it is correct!

The best answer is to have the competency to point at a certified and recognized algorithm you used. Associate the algorithm to your business terminology to accomplish success with your projects.

Machine Learning

The business world is buzzing with activities and ideas about machine learning and the application to numerous business environments. Machine learning is the capability of systems to learn without explicitly providing the rules, and software development may or may not be part of machine learning depending on what tool/technique is being used. It evolved from the study of pattern recognition and computational learning theory.

The impact is that with the appropriate processing and skills, you can augment your own data capabilities. Training enables a processing environment to complete several magnitudes of discoveries while you have a cup of coffee.

Work smarter … Not harder! Offload your data science to the machines; they are faster and more consistent in processing.

This skill is an essential part of achieving major gains in shortening the data-to-knowledge cycle, and this book will cover the essential practical ground rules.

Data Mining

Data mining is processing data to isolate patterns and establish relationships between data entities within the data lake. During later chapters, I will expand how you can mine your data for insights, which will help you discover new actionable knowledge. But first, there are some critical data mining theories you need to know about data patterns before you can be successful with data mining.

Statistics

Statistics is the study of the collection, analysis, interpretation, presentation, and organization of data. Statistics deals with all aspects of data including the planning of data collection in terms of the design of surveys and experimentations.

Data science and statistics are closely related. I will show how you run through such statistics models as data collections, populations, and samples to enhance your data science deliveries. The book will deliberate in later chapters on how you amalgamate it together into an effective and efficient process.

Algorithms

An algorithm is a self-contained, step-by-step set of processes to achieve a specific outcome. Algorithms execute calculations, data processing, or automated reasoning tasks with repeatable outcomes.

Algorithms are the backbone of the data science process. You should assemble a series of methods and procedures that will ease the complexity and processing of your specific data lake. In this book, I will discuss numerous algorithms and good practices for performing practical data science.

Data Visualization

Data visualization is your key communication channel with the business. It consists of the creation and study of the visual representation of business insights. The principal deliverable of data science is visualization. You will need to take your highly technical results and transform them into a format that you can show to non-data science people.

The successful transformation from data results to actionable knowledge is a skill set I will cover in detail in later chapters. If you master the visualization skill, you will be most successful in data science.

Storytelling

Data storytelling is the process of translating data analyses into layman's terms in order to influence a business decision or action. You can have the finest data science, but without the business story to translate your findings into business-relevant actions, you will not succeed. I will give details and practical insights on what to check for to ensure you have both the story and the actions.

What Next?

As we progress in this book, I will demonstrate with core theoretical knowledge of the underlining science how you can make a capable start to handle the transformation process of your data lake into actionable knowledge using machine learning. The sole requirement is to understand the data science of your own data lake and then use machine learning to turn it into a disruptor.

My advice is to apply the data science via machine learning on smaller-scale activities for insights from the data lake and then deploy it at scale to reap the complete benefits of machine learning.

Experiment. Push the boundaries of your own insights by training a machine.

I have included examples at the end of each chapter, along with code, which will enable you to generate machine learning methods to handle the requirements set at the beginning of each example.

But note that it is not required for you to complete all the examples in each chapter in order to understand the core concepts. I have provided a number of examples to clarify diverse industries and levels of experience.

I advise performing all the examples as this will give you more practice with the concepts and theories. Nothing improves a skill like practice!

CHAPTER 3

Classic Machine Learning

Machine learning is an application of artificial intelligence (AI) that provides business systems the ability to automatically learn and improve from experience without being explicitly programmed. Machine learning focuses on the development of processes that can access data and use it for learning to improve future processing.

I discuss the details of the specific machine learning algorithms later in the book in Chapters 4–10.

In this book, I will introduce you to the machine learning fields that we will turn to next.

Accuracy Testing of Machine Learning

The technique I regularly use is to take data from the history and split it into training and a test set. This way you can use real data to train the model and then predict the values you should get. As you already know, the outcome of your test data can now calculate the predicted values. This enables checking against this real data against the model prediction using test data from the past where you already know the outcome.

Supervised Learning

This type of machine learning acquires insight by creating a function that maps an input to an output based on example input-output pairs. It infers a function from labeled training data consisting of a set of training examples that are prepared or recorded by another source.

Remember that you need existing train and test data to enable the learning, hence the name supervised.

We will discuss this process in detail in Chapters 4 and 5.

A. F. Vermeulen, *Industrial Machine Learning*, https://doi.org/10.1007/978-1-4842-5316-8_3

Unsupervised Learning

This type of machine learning acquires insight by inferring a function to describe hidden structure and patterns from unlabeled data. The classification or categorization is not included in the training observations. There is therefore no right or wrong evaluation of the learner and no evaluation of the accuracy of the learned insights that is output by the relevant algorithm used.

We will discuss this process in detail in Chapters 6, 7, and 8.

Reinforcement Learning

In reinforcement learning, the algorithm gets to choose an action in response to each data point. The learning algorithm also receives a reward signal a short time later by indicating ... how good the decision was. Based on this reward, the algorithm then modifies its strategy in order to achieve the highest reward.

We will discuss this process in detail in Chapter 9.

Evolutionary Computing

Evolutionary computation is a set of algorithms for global optimization mimicking biological evolution and the subfield of AI and soft computing studying these algorithms. In technical terms, you evolve a generation of a specific population on a trial-and-error basis to solve a given problem with a metaheuristic or stochastic optimization character. The code that is generated is capable of adapting to changes to the data or the processing ecosystem.

Warning The adaptions done by evolutionary computing are theoretically capable of adapting to any changes, but my experience is that most ecosystems and implementations have hard limitations on the range of changes that it will investigate to adapt to the new requirements. I suggest you understand these clearly and precisely to ensure your solution can evolve properly.

We will discuss this process in detail in Chapter 10.

Basic Machine Learning Concepts

The processing of data using machine learning is a body of knowledge that is changing at an increasing rate. New techniques and methods are formulated, and current processes are evolving. It is an exhilarating time to be involved with this field.

I will now discuss several basic machine learning concepts you need to understand to progress with the book. So, please open Chapter 003 Examples 001.ipynb from the example directory.

Actually, Positive Samples (P)

The actual positive samples (P) are the numeric count of the "true" status of an observable state of the real world the machine learning is processing.

Example: In a group of one hundred, ten categorically contracted the flu, then P=10.

Actually, Negative Samples (N)

The actual negative samples (N) are the numeric count of the "false" status of an observable state of the real world that the machine learning is processing.

Example: In a group of one hundred people, ninety categorically contracted the flu, then N=90.

True Positives (TP)

The true positives (TP) are the numeric count of all the actual positive samples that were correctly classified as positive. Data scientists also refer to it as a "Hit."

Example: In a group of hundred people, seven are classified or projected to have the flu and are proven to have contracted the flu, then TP=7.

True Negatives (TN)

The true negatives (TN) are the numeric count of all the actual negative samples that were correctly classified as negative. Data scientists also refer to it as a "Correct rejection."

Example: In a group of one hundred people, eighty-five are classified or projected to not have the flu and are proven to not have contracted the flu, then TN=85.

False Positives (FP)

The false positives (FP) are the numeric count of all the actual negative samples that were incorrectly classified as positive. Data scientists also refer to it as a "False alarm" or "Type I error."

Example: In a group of one hundred people, five are classified or projected to have the flu and are proven to not have contracted the flu, then FP=5.

False Negatives (FN)

The false negatives (FN) are the numeric count of all the actual positive samples that were incorrectly classified as negative. Data science also refers to it as a "Miss" or "Type II error."

Example: In a group of one hundred people, three are classified or projected to not have the flu and are proven to have contracted the flu, then FN=3.

Sensitivity or True Positive Rate (TPR)

I will now show you how to calculate an indicator for your machine learning that is called a "sensitivity" or "true positive rate." It is also referred to as "Hit rate" or "Recall" or "probability of detection."

$$TPR = \frac{TP}{P} = \frac{TP}{TP + FN}$$

So, for our example:

$$TPR = \frac{TP}{P}$$

$$= \frac{7}{10}$$

$$= 0.7$$

Or calculate as follows:

$$TPR = \frac{TP}{TP + FN}$$

$$= \frac{7}{7+3}$$

$$= \frac{7}{10}$$

$$= 0.7$$

In general, this will be reported as 70% sensitivity.

Open Jupyter Notebook in example Chapter 003 called: Chapter 003 Examples 001.ipynb.

This simple example to show you how to use the sklearn's metrics sub-libraries recall_score function to calculate the recall score of a predicted result set from a machine learning model for everyday activities.

Example for Proof A - Predict which desk at the bank will serve the customer?

```
from sklearn.metrics import recall_score
y_true = [0, 1, 2, 3, 0, 1, 2, 3, 0, 1]
y_pred = [0, 1, 2, 3, 0, 1, 2, 3, 0, 2]
print(recall_score(y_true, y_pred, average=None))
```

Result:

```
[1. 0.66666667 1. 1.]
```

This indicates, for each of the types of classifications, what the machine learning scored:

```
Desk 0 : 3 values 3 correct = 1.00 or 100%
Desk 1 : 3 values 2 correct = 0.667 or 66.7%
Desk 2 : 2 values 2 correct = 1.00 or 100%
Desk 3 : 2 values 2 correct = 1.00 or 100%
```

So, if I introduce a new classification where Proof B – Predict which doctor at a clinic will serve the customer?

```
from sklearn.metrics import recall_score
y_true = [0, 1, 2, 3, 0, 1, 2, 3, 0, 4]
y_pred = [0, 1, 2, 3, 0, 1, 2, 3, 0, 1]
print(recall_score(y_true, y_pred, average=None))
```

What does the answer now show?

```
[1. 1. 1. 1. 0.]
```

The classification of Category 4 is not good at recall, or I suspect that the doctor is not in the clinic?

Specificity (SPC) or True Negative Rate (TNR)

I will now show you how to calculate an indicator for your machine learning that is called “specificity” or “true negative rate.”

$$TNR = \frac{TN}{N} = \frac{TN}{TN + FP}$$

So, for our example:

$$TNR = \frac{TN}{N}$$

$$= \frac{85}{90}$$

$$= 0.945$$

Or calculate as follows:

$$TNR = \frac{TN}{TN + FP}$$

$$= \frac{85}{85 + 5}$$

$$= \frac{85}{90}$$

$$= 0.945$$

In general, this will be reported as 94.5% specificity.

Example:

You are a doctor and you are predicting if somebody could get the flu. You evaluate data for ten people and predict six will have the flu. By end of flu season, you find only five had the flu.

Here is the calculation for your prediction as Proof C:

```
from sklearn.metrics import confusion_matrix
y_true = [0, 0, 0, 0, 0, 1, 1, 1, 1, 1]
y_pred = [0, 0, 0, 0, 1, 1, 1, 1, 1, 1]
tn, fp, fn, tp = confusion_matrix(y_true, y_pred).ravel()
sensitivity = tp / (tp+fn)
specificity = tn / (tn+fp)

print('sensitivity :', sensitivity)
print('specificity :', specificity)
```

Result:

```
Sensitivity: 1.0 or 100%
Specificity: 0.8 Or 80%
```

The result shows you can 100% correctly predict people with the flu, that is, Sensitivity is 100%. It also shows that you can correctly predict by 80% that people will not get the flu, that is, Specificity is 80%.

The means your Machine Learning (ML) misdiagnosed somebody with the flu, not a bad outcome you could say. The person was lucky not to get the flu.

But what if the flu shot that you incorrectly prescribed caused long-term damage to you patient? That is not good!

Remember Do no harm! So ensure you understand the impact of your decisions.

I will discuss the true real impact of machine learning as I get you to build them later in the book again.

Precision or Positive Predictive Value (PPV)

I will now show you how to calculate an indicator for your machine learning that is called "precision" or "positive predictive value."

$$PPV = \frac{TP}{TP + FP}$$

Example:

$$PPV = \frac{TP}{TP + FP}$$

$$= \frac{7}{7 + 5}$$

$$= \frac{7}{12}$$

$$= 0.583$$

In general, this will be reported as 58.3% precision.

Example:

Let's look at our Flu ML case study again: (See Proof D)

Precision is the ratio of properly predicted positive clarifications to the total predicted positive clarifications.

```
from sklearn.metrics import precision_score
y_true = [0, 0, 0, 0, 0, 1, 1, 1, 1, 1]
y_pred = [0, 0, 0, 0, 1, 1, 1, 1, 1, 1]
print(precision_score(y_true, y_pred, average=None))
```

The results:

```
[1. 0.83333333]
```

The ML can 100% predict 'No-Flu' but only 83.3% 'Flu' cases correctly.

Negative Predictive Value (NPV)

I will now show you how to calculate an indicator for your machine learning that is called "negative predictive value."

Negative predictive value is the probability that records with a negative predicted result truly should be negative.

$$NPV = \frac{TN}{TN + FN}$$

Example:

$$NPV = \frac{TN}{TN + FN}$$

$$NPV = \frac{85}{85 + 3}$$

$$= \frac{85}{88}$$

$$= 0.966$$

In general, this will be reported as 96.6% NPV.

Example:

Here is the calculation for your prediction: (See Proof E)

```
from sklearn.metrics import confusion_matrix
y_true = [0, 0, 0, 0, 0, 1, 1, 1, 1, 1]
y_pred = [0, 0, 0, 0, 0, 1, 1, 1, 1, 0]
tn, fp, fn, tp = confusion_matrix(y_true, y_pred).ravel()
npv = tn / (tn+fn)
print('Negative predictive value : %7.3f %%' % (npv*100))
```

Result:

```
Negative predictive value: 0.833 or 83.3%
```

This negative predictive value shows that the probability is 83.3% that the negative predictions should be negative, so this is a good result.

Miss Rate or False Negative Rate (FNR)

I will now show you how to calculate an indicator for your machine learning that is called the "Miss Rate" or "false negative rate." The false negative rate (FNR) is the proportion of positives that yield negative prediction outcomes with the specific model.

$$FNR = \frac{FN}{P} = \frac{FP}{FN + TP} = 1 - TPR$$

Example:

$$FNR = \frac{FN}{P}$$

$$= \frac{3}{90}$$

$$= 0.033$$

In general, this will be reported as 3.3% FNR.

Example:

Here is the calculation for your prediction: (See Proof F)

```
from sklearn.metrics import confusion_matrix
y_true = [0, 0, 0, 0, 0, 1, 1, 1, 1, 1]
y_pred = [0, 0, 0, 0, 1, 1, 1, 1, 1, 1]
tn, fp, fn, tp = confusion_matrix(y_true, y_pred).ravel()
fnr = fp / (fn+tp)
print('False negative : %7.3f %%' % (fnr*100))
```

Result:

```
False negative rate : 0.2 or 20%
```

This FNR shows that 20% of the predicted positives are negative.

This means 1 in 5 predicted negative outcomes are positive.

The practical implementation is that if you test for cancer, you will miss 1 in 5 patients.

Fall-Out or False Positive Rate (FPR)

I will now show you how to calculate an indicator for your machine learning that is called the "Fall-out" or "Miss Rate" or "false positive rate" (FPR):

$$FPR = \frac{FP}{N} = \frac{FP}{FP + TN} = 1 - TNR$$

Example:

$$FPR = \frac{FP}{N}$$

$$= \frac{5}{90}$$

$$= 0.056$$

In general, this will be reported as 5.6% FPR.

Example:

Here is the calculation for your prediction: (See Proof G)

```
from sklearn.metrics import confusion_matrix
y_true = [0, 0, 0, 0, 0, 0, 1, 1, 1, 1]
y_pred = [0, 0, 0, 0, 0, 1, 1, 1, 1, 1]
tn, fp, fn, tp = confusion_matrix(y_true, y_pred).ravel()
fpr = fp / (fp+tn)
print('False positive rate: %7.3f %%' % (fpr*100))
```

Result:

```
False positive rate: 0.16666666666666666
```

This false positive rate shows that nearly 17% of the predicted negative is positive. This means 17 in 100 predicted positive outcomes are negative.

False Discovery Rate (FDR)

I will now show you how to calculate an indicator for your machine learning that is called the "false discovery rate."

The false discovery rate (FDR) is a technique of conceptualizing the rate of type I errors in null hypothesis testing when conducting multiple comparisons. FDR-controlling procedures are designed to control the expected proportion of "discoveries" (rejected null hypotheses) that are false (incorrect rejections).

$$FDR = \frac{FP}{FP + TP} = 1 - PPV$$

Example:

$$FDR = \frac{FP}{FP + TP}$$

$$= \frac{5}{5 + 7}$$

$$= \frac{5}{12}$$

$$= 0.417$$

In general, this will be reported as 41.7% FDR.

Example:

Here is the calculation for your prediction: (See Proof H)

```
from sklearn.metrics import confusion_matrix
y_true = [0, 0, 0, 0, 0, 0, 1, 1, 1, 1]
y_pred = [0, 0, 0, 0, 0, 1, 1, 1, 1, 1]
tn, fp, fn, tp = confusion_matrix(y_true, y_pred).ravel()
fdr = fp / (fp+tp)
print('False discovery rate:', fdr)
```

Results:

```
False discovery rate: 0.2
```

This means you have a 20% probability that type I errors can occur.

False Omission Rate (FOR)

I will now show you how to calculate an indicator for your machine learning that is called the "false omission rate."

The false omission rate (FOR) is a statistical technique used in multiple hypotheses testing to correct for multiple comparisons, and it is the complement of the negative predictive value. It measures the proportion of false negatives that are incorrectly rejected.

$$FOR = \frac{FN}{FN + TN} = 1 - NPV$$

Example:

$$FOR = \frac{FN}{FN + TN}$$

$$= \frac{3}{3 + 85}$$

$$= \frac{3}{88}$$

$$= 0.034$$

In general, this will be reported as 3.4% FOR.

Example:

Here is the calculation for your prediction: (See Proof K)

```
from sklearn.metrics import confusion_matrix
y_true = [0, 0, 0, 0, 0, 1, 1, 1, 1, 1]
y_pred = [0, 0, 0, 0, 0, 0, 1, 1, 1, 1]
tn, fp, fn, tp = confusion_matrix(y_true, y_pred).ravel()
fomr = fn / (fn+tn)
print('False omission rate:', fomr)
```

Result:

```
False omission rate: 0.17 or 17%
```

This means that in 17 out of 100 cases, false negatives could be incorrectly rejected.

Accuracy (ACC)

I will now show you how to calculate an indicator for your machine learning that is called "accuracy" (ACC).

It is a measure of statistical bias, as these cause a difference between a result and a "true" value.

$$ACC = \frac{TP + TN}{P + N} = \frac{TP + TN}{TP + TN + FP + FN}$$

Example:

$$ACC = \frac{TP + TN}{P + N}$$

$$= \frac{7 + 85}{10 + 90}$$

$$= \frac{92}{100}$$

$$= 0.92$$

In general, this will be reported as 92% accuracy.

This is an important calculation as it shows the Bias. It has a standard called ISO 5725-1 that states the general term "accuracy" is used to refer to the closeness of a measurement to the true value.

Example:

Here is the calculation for your prediction: (See Proof L)

```
from sklearn.metrics import confusion_matrix
y_true = [0, 0, 0, 0, 0, 0, 1, 1, 1, 1]
y_pred = [0, 0, 0, 0, 0, 1, 1, 1, 1, 1]
tn, fp, fn, tp = confusion_matrix(y_true, y_pred).ravel()
acc = (tp + tn) / (tp + tn + fp + fn)
print('Accuracy %7.3f %%' % (acc*100))
```

Results:

```
Accuracy: 0.9 or 90%
```

Compare the two results: (See Proof M)

```
from sklearn.metrics import accuracy_score
y_true = [0, 0, 0, 0, 0, 0, 1, 1, 1, 1]
y_pred = [0, 0, 0, 0, 0, 1, 1, 1, 1, 1]
print('Accuracy:',accuracy_score(y_true, y_pred))
print('Accuracy Count:',accuracy_score(y_true, y_pred, normalize=False))
```

Result is:

```
Accuracy:  90.000 %
Accuracy Count:   9 of  10
```

The result is 90% accurate for 9 points. This is a good outcome for this Flu model.

Extra example:

```
from sklearn.metrics import accuracy_score
y_true = [0, 0, 0, 0, 0, 0, 0, 0, 0, 0, 0, 0, 1, 1, 1, 1, 1, 1, 1, 1]
y_pred = [0, 0, 0, 0, 0, 0, 0, 0, 0, 0, 0, 1, 1, 1, 1, 1, 1, 1, 1, 1]
print('Accuracy:',accuracy_score(y_true, y_pred))
print('Accuracy Count:',accuracy_score(y_true, y_pred, normalize=False))
```

Result:

```
Accuracy:  95.000 %
Accuracy Count:  19 of  20
```

The result is 95% accurate for 19 points. This is a good outcome for this Flu model.

F1 Score

I will now show you how to calculate an indicator for your machine learning that is called "The harmonic mean of precision" or "sensitivity" or "F1 Score."

The F1 score is defined as the harmonic mean between precision and recall. It is utilized as a statistical measure to rate performance. That results in an F1 score (from 0 to 9, 0 being the lowest and 9 being the highest) that is a mean of an individual's performance, based on two factors, that is, precision and recall.

$$F1 = 2.\frac{PPV.TPR}{PPV + TPR} = \frac{2.TP}{2.TP + FP + FN}$$

Example:

$$F1 = \frac{2.TP}{2.TP + FP + FN}$$

$$= \frac{2.7}{2.7 + 5 + 3}$$

$$= \frac{14}{22}$$

$$= 0.636$$

In general, this will be reported as a 63.6% F1 Score.

The F-measure is the harmonic mean of your precision and recall.

Example: (See Proof N)

```
from sklearn.metrics import f1_score
y_true = [0, 1, 2, 0, 1, 2 ,0, 1, 2, 0, 1, 2, 0, 1, 2, 0, 1, 2]
y_pred = [0, 2, 1, 0, 0, 1 ,0, 1, 2, 0, 1, 2, 0, 1, 2, 0, 1, 2]
print('F1 score:', f1_score(y_true, y_pred, average='macro'))
```

Results:

```
F1 score: 0.772 or 77.2%
```

This means that the F1 Score is 72.2%.

Matthews Correlation Coefficient (MCC)

I will now show you how to calculate an indicator for your machine learning that is called the "Matthews correlation coefficient" (MCC).

The MCC is the correlation coefficient between the observed and predicted classifications.

$$MCC = \frac{(TP.TN) - (FP.FN)}{\sqrt{(TP + FP)(TP + FN)(TN + FP)(TN + FN)}}$$

Example:

$$MCC = \frac{(TP.TN)-(FP.FN)}{\sqrt{(TP+FP)(TP+FN)(TN+FP)(TN+FN)}}$$

$$= \frac{(7.85)-(5.3)}{\sqrt{(7+5)(7+3)(85+5)(85+3)}}$$

$$= \frac{(595)-(15)}{\sqrt{(12)(10)(90)(88)}}$$

$$= \frac{580}{\sqrt{950400}}$$

$$= \frac{580}{974.885}$$

$$= 0.595$$

In general, this will be reported as a 59.5% MMC.

The MCC is utilized as a measure of the quality of binary classifications.

Example: (See Proof O)

```
from sklearn.metrics import matthews_corrcoef
y_true = [+1, +1, +1, +1, +1, -1, +1, -1]
y_pred = [-1, +1, +1, +1, +1, -1, +1, -1]
print(matthews_corrcoef(y_true, y_pred))
```

Results:

Matthews's correlation coefficient (MCC): 0.745

Cohen's Kappa

Cohen's kappa is a score that articulates the level of agreement between two annotators on a classification problem.

Example: (See Proof P)

```
from sklearn.metrics import cohen_kappa_score
y_true = [+1, +1, +1, +1, +1, -1, +1, -1, +1, -1, +1, -1, +1]
y_pred = [-1, +1, +1, +1, +1, -1, +1, -1, +1, -1, +1, -1, +1]
print("Cohen's Kappa:", cohen_kappa_score(y_true, y_pred))
```

Result:

```
Cohen's Kappa: 0.831
```

The kappa score is a number between -1 and 1.

Scores above 0.8 are generally considered good agreement; zero or less means no labeling agreement (basically just random labeling).

Let's investigate an earlier set of data also: (See Proof Q)

```
from sklearn.metrics import cohen_kappa_score
y_true = [0, 1, 2, 0, 1, 2 ,0, 1, 2, 0, 1, 2, 0, 1, 2, 0, 1, 2]
y_pred = [0, 2, 1, 0, 0, 1 ,0, 1, 2, 0, 1, 2, 0, 1, 2, 0, 1, 2]
print("Cohen's Kappa:", cohen_kappa_score(y_true, y_pred))
```

Results:

```
Cohen's Kappa: 0.6666666666666667
```

Informedness

I will now show you how to calculate an indicator for your machine learning that is called "Informedness" or "Bookmaker Informedness" (BM).

The BM is the probability of an informed decision.

This is an important calculation that you need to publish with the results of your machine learning model.

The formula is:

$$BM = TPR + TNR - 1 = sensitivity + specificity - 1$$

Example:

$$BM = 0.7 + 0.945 - 1$$

$$= 0.645$$

In general, this will be reported as 64.5% Informedness.

Example: (See Proof R)

```
from sklearn.metrics import confusion_matrix
y_true = [0, 0, 0, 0, 0, 0, 0, 0, 0,0, 1, 1, 1, 1, 1,1, 1, 1, 1, 1]
y_pred = [0, 0, 0, 0, 0, 0, 0, 0, 0,0, 1, 1, 1, 1, 1,1, 1, 1, 1, 0]
tn, fp, fn, tp = confusion_matrix(y_true, y_pred).ravel()
tpr = tp / (tp+fn)
print('True positive rate:', tpr)
tnr = tn / (tn+fp)
print('True negative rate :', tnr)
bm=tpr+tnr-1
print('Bookmaker Informedness:', bm)
```

Results:

```
True positive rate: 0.9
True negative rate : 1.0
Bookmaker Informedness: 0.8999999999999999
```

Markedness

I will now show you how to calculate an indicator for your machine learning that is called "Markedness" (MK). The markedness of the process is rate that a prediction can cause a specific result.

The formula is:

$$MK = PPV + NPV - 1$$

Example:

$$MK = 0.583 + 0.966 - 1$$

$$= 0.549$$

In general, this will be reported as 54.9% Markedness.

Example: (See Proof S)

```
from sklearn.metrics import confusion_matrix
y_true = [0, 0, 0, 0, 0,  0, 0, 0, 0, 0,  1, 1, 1, 1, 1, 1,  1, 1, 1, 1]
y_pred = [0, 0, 0, 0, 0,  0, 0, 0, 0, 0,  1, 1, 1, 1, 1, 1,  1, 1, 1, 0]
tn, fp, fn, tp = confusion_matrix(y_true, y_pred).ravel()
ppv = tp / (tp+fn)
print('Positive predictive value:', ppv)
npv = tn / (tn+fn)
print('Negative predictive value:', npv)
mk=ppv+npv-1
print('Markedness:', mk)
```

Results:

```
Positive predictive value: 0.9
Negative predictive value: 0.9090909090909091
Markedness: 0.8090909090909091
```

You can close your example code. I will discuss a few other general indicators you must know for reporting your machine learning performance.

Positive Likelihood Ratio (LR+)

I will now show you how to calculate an indicator for your machine learning that is called a "Positive likelihood ratio" (LR+)

The (LR+) is the likelihood for a positive result.

The formula is:

$$LR+ = \frac{TPR}{FPR} = \frac{sensitivity}{1 - specificity}$$

Example:

$$LR+ = \frac{TPR}{FPR}$$

$$= \frac{0.7}{0.056}$$

$$= 12.5$$

In general, this will be reported as an LR+ ratio of 12.5.

A good LR+, if > 10.

Negative Likelihood Ratio (LR-)

I will now show you how to calculate an indicator for your machine learning that is called a "Negative likelihood ratio" (LR-),

The LR- is the likelihood of a negative result.

The formula is:

$$LR- = \frac{FNR}{TNR} = \frac{1 - sensitivity}{specificity}$$

Example:

$$LR- = \frac{FNR}{TNR}$$

$$= \frac{0.033}{0.945}$$

$$= 0.035$$

In general, this will be reported as a LR- of 0.035.

A good LR-, if < 0.1.

If LR+ or LR- is close to 1, the model is questionable or not working.

The result you will get at 0.035 is good.

Diagnostic Odds Ratio (DOR)

I will now show you how to calculate an indicator for your machine learning that is called a "Diagnostic odds ratio" (DOR). The DOR is the measure of the effectiveness of a test.

The formula is:

$$DOR = \frac{LR+}{LR-}$$

Example:

$$DOR = \frac{LR+}{LR-}$$

$$= \frac{12.5}{0.035}$$

$$= 357.14$$

In general, this will be reported as Diagnostic odds of 357.14 to 1.

You are making good progress with the basics of machine learning.

Warning The ROC analysis is a complex concept and the description is complex. So, get a nice drink and then you can start again.

Receiver Operating Characteristic Curve (ROCC)

The Receiver Operating Characteristic (ROC) curve or the true positive rate (Sensitivity) is plotted as a function of the false positive rate (100-Specificity) for different cut-off points.

ROC curves typically feature a true positive rate on the Y-axis, and a false positive rate on the X-axis. This means that the top-left corner of the plot is the "ideal" point - a false positive rate of zero, and a true positive rate of one. This is not very realistic, but it does mean that a larger area under the curve (AUC) is usually better.

The "steepness" of ROC curves is also important, since it is ideal to maximize the true positive rate while minimizing the false positive rate.

Open Jupyter Notebook in example Chapter 003 called: Chapter 003 Examples 002.ipynb

Let's start the ROC example.

The real problem is my mother has planted several roses in same flowerbed, and my dog Angus pulled three different description labels off the rose trees. So now I need to match them back together. I did some measurements.

```
import numpy as np
import matplotlib.pyplot as plt
from itertools import cycle
```

```
from sklearn import svm, datasets
from sklearn.metrics import roc_curve, auc
from sklearn.model_selection import train_test_split
from sklearn.preprocessing import label_binarize
from sklearn.multiclass import OneVsRestClassifier
from scipy import interp

imagepath = os.path.join(*[os.path.dirname(os.path.dirname(os.
getcwd())),'Results','Chapter 03'])
print(imagepath)

if not os.path.exists(imagepath):
    os.makedirs(imagepath)

# Import Rose data to play with
xfile=os.path.abspath("../../Data/x-roses.csv")
yfile=os.path.abspath("../../Data/y-roses.csv")

X = np.loadtxtxfile, delimiter=",")
y = np.loadtxtyfile delimiter=",")

# Binarize the output
y = label_binarize(y, classes=[0, 1, 2])
n_classes = y.shape[1]

# Added noisy features to make the problem harder to solve
random_state = np.random.RandomState(0)
n_samples, n_features = X.shape
X = np.c_[X, random_state.randn(n_samples, 200 * n_features)]

# shuffle and split training and test sets
X_train, X_test, y_train, y_test = train_test_split(X, y, test_size=.5,
                                                    random_state=0)

# Learn to predict each class against the other
classifier = OneVsRestClassifier(svm.SVC(gamma='auto'))
y_score = classifier.fit(X_train, y_train).decision_function(X_test)
```

```
# Compute ROC curve and ROC area for each class
fpr = dict()
tpr = dict()
roc_auc = dict()
for i in range(n_classes):
    fpr[i], tpr[i], _ = roc_curve(y_test[:, i], y_score[:, i])
    roc_auc[i] = auc(fpr[i], tpr[i])

# Compute Micro-Average ROC curve and ROC area

fpr["micro"], tpr["micro"], _ = roc_curve(y_test.ravel(), y_score.ravel())
roc_auc["micro"] = auc(fpr["micro"], tpr["micro"])
fig1 = plt.figure(figsize=(10, 10))
lw = 2
plt.plot(fpr[2], tpr[2], color='darkred',
         lw=lw, label='ROC curve (area = %0.3f)' % roc_auc[2])
plt.plot([0, 1], [0, 1], color='navy', lw=lw, linestyle='--')
plt.xlim([0.0, 1.0])
plt.ylim([0.0, 1.1])
plt.xlabel('False Positive Rate')
plt.ylabel('True Positive Rate')
plt.title('Receiver Operating Characteristic for Roses')
plt.legend(loc="lower right")
plt.show()
```

See the results in Figure 3-1.

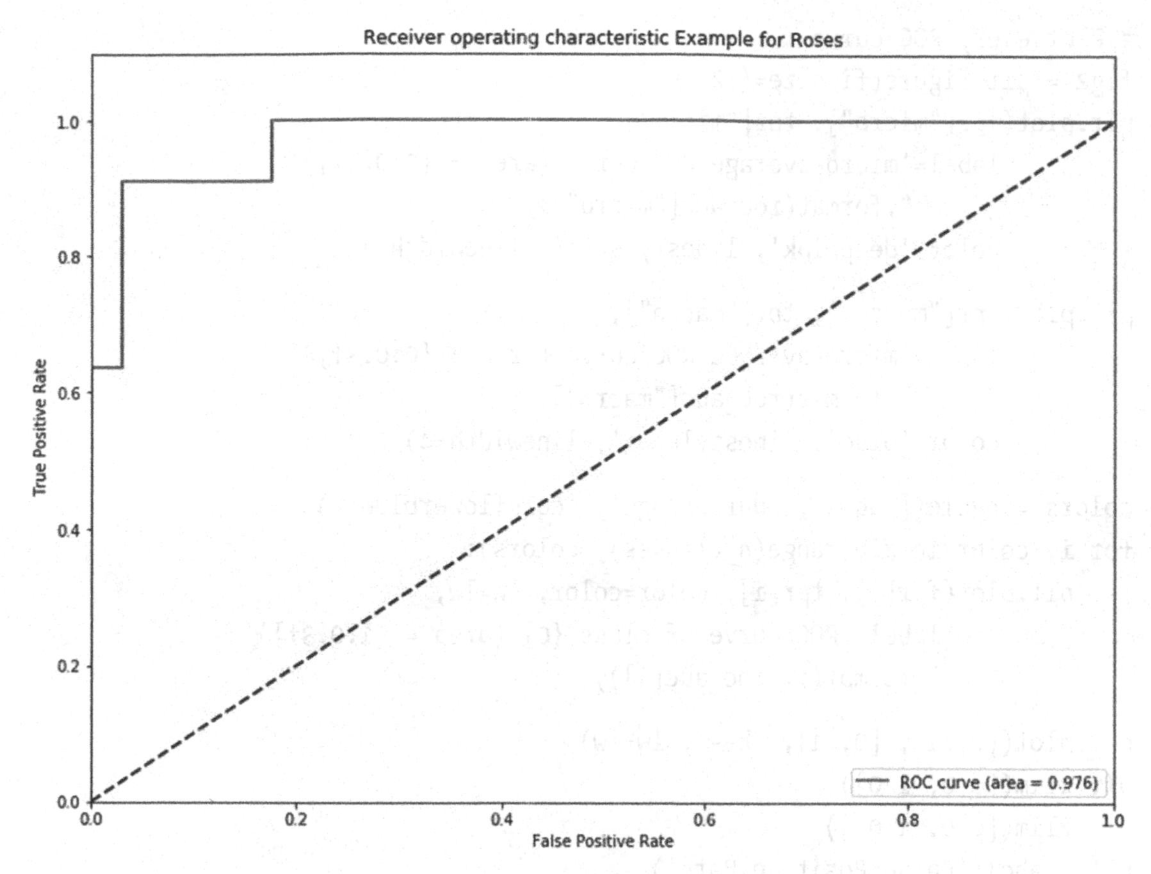

Figure 3-1. *ROC*

To check Plot ROC curves for the multiclass problem, just add the following:

```
# First aggregate all false positive rates
all_fpr = np.unique(np.concatenate([fpr[i] for i in range(n_classes)]))

# Then interpolate all ROC curves at this points
mean_tpr = np.zeros_like(all_fpr)
for i in range(n_classes):
    mean_tpr += interp(all_fpr, fpr[i], tpr[i])

# Finally average it and compute AUC
mean_tpr /= n_classes

fpr["macro"] = all_fpr
tpr["macro"] = mean_tpr
roc_auc["macro"] = auc(fpr["macro"], tpr["macro"])
```

```
# Plot every ROC curve
fig2 = plt.figure(figsize=(12, 9))
plt.plot(fpr["micro"], tpr["micro"],
         label='micro-average ROC curve (area = {0:0.3f})'
               ".format(roc_auc["micro"]),
         color='deeppink', linestyle=':', linewidth=4)

plt.plot(fpr["macro"], tpr["macro"],
         label='macro-average ROC curve (area = {0:0.3f})'
               ".format(roc_auc["macro"]),
         color='blue', linestyle=':', linewidth=4)

colors = cycle(['aqua', 'darkorange', 'cornflowerblue'])
for i, color in zip(range(n_classes), colors):
    plt.plot(fpr[i], tpr[i], color=color, lw=lw,
             label='ROC curve of class {0} (area = {1:0.3f})'
             ".format(i, roc_auc[i]))

plt.plot([0, 1], [0, 1], 'k--', lw=lw)
plt.xlim([0.0, 1.0])
plt.ylim([0.0, 1.05])
plt.xlabel('False Positive Rate')
plt.ylabel('True Positive Rate')
plt.title('Extension of Receiver operating characteristic to multi-class')
plt.legend(loc="lower right")
plt.show()
```

See the results in Figure 3-2.

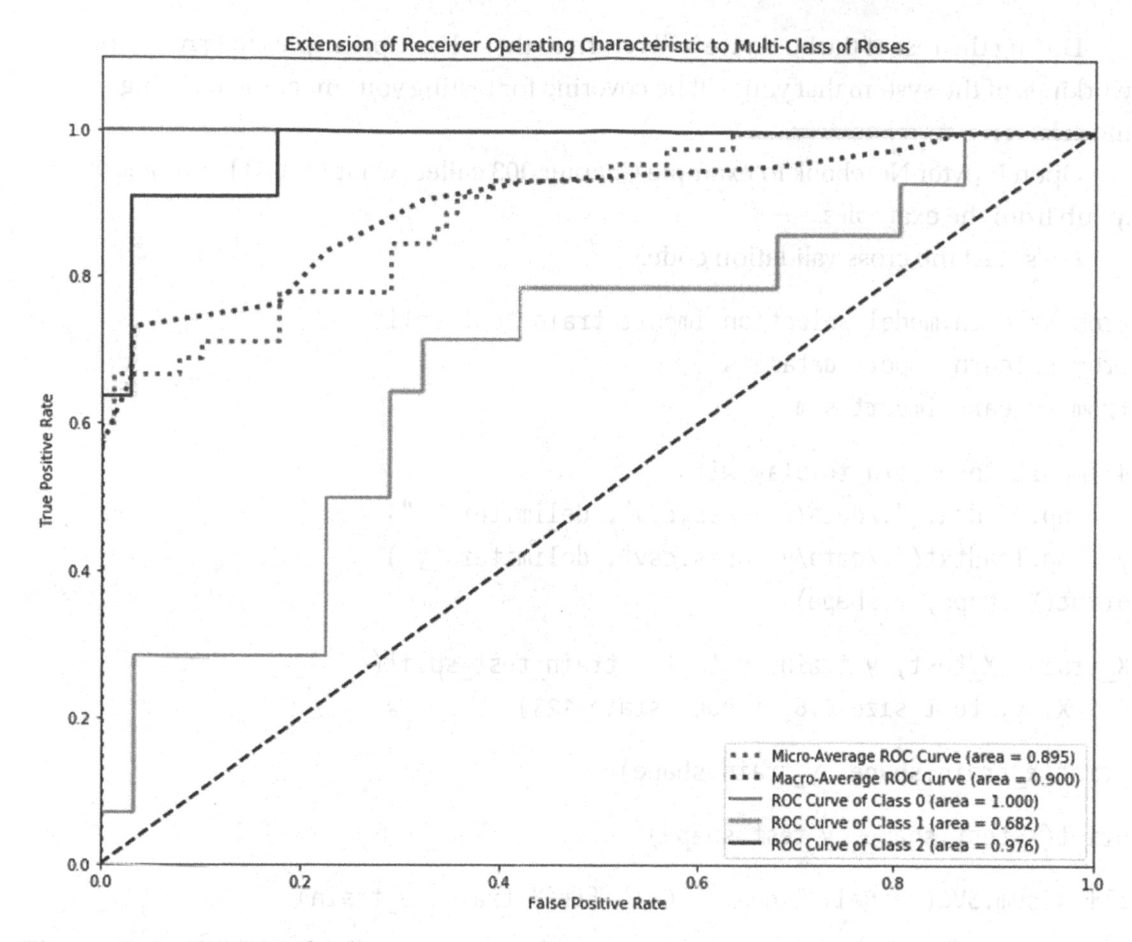

Figure 3-2. *ROC Multiclass*

You can close your Jupyter Notebook. You have made great progress. You can now perform basic machine learning and you have a way (ROCC) to visually validate your machine learning. Next, I will show you a more formal technique to validate the models.

Cross-Validation Testing

K-**fold cross-validation** is a technique for **cross-validation**: it splits the training data into K **folds**. It then builds the model based on the data from (K - 1) **folds** and tests the model on the remaining **fold** (the **validation** set). This enables improved test standards of your models.

During the rest of the book, we will return to these basic principles to explain the workings of the system that you will be covering for testing your machine learning models.

Open Jupyter Notebook in example Chapter 003 called: Chapter 003 Examples 003. ipynb from the examples.

Let's start the cross-validation code.

```
from sklearn.model_selection import train_test_split
from sklearn import datasets
from sklearn import svm

# Import Rose data to play with
X = np.loadtxt("./data/x-roses.csv", delimiter=",")
y = np.loadtxt("./data/y-roses.csv", delimiter=",")
print(X.shape, y.shape)

X_train, X_test, y_train, y_test = train_test_split(
    X, y, test_size=0.6, random_state=123)

print(X_train.shape, y_train.shape)

print(X_test.shape, y_test.shape)

clf = svm.SVC(kernel='linear', C=1).fit(X_train, y_train)

scoreResult=clf.score(X_test, y_test)
print(scoreResult)

from sklearn.model_selection import cross_val_score

clf = svm.SVC(kernel='linear', C=1)

scoreTypes = [
        'accuracy',
        'precision_macro',
        'precision_micro',
        'precision_weighted',
        'r2',
        'recall_macro',
        'recall_micro',
        'recall_weighted']
```

```
for scoreType in scoreTypes:
    print(scoreType)
    print('-------------------------------')
    scores = cross_val_score(clf, X, y, cv=3, scoring=scoreType)
    for i in range(scores.shape[0]):
        print('%0.4f' % scores[i])
    print('-------------------------------')
    print("Score: %0.4f (+/- %0.4f)" % (scores.mean(), scores.std() * 2))
    print('-------------------------------','\n')
```

The validation checks the results using several different cross-validation scores. The results:

Note This result is big … You could select a smaller set of tests if you want.

I will discuss each of these as we proceed through the results. Let's start:

```
accuracy
-------------------------------
1.0000
0.9216
1.0000
-------------------------------
Score: 0.9739 (+/- 0.0739)
--------------------------------

precision_macro
-------------------------------
1.0000
0.9365
1.0000
-------------------------------
Score: 0.9788 (+/- 0.0599)
--------------------------------

precision_micro
-------------------------------
```

```
1.0000
0.9216
1.0000
--------------------------------
Score: 0.9739 (+/- 0.0739)
---------------------------------

precision_weighted
--------------------------------
1.0000
0.9365
1.0000
--------------------------------
Score: 0.9788 (+/- 0.0599)
---------------------------------

r2
--------------------------------
1.0000
0.8824
1.0000
--------------------------------
Score: 0.9608 (+/- 0.1109)
---------------------------------

recall_macro
--------------------------------
1.0000
0.9216
1.0000
--------------------------------
Score: 0.9739 (+/- 0.0739)
---------------------------------

recall_micro
```

```
--------------------------------
1.0000
0.9216
1.0000
--------------------------------
Score: 0.9739 (+/- 0.0739)
--------------------------------
recall_weighted
--------------------------------
1.0000
0.9216
1.0000
--------------------------------
Score: 0.9739 (+/- 0.0739)
--------------------------------
```

This will supply you with a good accuracy indication as cross-validation is a statistical method used to estimate the accuracy of machine learning models.

The concept of generalization is when the machine learning engineer expands the findings against wider real-world data set to predict how it will behave against the same model.

This assessment indicates how the results of a statistical analysis will generalize to an independent data set or in general ... What is the probability that the model will work correctly against a new unknown set of data?

You can now close the example Jupyter Notebook.

You should by now understand how to test a model by using a series of cross-validation techniques.

Tip Constantly test with the unchanged set of scoring types as your testers will get used to signing off the models during the production readiness phase of your models.

Imputing Missing Values

I will show you how performing imputing of the missing values can give better results than simply discarding the samples containing any missing value.

Warning Imputing does not always improve the predictions, so please check via cross-validation before and after any imputing actions.

This shows that sometimes dropping rows or using marker values is more effective, but it can result in undesirable impacts that can cause the selection of an incorrect model and not the one truly required.

Open Jupyter Notebook in example Chapter 003 called: Chapter 003 Examples 004.ipynb from the examples.

Let's start by discovering the example code:

```
import matplotlib
matplotlib.use('TkAgg')
%matplotlib inline

import numpy as np
import matplotlib.pyplot as plt

from sklearn.datasets import load_diabetes
from sklearn.datasets import load_boston
from sklearn.ensemble import RandomForestRegressor
from sklearn.pipeline import make_pipeline, make_union
from sklearn.impute import SimpleImputer, MissingIndicator
from sklearn.model_selection import cross_val_score
import pandas as pd
import os

rng = np.random.RandomState(0)

sickdf= pd.read_csv(os.path.abspath('../../Data/sickness02.csv'),header=0)
print(sickdf.shape)

homesdf= pd.read_csv( os.path.abspath('../../Data/Homes02.csv'),header=0)
print(homesdf.shape)
```

```
def get_results(dataset):
    X_full_df = dataset.copy(deep=True)
    X_full_df.drop(['T'], axis=1, inplace=True)
    y_full_df = dataset[['T']]

    X_full=np.array(X_full_df)
    y_full=np.array(y_full_df['T'])

    n_samples = X_full.shape[0]
    n_features = X_full.shape[1]

    # Estimate the score on the entire dataset, with no missing values
    estimator = RandomForestRegressor(random_state=0, n_estimators=100)
    full_scores = cross_val_score(estimator, X_full, y_full,
                                  scoring='neg_mean_squared_error', cv=5)

    # Add missing values in 80% of the lines
    missing_rate = 0.8
    n_missing_samples = int(np.floor(n_samples * missing_rate))
    missing_samples = np.hstack((np.zeros(n_samples - n_missing_samples,
                                          dtype=np.bool), np.ones(n_missing_
                                          samples, dtype=np.bool)))
    rng.shuffle(missing_samples)
    missing_features = rng.randint(0, n_features, n_missing_samples)

    # Estimate the score after replacing missing values by 0
    X_missing = X_full.copy()
    X_missing[np.where(missing_samples)[0], missing_features] = 0
    y_missing = y_full.copy()
    estimator = RandomForestRegressor(random_state=0, n_estimators=100)
    zero_impute_scores = cross_val_score(estimator, X_missing, y_missing,
                                         scoring='neg_mean_squared_error',
                                         cv=5)

    # Estimate the score after imputation (mean strategy) of the missing
      values
    X_missing = X_full.copy()
    X_missing[np.where(missing_samples)[0], missing_features] = 0
```

```
    y_missing = y_full.copy()
    estimator = make_pipeline(
        make_union(SimpleImputer(missing_values=0, strategy="mean"),
                   MissingIndicator(missing_values=0)),
        RandomForestRegressor(random_state=0, n_estimators=100))
        mean_impute_scores = cross_val_score(estimator, X_missing, y_missing,
                                      scoring='neg_mean_squared_error',
                                      cv=5)

    return ((full_scores.mean(), full_scores.std()),
            (zero_impute_scores.mean(), zero_impute_scores.std()),
            (mean_impute_scores.mean(), mean_impute_scores.std()))

results_sick = np.array(get_results(sickdf))
mses_sick = results_sick[:, 0] * -1
stds_sick = results_sick[:, 1]

results_homes = np.array(get_results(homesdf))
mses_homes = results_homes[:, 0] * -1
stds_homes = results_homes[:, 1]

n_bars = len(mses_sick)
xval = np.arange(n_bars)

x_labels = ['Full Data',
            'Zero Imputation',
            'Mean Imputation']
colors = ['red', 'green', 'blue', 'orange']

# plot sickness results
plt.figure(figsize=(12, 6))
ax1 = plt.subplot(121)
for j in xval:
    ax1.barh(j, mses_sick[j], xerr=stds_sick[j],
             color=colors[j], alpha=0.6, align='center')

ax1.set_title('Imputation Techniques with Sickness Data')
ax1.set_xlim(left=np.min(mses_sick) * 0.9,
             right=np.max(mses_sick) * 1.1)
```

```
ax1.set_yticks(xval)
ax1.set_xlabel('MSE')
ax1.invert_yaxis()
ax1.set_yticklabels(x_labels)

# plot homes results
ax2 = plt.subplot(122)
for j in xval:
    ax2.barh(j, mses_homes[j], xerr=stds_homes[j],
             color=colors[j], alpha=0.6, align='center')

ax2.set_title('Imputation Techniques with Homes Data')
ax2.set_yticks(xval)
ax2.set_xlabel('MSE')
ax2.invert_yaxis()
ax2.set_yticklabels([''] * n_bars)

plt.show()

imagepath = os.path.join(*[os.path.dirname(os.path.dirname(os.
getcwd())),'Results','Chapter 03'])
print(imagepath)

if not os.path.exists(imagepath):
    os.makedirs(imagepath)

imagename = os.path.join(*[os.path.dirname(os.path.dirname(os.
getcwd())),'Results','Chapter 03','Chapter-003-Examples-004-01.jpg'])
fig.savefig(imagename, bbox_inches='tight')
```

See the results in Figure 3-3.

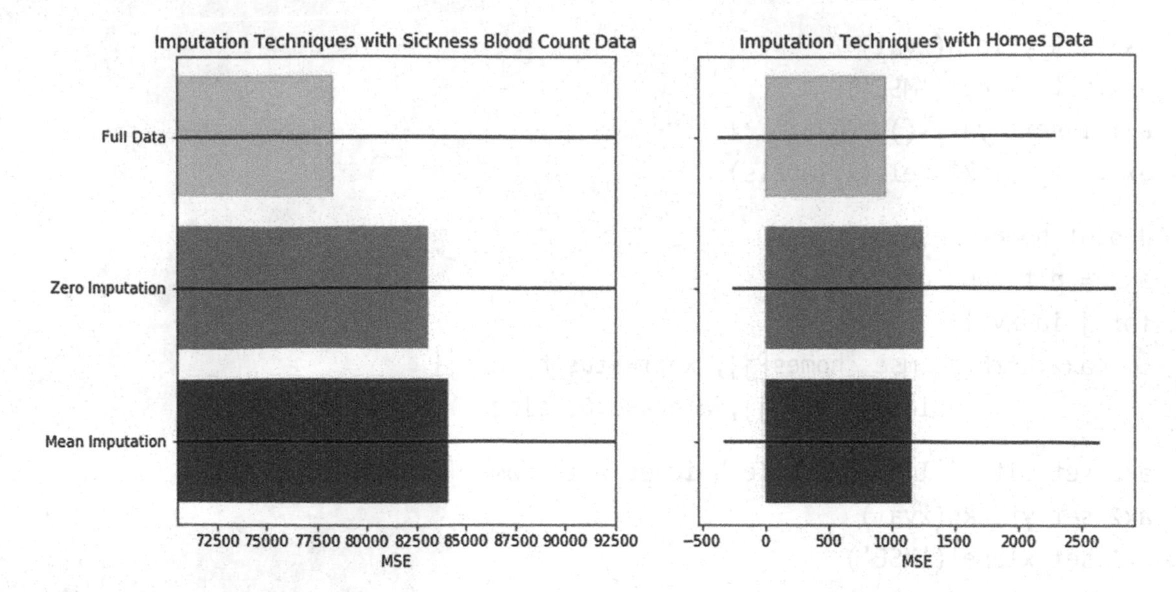

Figure 3-3. *Imputing Data*

The example clearly shows an improvement of the data science results as the imputing of the missing values changes the data set.

You can close your Jupyter Notebook.

You can now resolve missing data in your data sets, and this will help you with your machine learning.

I will now take you through several industrial knowledge fields to explain how they fit into the machine learning field.

Knowledge Fields

I will take you through the detailed processes to apply this machine learning knowledge on the world around you. The following areas will be discussed at the during your progression throughout this book.

Mechatronics

Mechatronics is a multidisciplinary field of science that includes a combination of mechanical engineering, electronics engineering, control engineering, telecommunications engineering, systems engineering, and computer engineering.

What Is Mechanical Engineering?

Mechanical engineering is a branch of engineering that applies the principles of mechanics and materials science for analysis, design, manufacturing, and maintenance of mechanical systems. It involves the production and usage of heat and mechanical power for the design, production, and operation of machines.

You create the physical components (arms, legs, wheels, tracks, and many more) in this portion of mechatronics. I will not explain in detail how this is done but will introduce some known constraints in this field that will pass into your machine learning processing requirements.

What Is Electronics Engineering?

Electronic engineering is an electrical engineering discipline that utilizes nonlinear and active electrical components (semiconductor devices, diodes, and integrated circuits) to design electronic circuits, devices, VLSI devices and their systems. The discipline typically also designs passive electrical components, usually based on printed circuit boards.

You will create the physical components (embedded integrated circuits) that interact with the mechanical and telecommunication components in this portion of mechatronics. I will not explain in detail how this is done but will introduce some specified constraints this field will pass into your machine learning processing requirements.

What Is Telecommunications Engineering?

Telecommunications engineering is an engineering discipline centered on electrical and computer engineering that seeks to support and enhance telecommunication systems. The work designs and produces telecommunications equipment and facilities like complex electronic switching systems, optical fiber cabling, IP networks, and wireless transmission systems.

You will create the physical communication components (embedded integrated circuits) that interact with the non-communication electronics components in this portion of mechatronics. I will not explain in detail how this is done but will introduce some specified constraints this field will pass into your machine learning processing requirements.

What Is Control Engineering?

Control engineering or control systems engineering is an engineering discipline that applies automatic control theory to design systems with desired behaviors in control environments.

You use sensors and detectors to measure the output performance of the process being controlled; these measurements are used to provide corrective feedback helping to achieve the desired performance in this portion of mechatronics. I will not explain in detail how this is prepared but will introduce numerous constraints this field will pass into your machine learning processing requirements.

This area is booming as the Internet of things (IoT) is the latest network of physical devices, vehicles, home appliances, and other items embedded with electronics, software, sensors, actuators, and connectivity that empowers this equipment to connect, collect, and exchange data.

The number of IoT devices increased 31% year over year to 8.4 billion in the year 2017, and it is estimated that there will be 30 billion devices by 2020. The global market value of IoT is projected to reach $7.1 trillion by 2020.

What Is Systems Engineering?

Systems engineering is an interdisciplinary field of engineering and engineering management that focuses on how to design and manage complex systems over their life cycles.

You use this design and interconnect different systems in this in this portion of mechatronics. I will not explain in detail how this is done but will introduce some specified constraints this field will pass into your machine learning processing requirements.

This area is booming as the IoT is now requiring complex systems to be designed using heterogeneous control systems.

Machine learning is a major role player in this process of developing the most cost-effective systems.

What Is Computer Engineering?

The computer engineering interdisciplinary field integrates electronic engineering and computer sciences to design and develop computer systems and other technological devices by programming them, using a specific form of programming language that enables you to tell the machine what to execute.

I will explain from a computer engineering point of view how you use machine learning to determine the next possibilities and results that are required to formulate the commands you need to hand off to the control engineering to use the mechanical engineering, electronics engineering, telecommunications engineering, and systems engineering to achieve your end goal.

At this point in the book, I will not discuss this subject in more detail, but we will revisit it numerous times during the rest of the book. This is the core of mechatronics you will use to generate your industrialized machine learning.

Robotics

Robotics is a branch of engineering that involves the conception, design, manufacture, and operation of robots. This field is at the intersections with electronics, computer science, AI, mechatronics, nanotechnology, and bioengineering.

Soft Robotics

Soft Robotics is the specific subfield of robotics dealing with constructing robots that have no physical manifestation. They are exclusively software that perform responsibilities in an autonomous technique following machine learning models or embedded guidelines.

Characteristic examples are robots that responses question on a website or publish content to customers guided by matching their profile with the marketing communication required.

Hard Robotics

Hard Robotics is an interdisciplinary branch of engineering and science that includes mechanical engineering, electrical engineering, computer science, and others. Robotics deals with the design, construction, operation, and use of robots, as well as computer systems for their control, sensory feedback, and information processing. These robots are capable of directly manipulating their surroundings.

I will demonstrate with simulations for aerial, land, and underwater drones how your machine learning needs to adapt to work in these environments.

Fourth Industrial Revolution

The Fourth Industrial Revolution is the concept of amalgamation of the real world with the technological world. The uses machine learning has at present starts to influence the everyday environment around us, and in many ways, we have accepted these changes without noticing that we took them, even though they have been assisting us to have a better quality of life.

I will demonstrate how you handle machine learning for a small just-in-time 3D printing factory to manufacture a simple six-wheel vehicle.

Challenges

Look at the environment around you where you spend your daily life and write down three things you believe are the result of machine learning.

Description	Yes/No

Then write down three things that would be better for you if they were automatically done for you.

Description	Yes/No

Later in the book, I will come back to these six items to investigate how you could achieve the goal of applying Industrialized Machine Learning (IML) (in to these requirements.

Disruptors

The news and the lecture circuits are bursting with discussions of the "disruptors" using machines. The world is shifting, and we are not the primary change agent within this new world. Humans have allowed machines and other soft agents to take control of our lives totally enthusiastically or unknowingly to us.

The impact of machine learning will increase as the agents that are assisting us in our daily lives take more control and get more attuned with individuals' behaviors. These disruptors create major opportunities for applying machine learning to amplify the process.

I will discuss several practical demonstrations of how a simple machine learning model can be industrialized to meet bigger real work challenges. Here's a small sample of the projects I handled over the last year.

Revenue-Enriched Shopping

The use of machine learning engages customers with a rich shopping experience by customizing the experience with recommendations of products and bundling products together that normally in the past would not have appeared in one shop together.

Examples:

A major retail store hired a dietitian firm to assist with altering their customers' nutrition food purchases based upon their specific medical conditions and individual needs. They then offer an option to get a specialist cooking school in the area, closest to your home, to deliver the ready-made food to your house.

They also staged full-sized mock-up of rooms in their shops that the customer can access via physically by visiting and then simply scanning the QR code, and the complete as-is room gets ordered and delivered. You can then simply scan the QR for the designer and get them to complete the interior decorating as a turnkey option. The same service is available as an online website, virtual reality shopping experience.

Their latest venture is to use a smartphone to upload your current room image and the machine learning designs you a new room with use of profiling at the customer level. Shopping experiences as we know are disrupted by machine learning to a level that has become the acceptable norm.

Re-shopping on Customers' Requests

Shopping for flight alternatives and pricing for all your customers' ticket changes and refunds can now be outsourced to companies that use machine learning and humans-in-the-loop technology to preplan and then rebook the required travel arrangements without your involvement.

You can file a request to buy a specific car and the system will collect the required collection of available cars together and simply display them as a basket option in an online website, virtual reality shopping experience.

The introduction of these new personal buyers is now changing the retail demographics for traditional shops. Your customer is no longer a person; you are now doing business with a middle agent that is a machine learning algorithm seeking the optimum profit.

Traditional customer profiling by shops is simply not valid anymore. The process is now disordered retail demographics as a primary selection condition.

Reward Shopping

Help your loyal passengers find reward flights and hotels while preserving your revenue objectives. The new process is the concept of a loyalty program where a financial institute pre-agrees to specific awards for their customers that are not directly part of their traditional business model.

The prime example is a disruptor bank that has a mobile application that tracks the behaviors of their customers and then learns, from a previous purchase, to supply an option list for places where their customers can buy lunch and receive a 2% discount or cash back because the merchant accepts the new bank's payment card or the merchant is a business customer of the new bank, so you get an extra discount or cash back at 3%.

The geo-spatial machine learning capabilities of the bank enable a shift in the customer's behavior to preapproved merchants. Traditional customer profiling by shops are overall not valid anymore. The process is interrupted by reward cards and coupons from the bank on behalf of their business clients.

Merchandising

Machine learning profiling enables increased revenue per customer through the sale of ancillary products that a specific customer requires. A supply chain model reacts to the change in the demand from the customers and then adapts the merchandising of the shops.

The new robotic shelving, automated storage, and retrieval system of products are enabling machine learning to predict the demand and then organize the shop in a manner that reflects the demand with minimal human interaction. Soft robots calculate the optimum solution for the shop and the hard robots pack the shelves in the warehouse and then deliver it to the shop floor. When the shelf is empty, it returns itself to the warehouse for re-merchandising.

The future target is to move stock between shops with autonomous drones on demand to ensure that the correct merchandising is achieved on a just-in-time delivery model.

Traditional shop floors, merchandising, and supply chains are changing, and the status quo is not effective anymore. The shopping process is seriously interrupted by only offering stock that matches the demand.

Meta-Search Offload

Simplifying an airline's website operations with a single interface for pricing and shopping content for metasearch engines enables the change of a simple search into a money transfer transaction.

Open banking as a result of the Second Payment Services Directive (PSD2) creates a new Account Information Service Provider (AISP). With the use of machine learning, this new business agent will consolidate or aggregate baking data from a variety of banks for the customer and create new propositions, dashboards presenting all customer account information in one place, and enable the customer to supply an metasearch request like "Find me a mortgage for buying 1600 Pennsylvania Avenue, NW, Washington, DC 20500," or "Want to save £2000.00 this year."

The result is a customer page with a complete portfolio on the property with options from the bank's preapproved mortgages. It also enables the machine learning to pick the correct banking account to perform your banking business. It can even book you the required flights or travel arrangements to visit the property.

Book holidays from a search, "Want to visit Tour Eiffel, Parc du Champs de Mars, 5 Avenue Anatole France 75007 Paris, France," and machine learning will then act on your behalf and start setting up a complete itinerary and then complete the transactions to book the holiday by a simple click from the customer on a mobile application.

These metasearch offload solutions will disrupt the way we perform our daily business in the future all because machine learning can learn what our requirements are and then handle the required solutions without complex involvement by the end user of the solution.

Internet-of-Things Sensors

The Internet of Things is the global network of physical devices, vehicles, home appliances, and numerous commonplace items embedded with electronics, software, sensors, actuators, and connectivity. This enables these objects to connect and exchange data while everything is uniquely identifiable through its embedded computing system but can interoperate within the prevailing internet infrastructure.

Billions of wearable electronic products are already sold on that track and record the world around us to the greatest details. You connect this global network of IoT devices with the correct level of machine learning and mechatronics to form a new entity that is capable of sensing the world in a detail that we have never seen before. This machine learning system has the capability to start true interaction with humans and their needs. This will enable the ultimate disruptor of current lives on the planet Earth and the future it will generate.

Autonomous Farming

The enablement of farming machine learning is changing the way we will grow future crops. The introduction of Internet-of-Things sensors that can monitor the farming activities at every step of the way from seedlings to harvest to ensure an optimum process is achieved. The new introduction of industrial-level containerized aquaponics now enables the growth of food at a scale that enables machine learning to predict the required harvests and then grow them when required.

This enables the agricultural activities in places that were not used before. It is using machine learning and robotics to deliver a harvest that was not possible before. The introduction of robotic agricultural equipment nowadays empowers the farming of traditional crops with minimum interaction by humans disturbs the traditional farming concept in favor of a machine learning-driven solution that enhances the volume and the diversity of farming to an extent never seen before in the history of the human race.

Autonomous Mining

The introduction of machine learning into the mining industry enables the use of mining robots to perform work in conditions where people cannot safely work. The process of mining is also more predictive as the process is planned and executed as an end-to-end process fully controlled by machine learning.

The mining of raw material in an autonomous manner also opens the future options of mining raw materials from space or other planets.

I will cover this industrial challenge as an example later in the book. We will model the environment and then simulate the mining operations of a mining company.

Autonomous Railway Repairs

The autonomous railway repair is a machine learning solution that assists a highly repetitive task that requires high-tech scanning of the tracks with a mobile system that can travel the tracks and detect the issues and then use 3D printing with industrial materials to perform minor repairs. The system can also send out a repair unit to replace tracks, forming a central track repair facility without the assistance of the people.

This is a major bonus for the rail organization to keep the railway going.

Predictive Maintenance on Machines

The introduction of the IoT and the use of supervisory control and data acquisition (SCADA) is a control system architecture that feeds the complete state of the system under control to a machine learning environment. The machine learning has direct control of the system and can perform corrections on a near real-time basis to ensure the system has a high Mean time between failures (MTBF). Mean time between failures is the predicted elapsed time between inherent failures of a mechanical or electronic system, during normal system operation.

The other factor it controls is Mean Time to Repair (MTTR), which is a basic measure of the maintainability of repairable components. It represents the average time required to repair a failed component or device. By predicting the failure, it is possible to prevent the failure and keep the repair time to a controlled and planned schedule.

The checking of industrial and even household equipment is common as most equipment has a common boot-up sequence that performs a diagnostic on the state of the environment and working state of the equipment.

Enhanced Health Care

The deployment of machine learning into the health care system is a major game changer for the future of health care. The capabilities of the machine learning process to analyze the volumes and complexities of the health-care environment's data is performed at accuracies and speeds not possible to be achieved by the average health care staff member.

The introduction of wearable electronic health products enhances the health-care capability of a health provider to monitor their customers and intervene with their health care when issues arise. The patient is connected 24/7 and monitored to predict problems before they become life-threatening ones. The machine learning can detect an issue with a patient's heart and then a system like Auto Pulse Resuscitation System can then activate to save the person's life.

Machine learning can monitor millions of measurements per second and determine who requires what medical response. This always-on medical service will be the norm for medical care in the future. The current medical staff will become the human in the loop that deals with the complex or unknown medical emergencies. These medical interventions will directly be a disruptor of the current lifestyles of the people in the world.

These are only a sample of systems that has changed because of successful industrialized machine learning that has been deployed into the world around us.

If you look around yourself and perform research, you could easily find a minimum of three systems that are already changed or that you can easily change by integrating them with machine learning knowledge, that you will have by completing this book and its practical examples.

There is a basic set of systems that any industrialized machine learning solution will require to operate in the real world around us.

I will now discuss a few of these core requirements.

Data Lakes

A data lake allows you to store your structured and unstructured data, in a unique centralized repository, and at any scale. With a data lake, you can store your data as is, without having to first structure the data, based on prospective questions you may have in the future. Data lakes also allow you to run different types of analytics on your data like SQL queries, big data analytics, full text search, real-time analytics, and machine learning to guide better insights and support better-informed decisions.

Data Lake Zones

The industrialized data lake consists of six core zones.

Workspace Zone

This workspace zone is where the machine learning tools store any data that they require to work within the data lake. It is a machine-only access zone, and no end users are allowed in this zone.

Raw Zone

The raw zone is the entry point of all outside-the-data-lake-data sources. The raw zone is an endpoint for several feeder systems like Enterprise Resource Planning (ERP), Supply Chain Management (SCM), Human Resource Management System (HRIS), Accounting Information System (AIS), and Internet of Things (IoT). In general, every aspect of our lives is generating s massive amount of useful raw data that can be piped to a data lake's raw zone ready for analysis.

Structured Zone

The structured zone is used to convert the raw data into enhanced data sources. This zone will contain data in harmonized formats to assist the next zone's processing to happen more effectively and efficiently.

Curated Zone

The curated zone is the single truth across the data lake. I always place my data vault and data warehouse in this zone to enable the consolidation and amalgamation of data sources from the structured zone. I will discuss the use of sun models, data vaults, and data warehouse structures within this zone with more details later in the book.

Consumer Zone

The consumer zone is the area where different data marts for the business's insights are stored ready for the end users to visualize their business insights. This is the primary zone for a majority of businesspeople. This is the Industrialized Machine Learning (IML) solutions' primary delivery area.

Analytics Zone

The analytics zone is the "sandbox" of the data lake. This zone is used to design, develop, and train new and novel machine learning solutions.

I will discuss these data lake zones with supplementary details later in the book in Chapters 13, 14, and 15.

Data Engineering

Data Lakes allow you to import any amount of data that can come in real time. Data is collected from multiple sources and moved into the data lake's raw zone in its original format. This process allows you to scale to data of any size, while saving time of having to define data structures, schema, and transformations.

Cautionary I have found that on average for each hour you spend on your machine learning model's creation, you can spend three hours on preparing the data the model requires. So, 25% machine learning needs 75% Data engineering.

My personal rule is that for each data scientist, you need two to three to perform the data engineering, that is, the pre-machine learning data preparation.

Securely Store, and Catalog Data

Data lakes allow you to store relational data -operational databases and data from line of business applications, and non-relational data - mobile apps, IoT devices, and social media. They also give you the ability to understand what data is in the lake through crawling, cataloging, and indexing of data. Finally, data must be secured to ensure your data assets are protected.

Analytics

Data lakes allow several business roles in your organization like data scientists, data developers, and business analysts to access data with their choice of analytic tools and frameworks. This includes open source frameworks such as Apache Hadoop, Presto, and Apache Spark, and commercial offerings from data warehouse and business intelligence vendors. A data lake's analytic zone allows you to run analytics without the need to move your data to a separate analytics system.

Tip The use of the analytic zone as a sandbox, saves you hours of development by allowing the rapid cycling of numerous ideas and concepts for solving the business solutions.

Autonomous Machine Learning

Data lakes will allow organizations to generate different types of insights including reporting on historical data and doing machine learning where models are built to forecast likely outcomes and suggest a range of prescribed actions to achieve the optimal result. The use of autonomous machine learning is causing major disruptions in the perception people have of future job prospects. News headline like ... "The robots are taking our jobs!" is causing prejudiced resentment of the proficiencies of machine learning.

I am assuming that with you reading this book, you are of the same belief that I am: machine learning at an industrialized scale will change the world around us. I suggest we are evolving, learning, and adapting to the changes and have no need to fear the changes it brings.

The process has just started, and we will still have numerous evolutions before we are 100% ready for the impact of IML.

Well done ... You now have the basic knowledge I believe you need to progress.

Enjoy your journey into the knowledge I will share with you over the rest of this book.

What Should You Know?

- You should now know how to load the examples as in this chapter as I supplied the complete code in the book. Later in this book, I will only supply snippets of code with the complete code only in the example Jupyter Notebooks that I have indicated for uploading before each section in the book.
- You should understand that the basic test for a quality evaluation of your machine learning is how accurate your predictions are on supervised learning when you apply it against the test data set.
- I would also advise you to look at your already deployed models to determine if they are still working as designed and deployed.

Warning Ensure your training and test data sets are not suffering from bias or variance that makes the data sets not true samples of the full population.

Next Steps

I suggest you take some time to understand the building blocks I have supplied to you.

Note These are only the basic building blocks and concepts you will use during the rest of the book to build the advanced IMLs; there is an endless collection of knowledge around us on the subjects covered.

Tip Please research what you are fascinated by in this chapter as the more knowledge you gain, the better your IML. Read Appendix A for some guidance on sources of extra knowledge.

In the next chapter, I will demonstrate how to use Supervised Learning to support IML.

CHAPTER 4

Supervised Learning: Using Labeled Data for Insights

Supervised Learning is a type of machine learning that learns by creating a function that maps an input to an output based on example input-output pairs. It infers a learned function from labeled training data consisting of a set of training examples, which are prepared or recorded by another source.

This method is only learning what was already agreed as the correct outcome or existing previous outcome for a selected set of features of the subject area that is being learned.

Solving Steps

To solve a given problem of supervised learning, you must perform the following steps:

1. Determine the type of training examples. Before executing any new model's function, the data scientist must choose what type of data is used as a training set. In case of handwriting analysis, for instance, this could be a single handwritten character, an entire handwritten word, or an entire line of handwriting to validate the model.

2. Gather a training set. The training set requirements are that the data must be representative of the real-world use of the function. Therefore, the set of input values are assembled, and corresponding outputs are also collected, either from human experts or from measurements in the real world.

A. F. Vermeulen, *Industrial Machine Learning*, https://doi.org/10.1007/978-1-4842-5316-8_4

3. Determine the input feature representation of the learned function. The accuracy of the learned function depends strongly on how the input object is represented. Typically, the input object is transformed into a feature vector, which encompasses several features that describe the object. The number of features should not be too high, because of the curse of dimensionality; but it should contain adequate information to accurately predict the output.
4. Determine the structure of the learned function and corresponding learning algorithm. For example, the engineer may choose to use support vector machines or decision trees.
5. Complete the design. Run the learning algorithm on the gathered training set. Supervised learning algorithms require the user to determine certain control parameters. These parameters should adjust the optimizing performance on a subset (called a validation set) of the training set, or via cross-validation.
6. Evaluate the accuracy of the learned function.
7. You should iterate steps 2 to 6 until the accuracy is acceptable.

Note 100% is not an option. Pick a % that is achievable.

After parameter adjustment and learning, the performance of the resulting function should be measured on a test set that is separate from the training set.

Concepts to Consider

The following concepts can help you understand the machine learning ecosystem.

Bias–Variance Trade-Off

The bias-variance trade-off is the property of a set of predictive models when models with a lower bias in parameter estimation have a higher variance of the parameter estimates across samples, and vice versa. The bias-variance dilemma or problem is the conflict in trying to simultaneously minimize these two sources of errors that prevent supervised learning algorithms from generalizing beyond their training set.

I have spent many hours turning systems to balance these two influencers to achieve effective and efficient running Industrialized Machine Learning (IML) solutions.

The bias is an error from erroneous assumptions in the learning algorithm. High bias can cause an algorithm to miss the relevant relations between features and target outputs; this is called underfitting.

Underfitting indicates that a model can neither model the training data nor generalize to new data. An underfitting machine learning model is not a suitable model as most predictions will inaccurately detect or predict outcomes obtained from poor performance on the training data.

The variance is an error from sensitivity to fluctuations in the training set.

High variance can cause an algorithm to model the random noise in the training data, rather than the intended outputs; this is called overfitting. Overfitting indicates that a model is too specific to the features of the training data. This happens when a model learns the detail and noise in the training data to the extent that it negatively impacts the performance of the model on new data. The noise or random fluctuations in the training data is picked up and learned as perceptions by the model. The problem is that these perceptions do not apply to new data and negatively impact the model's ability to generalize.

The bias-variance trade-off is a major impact factor on the long-term viability and use of an IML in the real-world applications.

Note that in the world where trained models are embedded into physical flash to drive integrated circuit or monolithic integrated circuit the bias-variance trade-off can extend or limit the life expectancy of a specific version of the model deployment.

Warning The bias–variance trade-off is one of the major issues in supervised learning as the data sets that you are training with should not include bias or variance.

A practical example of this trade-off is in the application of sorting of objects on a conveyor belt (see Figure 4-1 and Figure 4-2) in a factory by a robotic arm.

***Figure 4-1.** Real view*

***Figure 4-2.** Robot view*

The issue was happening in two ways after a model was trained with dark round and square sweets for two hours.

- During the underfitting the robot was required to pick the round sweets but kept on picking all of them.
- During the overfitting the robot only picks the dark round sweets.

The bias-variance trade-off was to train for less noise on the shape and accept more noise on the color.

Function Complexity

The complexity function of a data set is a finite or infinite sequence of samples that must be generated to cover every distinct feature combination. The higher the function complexity, the more sample data is required.

Example: If our sweets are only dark square or dark round, it will require fewer samples than having ten shapes and five colors.

The curse of dimensionality shows the various phenomena that arise when analyzing and organizing data in high-dimensional spaces (commonly with thousands of dimensions) as the high-dimensional spaces mean higher complexity function and larger training samples requirements.

Note that generally a high-dimensional space is prone to underfitting.

Amount of Training Data

The amount of data required for machine learning depends on two factors:

- The complexity of the problem, nominally the unknown underlying function that best relates your input variables to the output variable is more complex dimensional space.
- The complexity of the learning algorithm, nominally the algorithm used to inductively learn the unknown underlying mapping function from specific examples.

Note that more training data is always the furthermost preference as an absence of suitably qualified data in the supervised learning area is a major concern. There are also numerous recognized issues with bias in the prevailing, generally accessible training sets.

Dimensionality of the Input Space

The dimensionality of the input space of the data under investigation is an important factor to investigate. The curse of dimensionality shows the complexity phenomena that arise when analyzing and organizing data in high-dimensional spaces. As dimensionality becomes higher, additional input data is required to show all the combinations and permutations of the diverse dimensions.

Noise in the Output Values

The amount of noise in the output values can impact your training in a negative way. If the noise in your output due to unknown or uncontrollable factors is too intense in impact. it will result in rendering the modeling ineffective or inefficient because the model will underfit too easily.

Example: The sweets those are white on a white conveyor causes noise in the image detection model of the robot arm and results in the robot not detecting the sweets in Figure 4-2.

Heterogeneity of Data

Heterogeneity of data is incrementally more complex to handle in the same machine learning process. Heterogeneity in statistics means that the data populations, samples, or results are different. It is the opposite of homogeneity, which means that the population/data/results are the same.

The proverb in statistics of comparing apples with apples and oranges with oranges simplifies it.

An increase in heterogeneity results leads to significant intensification in model complexity and ultimately decreases the efficiency of the model outcomes. Diversity in the population is your biggest challenge. The more the variety, the more feature engineering is required to separate the variety in samples that are possible.

Example: The round sweet can be between one and five centimeters in diameter: this would make them heterogeneous and all being two centimeters would make them homogeneous. It is easier to pick sweets that are all round and the same size as machine learning only needs a binary result of detecting the sweet.

Note that most data sets would contain dimensions in the real world that is characteristically made up from an exceedingly heterogeneous population.

Curse of Dimensionality

Dimensionality is the result of the variance characteristics required to track or describe a specific object in the ecosystem. The more complex I have found the object, the more dimensionality the model needs to track the object's true features.

Beware the curse of dimensionality!!

The curse of dimensionality refers to a phenomenon that appears when analyzing and organizing data in high-dimensional spaces, which causes a dynamic optimization problem as the processing no longer grows in a workable fashion. The model processing does not complete because the number of dimensions makes it too complex and expensive to process due to the lack of common features. This phenomenon is seen in numerical analysis, sampling, combinatorics, machine learning, data mining, and database queries. The common theme of these problems is that when the dimensionality increases, the volume of the essential dimensions increases so fast that the available data becomes sparse.

This sparsity is challenging for any method that requires statistical significance. In order to obtain a statistically sound and reliable result, the amount of data needed to support the result often grows exponentially with the dimensionality.

In simple terms, even in the simplest case of d binary variables, the number of potential combinations is $O(2^d)$, that is, an exponential in the dimensionality.

Simply, each additional dimension doubles the effort needed to try all combinations.

For a d variable with v value combinations, it becomes $O(v^d)$ that is a major issue in high-dimensional solution spaces.

Data Redundancy

Redundancy of data causes invalid cause-and-effect relationships. Data redundancy is a condition generated within a data set when the identical piece of data is held in two separate dimensions. This increases the dimensionality of the data plus causes erroneous correlations between data dimensions.

Example: The sweets saving the size as centimeters and also meters in the same data record will cause a correlation between the two dimensions. Only use one of them to prevent these erroneous correlations.

Presence of Complex Interactions and Nonlinearities

Machine learning in a supervised space is heavily based on finding correlations in the relationship between the features and the outcomes required. The interactions are the main factors driving a successful model and an appropriate effective solution to the supervised learning process.

Presence of Multi-level Interactions

The presence of multi-level interactions has significant implications for the interpretation of machine learning models. If two variables of interest interact, the relationship between each of the interacting variables and a third "dependent variable" depends on the value of the other interacting variable. In practice, this creates a problematic relationship to predict the consequences of changing the value of a variable, particularly if the variables it interacts with are hard to measure or difficult to control in the machine learning space.

The solution is to only use multi-level interactions if no other features and interactions are possible.

Nonlinearities

Nonlinearity is a common concern when examining cause-effect relations. These data sets require complex modeling and hypotheses to offer explanations to nonlinear events. Nonlinearity without explanation can lead to random, badly forecasted outcomes that result in chaos. This happens because nonlinear in a general situation has a disproportionate cause and effect between features that the model cannot align as a model.

The most common solution is the application of a complex weighting factor across the data sets range.

Algorithms

Standard algorithms are the data processing backbone and the principal driving force for a supervised learning ecosystem. Here is the main grouping for machine learning methodologies available to empower IML.

Here also are the specific algorithms and a description for each of them. I will demonstrate an application for each of these industrialized solutions.

Warning I will start using code examples to explain the basic workings of machine learning. Please load the examples to enhance your interaction with the book. I will guide you by referencing to the part in the code I am discussing at that moment.

Support Vector Machines

A Support Vector Machine (SVM) is a specific, supervised machine learning algorithm that could be used for both classification and regression solutions. I personally use it mostly in classification solutions. This algorithm allows you to plot each data item as a point in n-dimensional space (the n is the number of features) with the value of each feature being the value of a coordinate. You perform classification by finding the hyperplane that differentiates the two classes best. It is simply a binary decision of 0 and 1.

I will guide you through a basic introduction to an SVN model using a Python example.

Open the sample example **`Jupyter Notebook: Chapter 004 Example 001.ipynb`** in your Jupyter software.

In this example's Part A, you will see all the libraries you need to process the SVM model. I will discuss the libraries quickly as these are the core of many of the examples in this book.

sklearn is scikit-learn library (`https://scikit-learn.org/stable/index.html`) is machine learning in Python.

matplotlib (`https://matplotlib.org/`) is a Python 2D plotting library that produces publication-quality figures in a variety of hard-copy formats and interactive environments across platforms.

Numpy (`http://www.numpy.org`) is the fundamental package for scientific computing within Python. Besides the scientific uses, its library can be used as an efficient multidimensional container of generic data.

Tip These three libraries are the backbone of most Python-based machine learning solutions.

Part B is a function to create a mesh grid to use for plotting.

Part C is a function to handle contours plotting. This function helps with visualization of the data.

Part D loads a sample test set for your experiment. You will observe that the data has 2,500 samples and is spread over 2 classes.

Using the pandas method: `ClusterDF.describe()` you get a general insight into the data's characteristics.

	F01	F02	T
count	2500	2500	2500
mean	-0.357842	48.737365	0.502
std	62.145431	46.793813	0.500096
min	-138.533496	-122.941273	0
25%	-50.59668	17.424954	0
50%	-14.791261	47.037317	1
75%	48.30878	81.998086	1
max	209.938456	194.04594	1

Part E will help you create the linear SVC (Support Vector Classifier) using four iterations.

You will observe that this experiment fails, as the lack of enough iteration has an impact on the results of your machine learning.

```
ConvergenceWarning: Liblinear failed to converge, increase the number of
iterations.
```

Part F trains the SVC machine learning model with more iterations.

```
clf = LinearSVC(random_state=123, max_iter=5, dual=False, verbose=True ,
class_weight='balanced' , multi_class='ovr')
```

You will get the following result:

```
LinearSVC(C=1.0, class_weight=None, dual=False, fit_intercept=True,
     intercept_scaling=1, loss='squared_hinge', max_iter=5,
     multi_class='ovr', penalty='l2', random_state=123, tol=0.0001,
     verbose=True)
```

So, what does this tell you?

C=1: Penalty parameter C of the error term is set to default = 1.0. This is a standard for this model.

```
Class_weight='balanced' : class weight is set to be n_samples / (n_classes
* np.bincount(y))
Class  0: count:  1245 weight:  1.0040
Class  1: count:  1255 weight:  0.9960
```

This enables the machine learning to adapt for an imbalance in the training data between class 0 and class 1.

> dual=False: Select the algorithm to use the dual or primal optimization problem. Choose dual=False when n_samples > n_features.

So, you are using the primal optimization for this model.

> fit_intercept=True: This is a default that indicates that by setting to true, an intercept will be used to calculate the model. fit_intercept=False sets the y-intercept to 0 and fit_intercept=True, the y-intercept will be determined by the line of best fit. I normally use the fit_intercept=True because it generally gives a better generalized fitting model.

> intercept_scaling=1: This is a default and indicates there is no scaling introduced into the model. I normally change the scale of the model if I have issues getting the model to train correctly within the time requirements. I suggest you experiment with scale to check which scale gives the best result.

> loss='squared_hinge' : Specifies the loss function. Your options are 'hinge' or 'squared_hinge'

That is, 'hinge' is the standard SVM loss and 'squared_hinge' is the square of the hinge loss.

max_iter=5 : The maximum number of iterations to be run to train the model. I have set the iterations to minimum. In a real-life model, I normally test for the minimum and then calculate a 1.5 x minimum safety margin; that is, 7.5 so I would set it to 8 for below 100 iterations. At > 100 iterations I use a 1.1 ratio, that is, 100 -> 110 or 1000 -> 1100.

```
multi_class=' ovr ' :  Your options is:
"ovr" trains n_classes one-vs-rest classifiers
"crammer_singer" optimizes a joint objective over all classes
```

(Note: crammer_singer option does not converge for this data set.)

penalty='l2' : The 'l2' penalty is the standard used by SVC. The other option is 'l1'.

The L1 regularization technique is known as LASSO Regression and L2 is known as Ridge Regression.

I pick the '12' or L2 technique as it is a mechanism that avoid easily overfitting of the model.

random_state=123 : random_state is the seed used by the random number generator but the dual=False causes this to have no effect on the model.

Warning Always check what combinations of hyperparameters activate and deactivate in the model.

---

tol=0.0001 : Tolerance for stopping criteria to supply a converge if the value is within the tolerance.

verbose=True : helps with debugging.

Part G scores the training with three indications.

```
-----------------------------------------------
Coefficient: [[-0.02296428  0.01051741]]
-----------------------------------------------
Intercept [-0.37975513]
-----------------------------------------------
Score 0.9132
-----------------------------------------------
```

This machine learning score is 91.32%.

Tip A score above 90% is normally acceptable unless you need more accurate predictions.

I have, however, found that level rare.

Part H enables plots of the results machine learning showing predicted vs. true. You would yield a graphic (Figure 4-3).

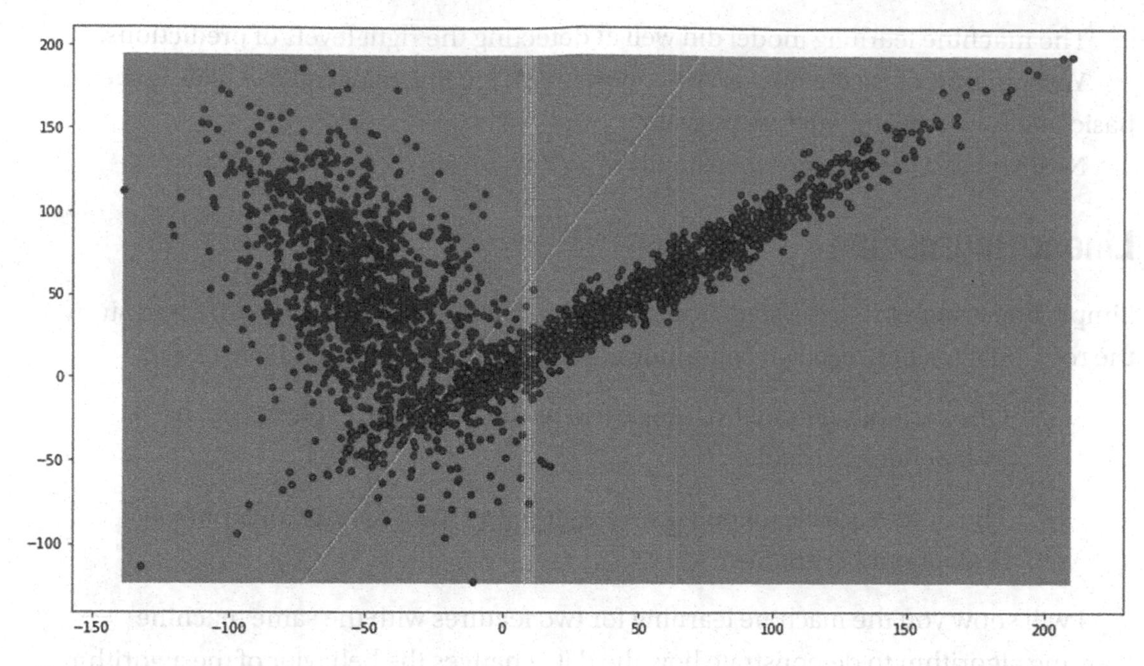

***Figure 4-3.** SVC Result*

Part J prepares the data for an F1 score test using a macro type averaging technique.

Your Results:

```
F1 Score (binary      ):  0.91190
F1 Score (micro       ):  0.91320
F1 Score (macro      ):   0.91318
F1 Score (weighted):      0.91318
```

This means the machine learning has a F1-Score of 91.1% minimum.

Part K prepares the Matthews correlation coefficient (MCC).

Your Results:

```
Matthews correlation coefficient (MCC): 0.82701
```

Part L calculates the confusion matrix (CM).

Your Results:

```
Confusion Matrix (CM)
=======================
TN =   1160 FP =     85
FN =    132 TP =   1123
```

The machine learning model did well at detecting the right levels of predictions.

You can now close the Jupyter Notebook as you are finished with calculating the basic machine learning success measures.

Next you can perform another standard SVN solution.

Linear Regression

Simple linear regression is a statistical method that allows you to summarize and study the relationships between two continuous (quantitative) variables:

- One variable, denoted x, is regarded as the predictor, explanatory, or independent variable.
- The other variable, denoted y, is regarded as the response, outcome, or dependent variable.

I will show you the machine learning for two features with the same machine learning algorithm to demonstrate how the data changes the behavior of the algorithm.

Open the sample example Jupyter Notebook: Chapter 004 Example 002A.ipynb in your Jupyter software.

Part A has the libraries you need. Have a look at each library and take note of the algorithm:

```
from sklearn import linear_model
```

This single import command loads the linear regression algorithm that is required for performing your machine learning.

Part B loads a Water Sickness data set that shows the presence of twelve kinds of germs features ('F01', 'F02', 'F03', 'F04', 'F05', 'F06', 'F07', 'F08', 'F09', 'F10', 'F11', 'F12') and four facts ('T01', 'T02', 'T03', 'T') identifying that person got sick from the water.

You need to extract only one feature (F03) by using Part C.

You should get `sickdf` with shape of (442, 1).

You will also see a data description of:

```
                  F
count  4.420000e+02
mean   1.131221e-11
std    4.761905e-02
min   -9.027530e-02
```

```
25%    -3.422907e-02
50%    -7.283766e-03
75%     3.124802e-02
max     1.705552e-01
```

Now that you have your feature (F). you need to use Part D to get your training feature set and Part E for the test feature set. Part F will provide the training target set and the test target set.

Part G is the algorithm activation. I am suggesting eight concurrent jobs. You can experiment with these to see what the performance impact is, but I am not going to handle this in the books as the diversity of ecosystem the readers have would make it practically impossible to predict your outcome.

Part H trains the model using the training sets from Part E.

Part I drives the predictions by using the testing set.

Part J and K explain the Coefficients and Mean squared error (MSE) for the prediction.

F01 feature delivers:

```
Coefficients = 1.15740949
Mean squared error (MSE) = 0.2473
Root Mean Square Error (RMSE): 0.4973
```

This is a really decent result as the model is processing the test data with great precision.

Part M calculates a Variance score of 0.0092 and Average Precision-Recall Score = 0.6006

The result is that the Precision-Recall Curve is as shown in Figure 4-4.

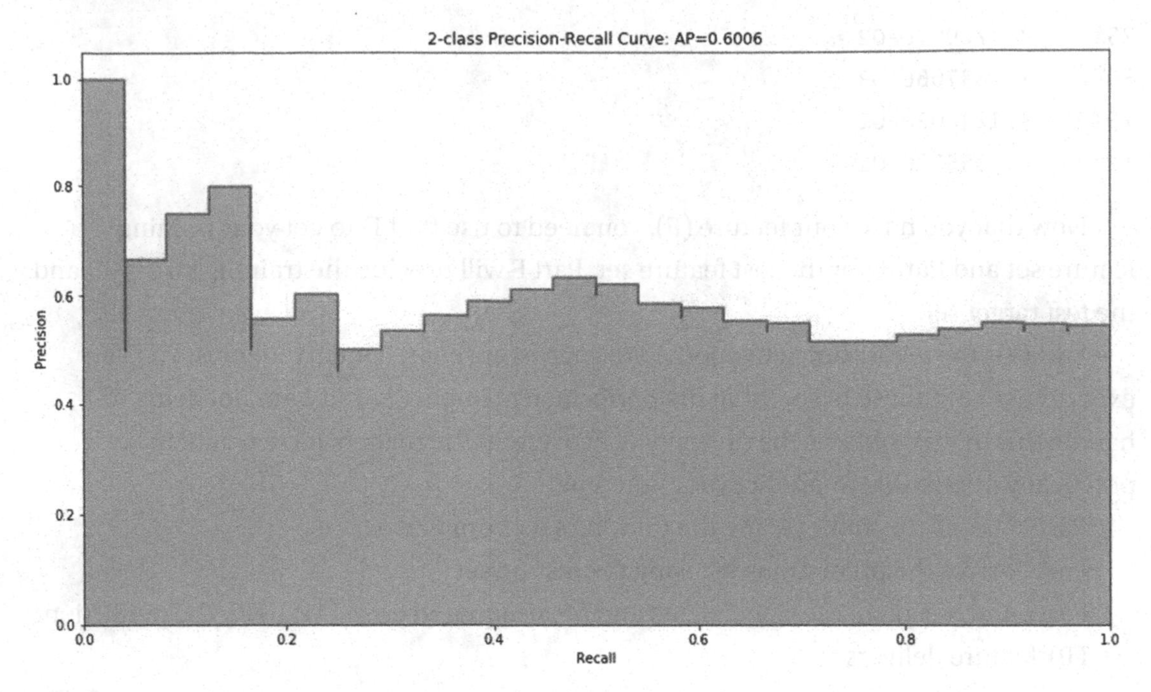

***Figure 4-4.** Precision-Recall Curve (F03)*

This graph shows that the model is above average on both Precision and Recall. For the given data set and the selected features, this model works 60.06%.

Warning This type of result is typical in real-world applications!

If you look at the data distribution, you can see why the specific data set is so easily learned.

Your machine learning results for feature F03 are shown in Figure 4-5.

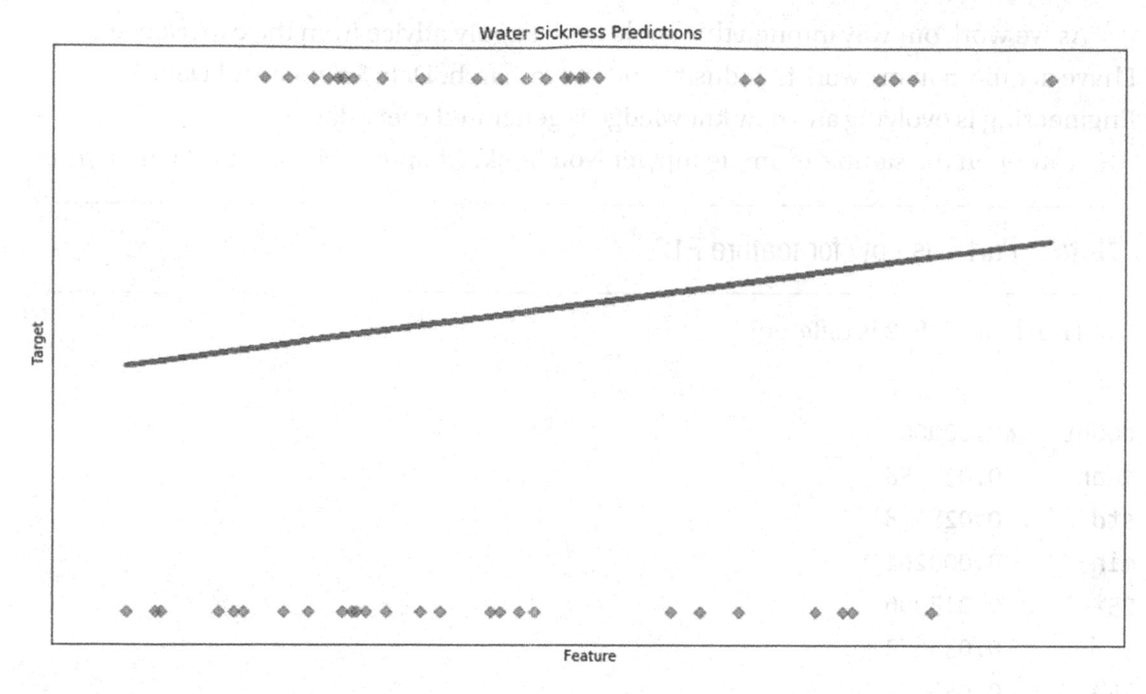

Figure 4-5. Linear Regression Result for F03

As you can witness, this data set is a neat bipolar cluster that is easily divided by the linear regression.

You can now close the Jupyter Notebook for Feature F03.

Well done; you have completed the first part of the analysis.

What have you learned?

- You can load data.
- You can select a single feature.
- You can train a machine learning model.
- You can predict a new outcome from unseen test data.
- You can create a graphic output of your findings.

If you understand these steps, you have done well already.

Note You will discover that the majority of the models are a basic set of data engineering to format the data ready for the training action by the selected machine learning.

As we work our way through this book, I will supply advice from the experience I have acquired in my work in industry and my research. Data Science and Data Engineering is evolving and new knowledge is generated every day.

Now open the sample example Jupyter Notebook: Chapter 004 Example 002B.ipynb.

Note Part C is now for feature F12.

The data for F12 is different:

```
                F
count  442.000000
mean     0.038358
std      0.028158
min      0.000261
25%      0.015906
50%      0.033673
75%      0.056724
max      0.170555
```

The result of the Precision-Recall Curve for F12 is as shown in Figure 4-6.

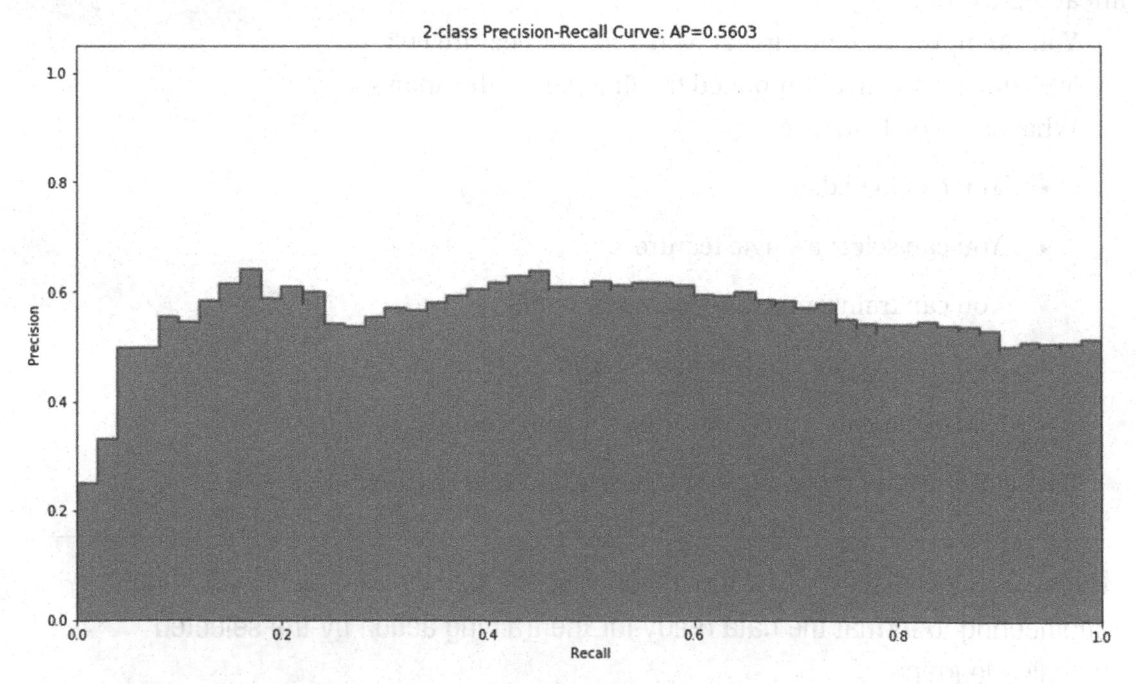

Figure 4-6. *Precision-Recall Curve (F12)*

This shows the second feature is not so good at indicating the outcome of the target.

If you look at the distribution, you will observe that the machine learning is having difficulty discovering the boundary for the data using linear regression.

Your machine learning results for feature F12 are shown in Figure 4-7.

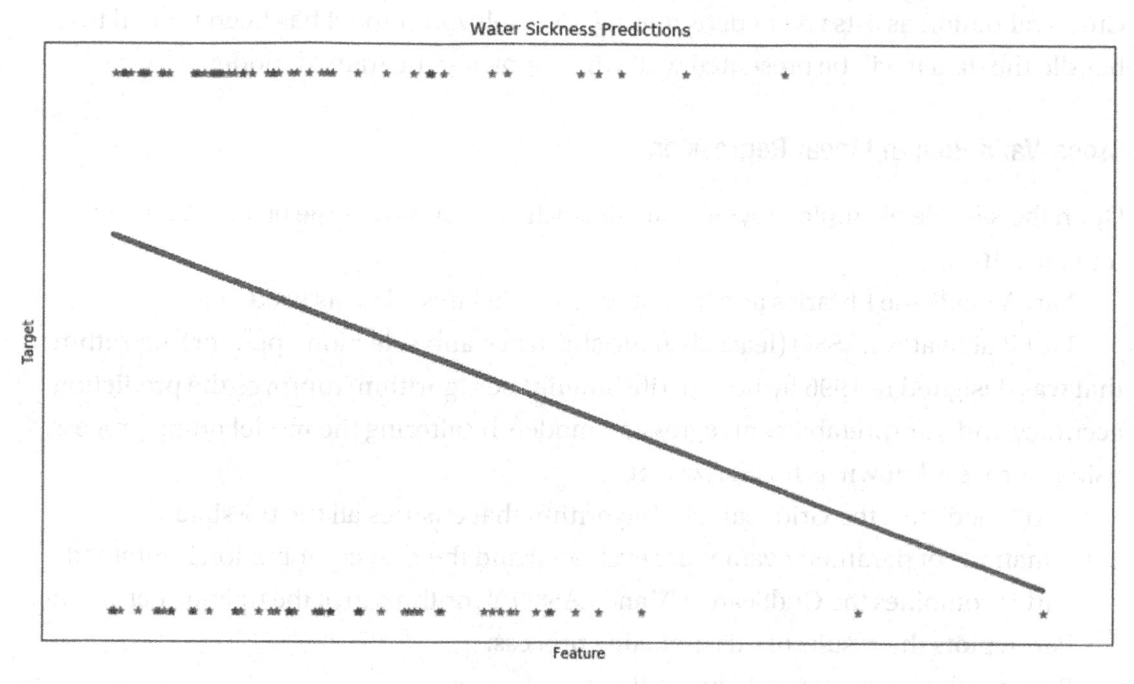

Figure 4-7. *Linear Regression Result for F12*

Warning The selection of machine learning model and features from the data set can extremely impact your outcome.

You can now close the Jupyter Notebook for Feature F12.

Extra training Copy the Notebook and compare the features of 1 to 12 for what their results are if you process them.

If you want to look at my notebooks (Chapter 004 Example 002B-F01 through to Chapter 004 Example 002B-F12), they will guide you with a possible solution/result for your training.

These will also show you the outcomes of each feature in the result directory.

Well done; you have completed the second part of the analysis.

Cross-Validation

Cross-validation assists you in determining how well your model has been trained to handle the data it will be presented with when applying the trained model.

Cross-Validation of Linear Regression

Open the sample example Jupyter Notebook: Chapter 004 Example 003.ipynb in your Jupyter software.

Part A loads the libraries and loads the Water Sickness data as used before.

Part B activates LASSO (least absolute shrinkage and selection operator) algorithm that was designed in 1996 by Robert Tibshirani. The algorithm improves the prediction accuracy and interpretability of regression models by altering the model fitting process, using a process known as regularization.

Part C activates the GridSearchCV algorithm that ensures all the possible combinations of parameter values are evaluated and the best combination is retained.

Part D combines the GridSearchCV and LASSO algorithms to fit the training data model.

Part E plots the results of this modeling process.

Your results are shown in Figure 4-8.

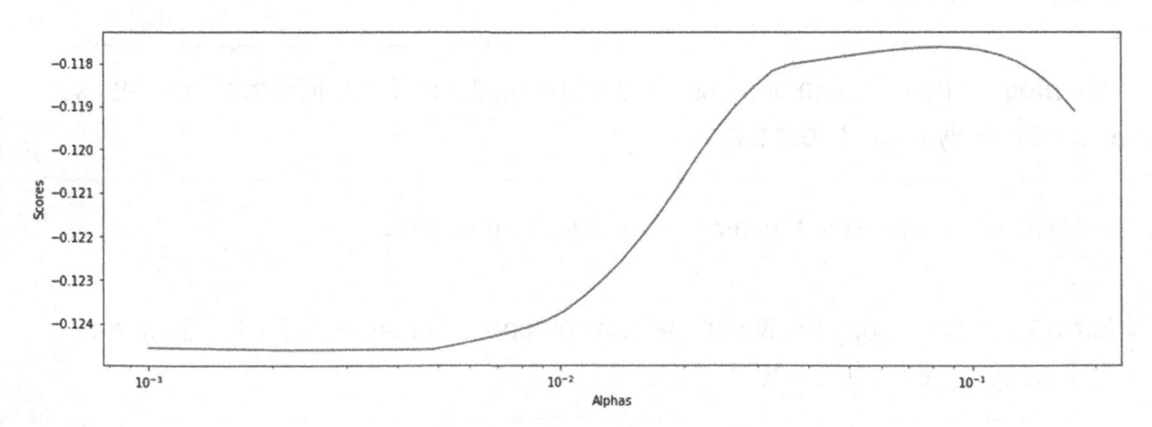

Figure 4-8. *Alphas vs. Scores*

Part F will assist you in plotting the error lines showing +/- std. errors of the scores.

Your Results are shown in Figure 4-9.

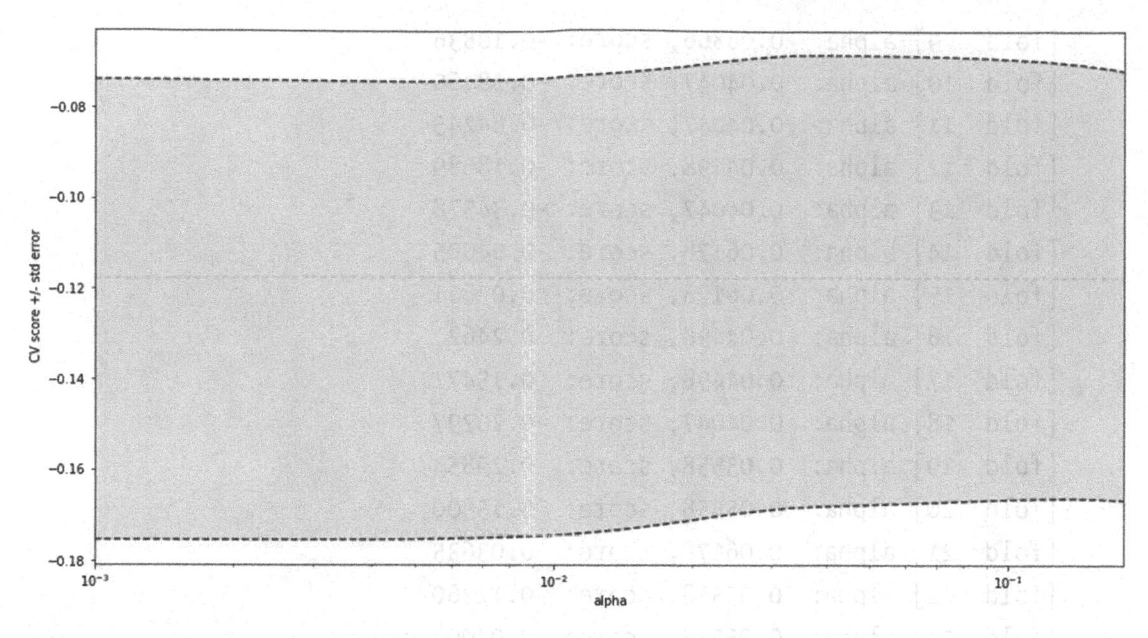

Figure 4-9. *CV Score*

Important Question:

How can you trust the selection of these alphas?

To answer this question, you use the LassoCV object that sets the alpha parameter automatically from the data by internal cross-validation (i.e., the process performs a cross-validation on each of the training data sets it receives).

You use external cross-validation to see how much the automatically obtained alphas differ across different cross-validation folds.

Part G will calculate the required results to answer this question.

Results to answer the important question: ***How can you trust the selection of alpha?***

Alpha parameters maximizing the generalization score on different subsets of the data:

```
[fold 0] alpha: 0.02947, score: -0.28715
[fold 1] alpha: 0.03276, score:  0.05880
[fold 2] alpha: 0.01931, score:  0.11514
[fold 3] alpha: 0.06866, score: -0.00656
[fold 4] alpha: 0.05000, score: -0.22933
[fold 5] alpha: 0.05558, score:  0.05195
[fold 6] alpha: 0.05558, score: -0.28285
[fold 7] alpha: 0.05000, score: -0.09796
[fold 8] alpha: 0.05000, score: -0.89133
```

```
[fold  9] alpha: 0.06866, score: -0.10836
[fold 10] alpha: 0.04047, score: -0.08460
[fold 11] alpha: 0.04047, score: -0.64245
[fold 12] alpha: 0.04498, score: -0.18689
[fold 13] alpha: 0.04047, score: -0.34578
[fold 14] alpha: 0.06178, score: -0.04005
[fold 15] alpha: 0.06178, score:  0.03641
[fold 16] alpha: 0.04498, score:  0.24627
[fold 17] alpha: 0.04498, score:  0.15471
[fold 18] alpha: 0.04047, score: -0.20797
[fold 19] alpha: 0.05558, score: -0.29854
[fold 20] alpha: 0.05558, score:  0.15500
[fold 21] alpha: 0.06178, score: -0.03635
[fold 22] alpha: 0.05558, score: -0.12260
[fold 23] alpha: 0.05558, score: -0.04003
[fold 24] alpha: 0.05558, score:  0.12859
```

Simple Answer:

For this example, I would not trust alpha since the solution acquires various alphas for different subsets of the data, and additionally the scores for your alphas vary significantly.

You can now save the notebook.

Well done; you can now evaluate parameters and make decisions on how you can set them to achieve success.

Tip Always test your hyperparameters, as most of the time the data has multidimensional behavior in real-world applications.

Sparsity

The concept of sparsity is common in numerous real-life data scenarios as the data is thinly dispersed or scattered and is missing across the given features. That data is missing or simply not present due to processing errors is the norm in most new IML ecosystems. In a mature IML ecosystem, the introduction of industrialized data engineering and improvements in the data provenance and lineage reduce this sparsity. I will demonstrate the concept of sparsity by observing at features 1 and 2 of the diabetes data set. It illustrates that even if feature 2 has a strong coefficient on the full model, it does not give you much regarding y when compared to just feature 1 due to the sparsity.

Open the sample example Jupyter Notebook: Chapter 004 Example 004A.ipynb in your Jupyter software.

Part A loads all the libraries you need for this investigation.

The *mpl_toolkits.mplot3d* supplies you with interactive 3D graphs.

Parts B and C generate the data you require for the example from the water sickness data.

I suggest you look at the data set from different angles to understand the true nature of the data distribution. You will achieve this by plotting the data plus your machine learning predictions on a 3D plane and then turning the plane to inspect the different angles or views of the data.

I suggest we look at the data on a 45-degree corner view, 90 degrees on the X-axis and 90 degrees on the Y-axis.

Your first angle of the data set and results are shown in Figure 4-10.

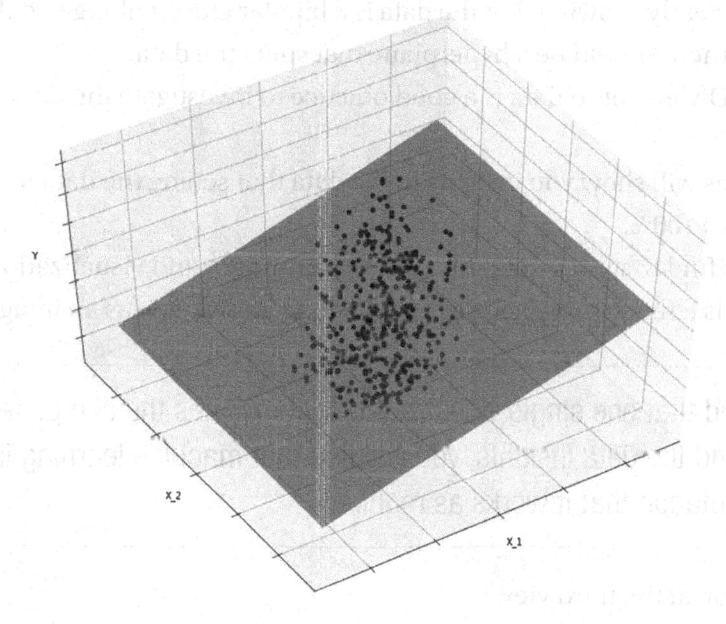

Figure 4-10. *3D Sparsity view of results*

You can agree this view is not showing you the true nature of the data distribution of the water sickness data.

So let's progress to the next view to look at this data from another angle.

The second angle of the data set and results are shown in Figure 4-11.

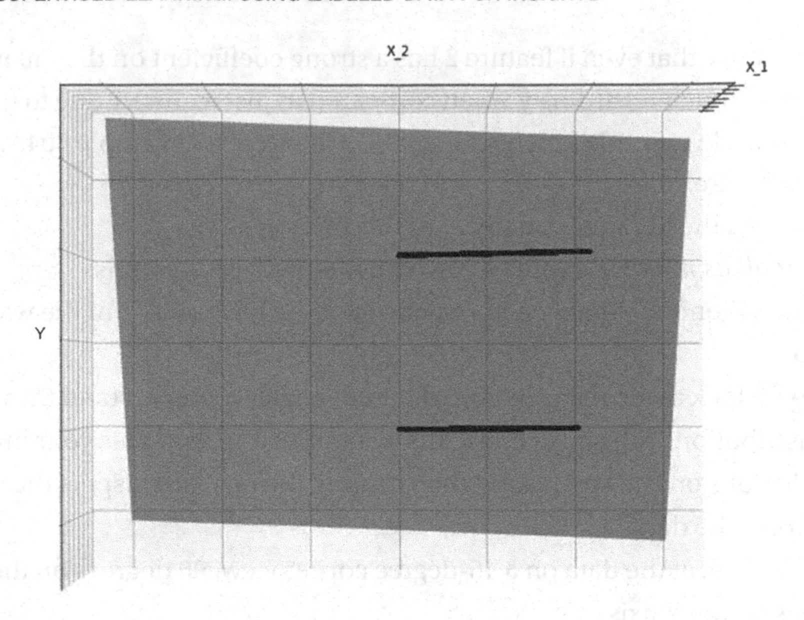

Figure 4-11. *3D sparsity - Top View on X*

This view is clearly showing that the data is a bipolar cluster along two distinct values. This means there should be a hyperplane that splits the data.

The use of 3D views onto data is a good practice to investigate the data's characteristics.

Visualizations will show you insights to the data that seeing the data in table format can never achieve 100%.

You will also find that non-data science people understand visualizations easier as the human race is a remarkable learner when seeing knowledge as an image.

Tip If you find that one single visualization that explains the complete business requirement and the data insights, you will find that machine learning indirectly will gain a reputation that it works as required.

Let's now look at the third view.

The third and final view of the data set is shown in Figure 4-12.

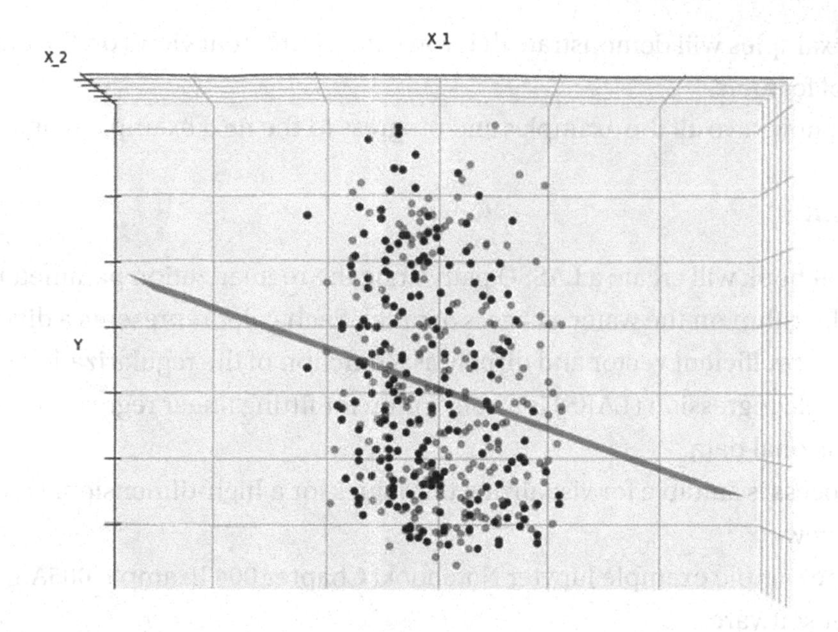

Figure 4-12. *3D sparsity – Top View on Y*

Now we have the complete insight. Sadly, this view clearly shows that the data is not 100% a clear bipolar cluster. There are several data points that do not split along the hyperplane.

Warning The issue is that the Linear Hyperplane is not the best model to split this data set.

You have completed the Jupyter Notebook and can save it.

You can now view data in multidimensions to investigate the true view of the data from different angles in the 3D space to discover insights that you would normally not observe by only investigating along one view angle.

If you want to investigate a few other feature combinations, do the following:

- Open the sample example Jupyter Notebook: Chapter 004 Example 004B.ipynb.
- Open the sample example Jupyter Notebook: Chapter 004 Example 004C.ipynb.
- Open the sample example Jupyter Notebook: Chapter 004 Example 004D.ipynb.

These examples will demonstrate that you can get different views on the data set by selections of features.

You can now save all the examples and progress to the next example type.

LASSO Path

The next notebook will create a LASSO path along the regularization parameter using the LARS algorithm on the water sickness data set. Each color represents a diverse feature of the coefficient vector and display as a function of the regularization parameter.

Least-angle regression (LARS) is an algorithm for fitting linear regression models to high-dimensional data.

This process is suitable for visualizing the values for a high-dimensional data set, in one single view.

Open the sample example Jupyter Notebook: Chapter 004 Example 005A.ipynb in your Jupyter software.

Run the complete notebook.

You will get following the result shown in Figure 4-13.

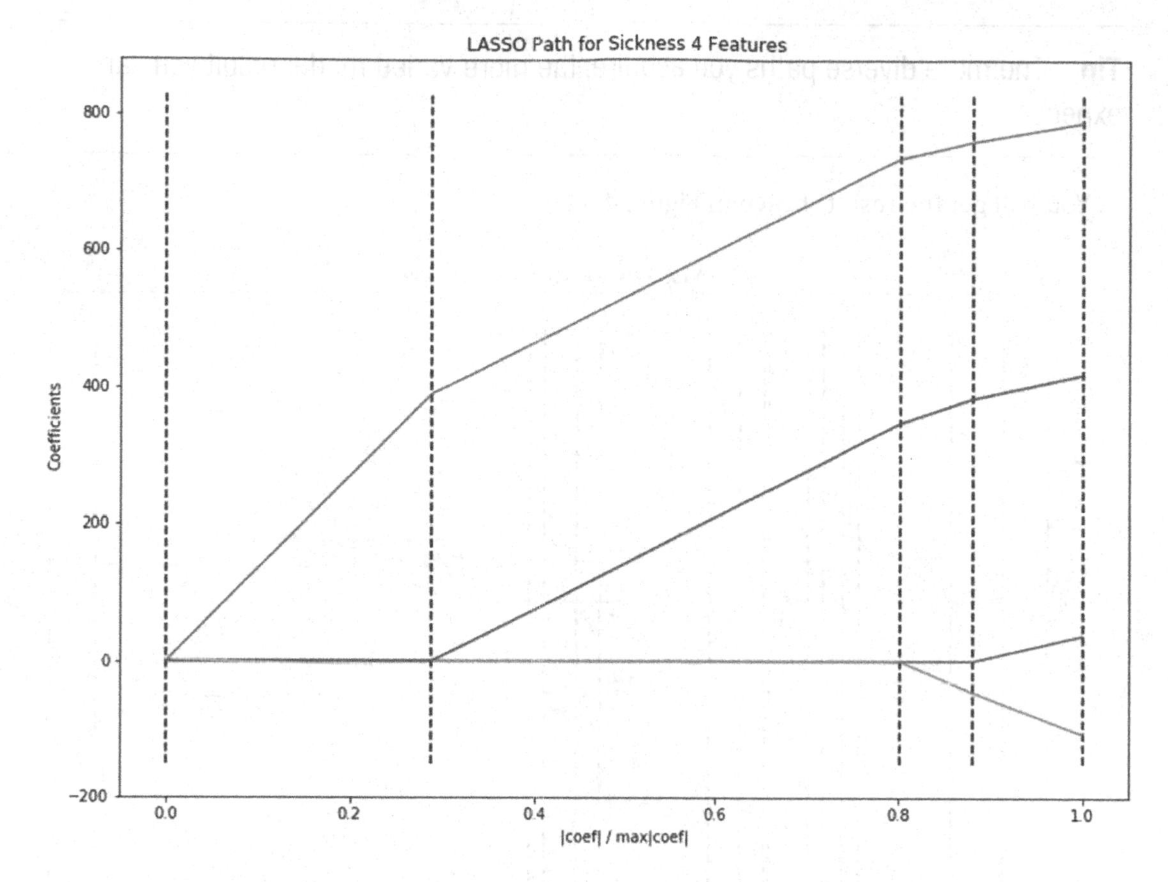

***Figure 4-13.** LASSO Path for four Features*

This result only shows four features, but it does clearly show how diverse the paths are for each feature.

Tip The diverse paths confirm what we have seen before: that this data set is complex and will require a nonlinear hyperplane to create the best boundaries.

Close the Jupyter Notebook now.

Open the sample example Jupyter Notebook: Chapter 004 Example 005B.ipynb in your Jupyter software.

Run the complete notebook.

Let's investigate what happens when we add more features and therefore more complexity.

Tip The more diverse paths you acquire, the more varied model result you can expect.

You will get the result shown in Figure 4-14.

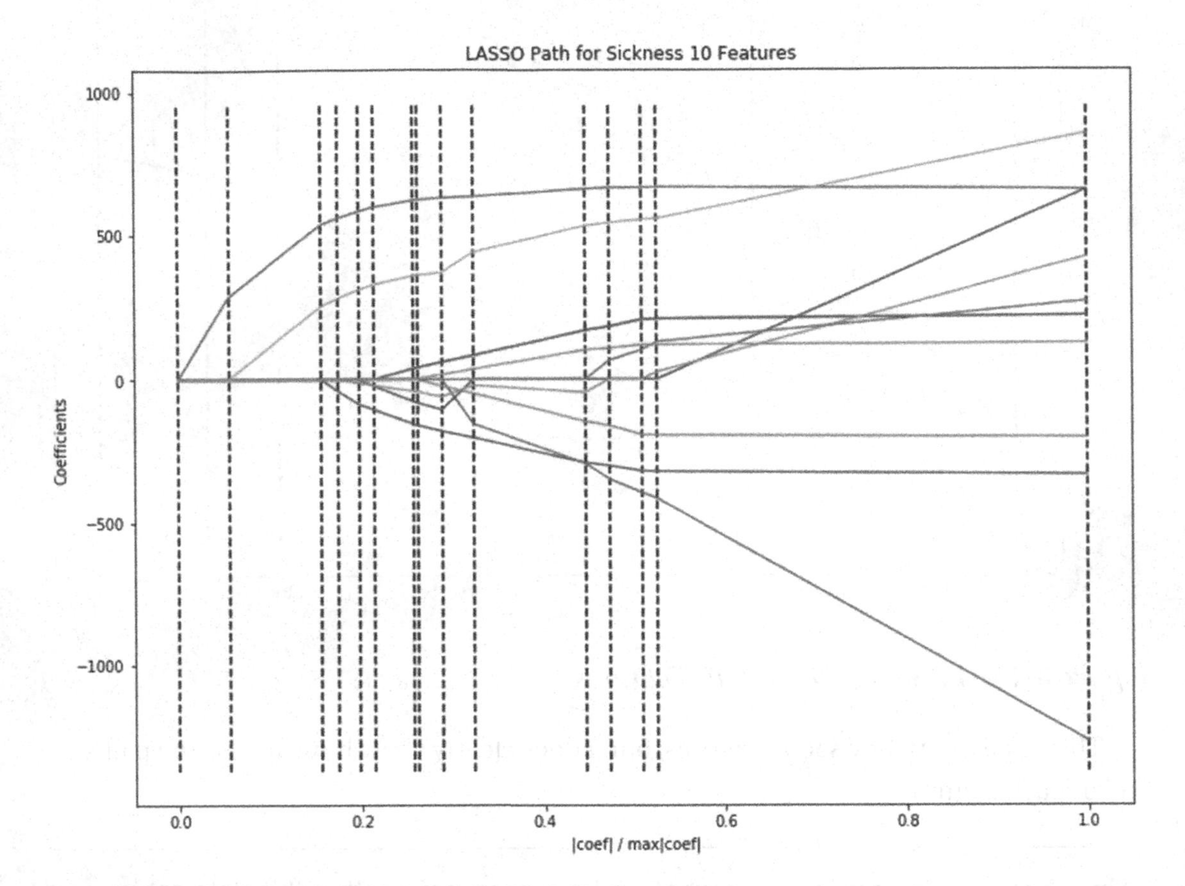

***Figure 4-14.** LASSO Path for ten Features*

This path analysis shows how there are a few features that are clearly outliers and a few that are clustered together.

I suggest you look at Part B's feature selection and experiment with which pairs work well together.

Once you have a set, you have an insight on the possible features that will work.

You are finished with this notebook and can now save it.

The LASSO path visualization will guide you to investigate the path of your different features. I suggest you map a LASSO path for every data set you want to investigate.

You are now finished with linear regression. Well done; this is good progress.You should now be able to perform linear regression on data sets.

I will now move onto another type of regression model.

Logistic Regression

Logistic regression is the appropriate regression analysis to conduct when the dependent variable is dichotomous (binary). Regression analysis is generally using the logistic regression as a predictive analysis.

Logistic regression labels data and explains the relationship between one dependent binary variable and one or more nominal, ordinal, interval, or ratio-level independent variables.

Open the sample example Jupyter Notebook: Chapter 004 Example 006A.ipynb in your Jupyter software.

Parts A, B, and C will enable you to load the data.

Part D is the new Logistic regression model.

```
logreg = linear_model.LogisticRegression(
C=1e6,
solver='lbfgs',
multi_class='ovr',
n_jobs=8,
verbose=True,
max_iter=1968,
dual=False,
class_weight='balanced',
warm_start=False
)
```

This core model's hyperparameters mean the following:

> C=1000000 -> Inverse of regularization strength and smaller values specify stronger regularization.

So this is a weaker regularization.

> solver='lbfgs' -> the options are 'newton-cg', 'lbfgs', 'liblinear', 'sag', 'saga' with default = 'liblinear'

Tip Algorithm usage for the different optimization problems are the following:

For small data sets, 'liblinear' is a decent choice, whereas 'sag' and 'saga' are faster for large data sets.

For multiclass problems, only 'newton-cg', 'sag', 'saga' and 'lbfgs' handle multinomial loss; 'liblinear' is limited to one-versus-rest schemes.

'newton-cg', 'lbfgs' and 'sag' only handle L2 penalty, whereas 'liblinear' and 'saga' handle an L1 penalty.

---

multi_class='ovr,' that is, a binary problem is activated.

The options are 'ovr,' 'multinomial,' 'auto'

For 'ovr,' the model performs a binary problem to fit for each label.

For 'multinomial,' the loss minimized is the multinomial loss fit across the whole probability distribution.

These hyperparameters are the controls you use to influence the model's processing and training methods.

Logistic regression uses that linear regression concept but performs a classification problem with a logistic regression that consists of transformation cooperation between an exponential function of x and a ratio.

This enables the hyperplane to create boundaries that a simple, flat linear plane cannot achieve.

Tip The logistic regression can also act like a linear regression with specific parameters. So, the logistics regression is more generally capable of finding the better boundaries in a data set rather than a linear regression.

Your results are shown in Figure 4-15.

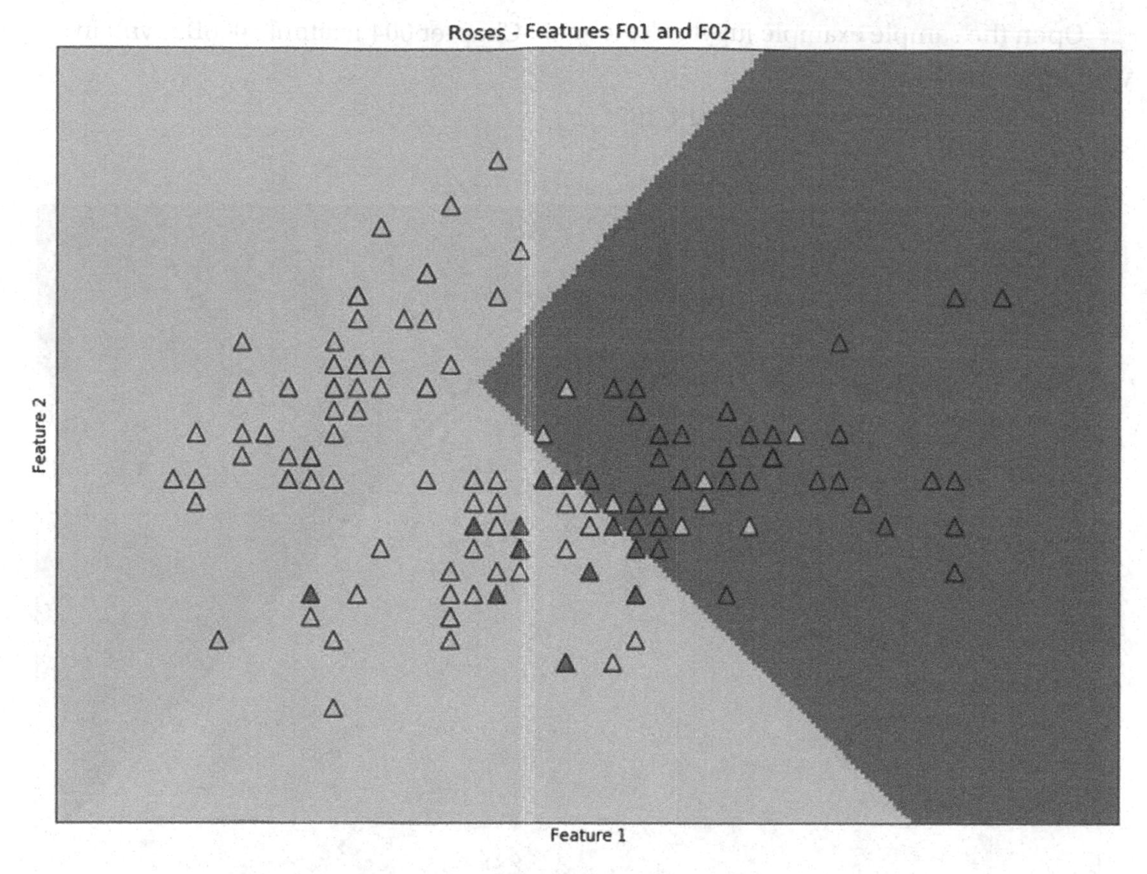

Figure 4-15. *Logistic regression - F01 and F02*

The two features have clear boundaries that identify the three different roses, but there are some issues between the yellow and orange roses as there are several data points that are close to the boundary.

Can you identify that the model can split the blue roses from the rest of the roses with ease? It is, however, less effective between the yellow and red roses. This is common in real-world applications of models.

Tip The features are not always a clear-cut boundary of values, as there are some values that belong to both classes. You should find these overlapping boundaries as soon as possible to prevent unwanted outcomes.

You're now finished with this notebook. Please save the notebook.

Now we change the features to features 2 and 3.

Open the sample example Jupyter Notebook: Chapter 004 Example 006B.ipynb in your Jupyter software.

Your results are shown in Figure 4-16.

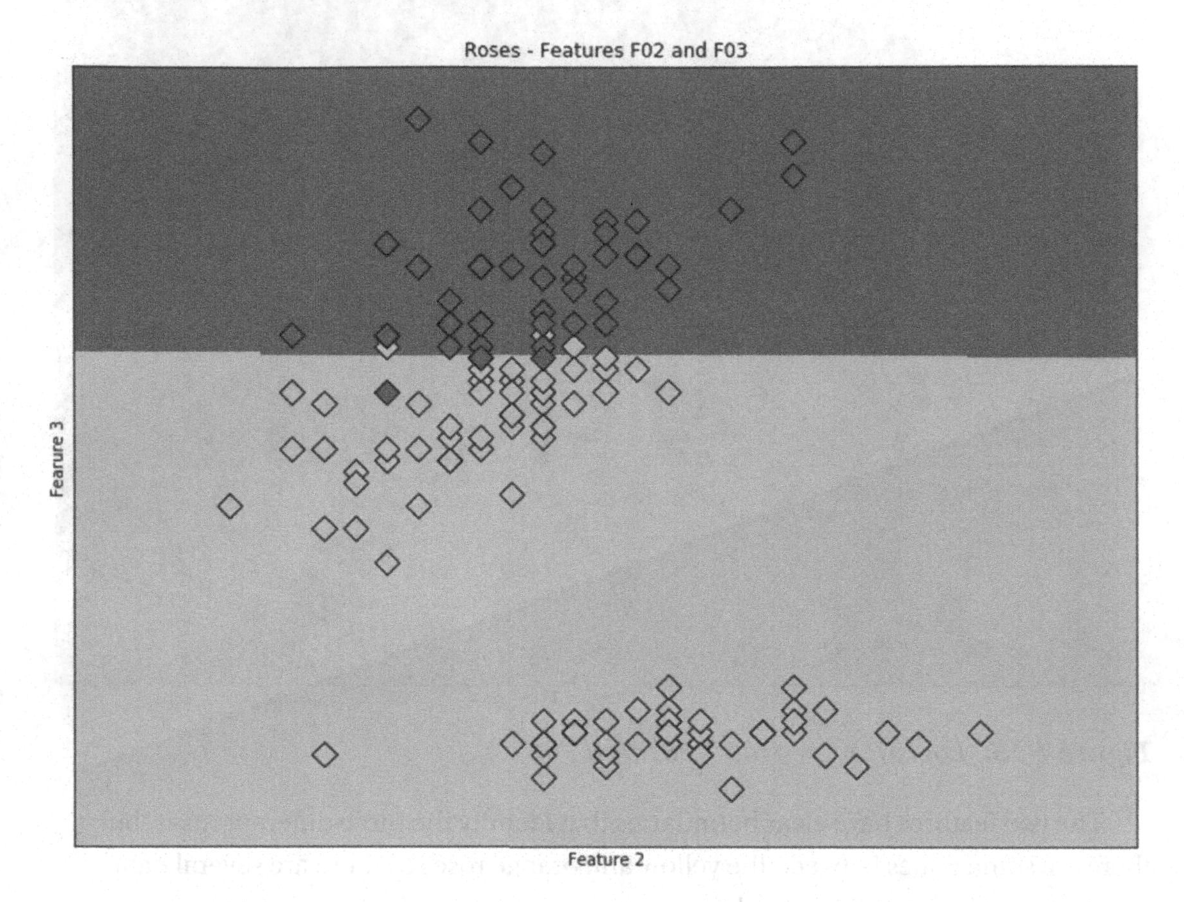

Figure 4-16. *Logistic regression - F02 and F03*

The two features have clear boundaries that identify the three different roses, but there are some issues between the yellow and red roses as there are several data points that are close to the boundary.

You're now finished with this notebook.

Please save the notebook as you can later practice with it.

Now we change the features to features 1 and 3.

Open the sample example Jupyter Notebook: Chapter 004 Example 006C.ipynb in your Jupyter software.

Your results are shown in Figure 4-17.

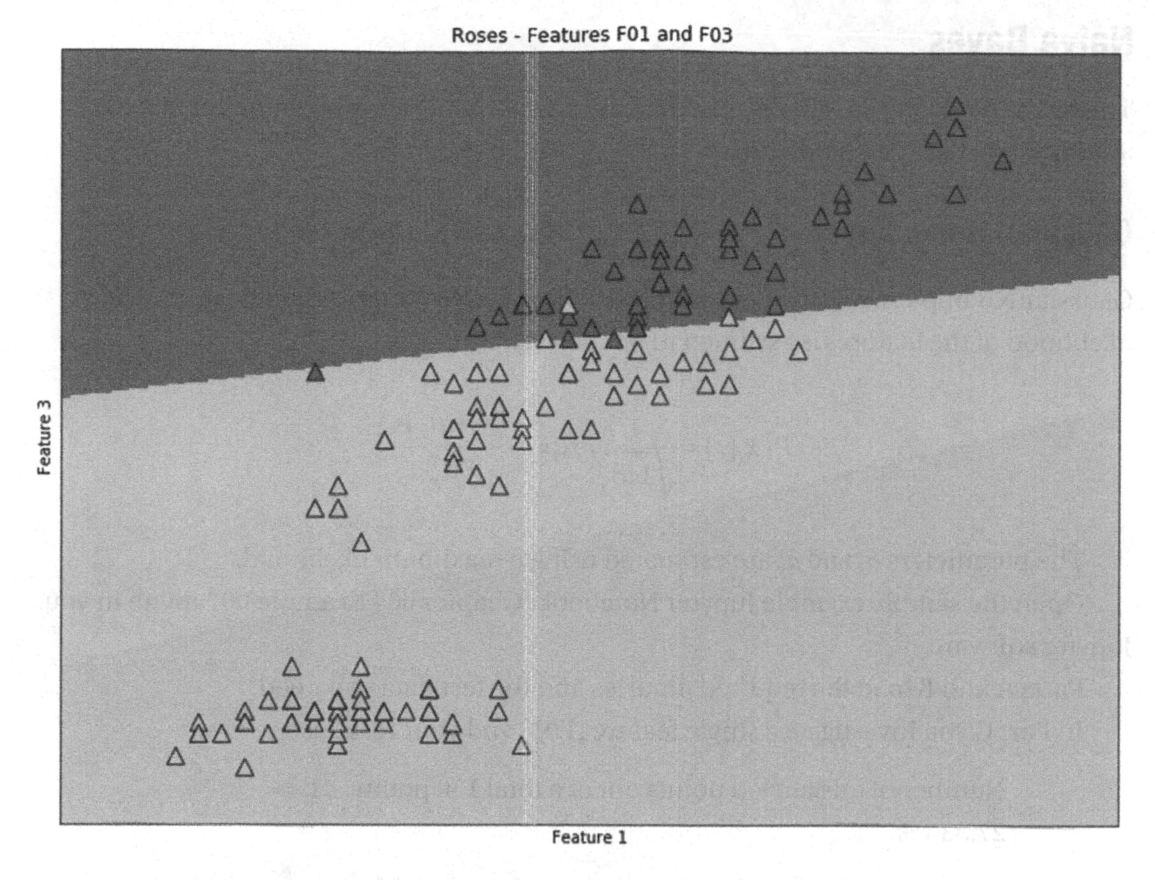

Figure 4-17. Logistic regression - F01 and F03

The two features have clear boundaries that identify the three different roses, but there are some issues between the yellow and orange roses as there are several data points that are close to the boundary.

You're now finished with this notebook.

Please save the notebook.

This investigation shows that the logistic regression improves the boundaries but not for all data points.

Tip Investigating different models will assist you in picking an appropriate ML model and feature discovery process to achieve success in IML.

Well done; you have now mastered boundaries discovery using logistic regression.

I will now introduce you to a new part of the model classification set of tools you can use to assist your work in machine learning.

Naive Bayes

The Naive Bayesian classifier is based on the Bayes' theorem with the independence assumptions between predictors.

Gaussian Naive Bayes

GaussianNB implements the Gaussian Naive Bayes algorithm for classification. The likelihood of the features is assumed to be Gaussian:

$$P(x_i|y) = \frac{1}{\sqrt{2\pi\sigma_y^2}} \exp\left(-\frac{(x_i - \mu_y)^2}{2\sigma_y^2}\right)$$

The parameters σ_y and μ_y are estimated using a maximum likelihood.

Open the sample example Jupyter Notebook: Chapter 004 Example 007.ipynb in your Jupyter software

Parts A and B load the required libraries and the test data you need.

In Part C you investigate a single feature (F01) and your result is:

> Number of mislabeled points out of a total 150 points: 41 => 27.333 %
>
> That means that this feature has over a 1 in 4 risk of been erroneously labeled.
>
> Let's investigate further by looking at other features.

Part D activates the model:

```
gnb = GaussianNB()
```

You investigate a single feature (F02) and your result is:

> Number of mislabeled points out of a total 150 points: 67 => 44.667 %
>
> Wow, this feature has nearly a 1 in 2 risk of been erroneously labeled.

In Part E you investigate a single feature (F03) and your result is:

> Number of mislabeled points out of a total 150 points: 7 => 4.667 %

That looks better; this feature has a nearly 5 in 100 risk of been erroneously labeled.

In Part F you investigate a single feature (F04) and your result is:

Number of mislabeled points out of a total 150 points: 6 => 4.000 %

That looks even better; this feature has a 4 in 100 risk of been erroneously labeled.

So let's investigate if we use them all if it changes the risk?

In Part G you investigate a single feature (F01, F02, F03. and F04) and your result is:

Number of mislabeled points out of a total 150 points: 6 => 4.000 %

That looks similar; these four features together have a 4 in 100 risk of been erroneously labeled.

I want to suggest that feature F04 is driving the core classification of this model.

You can save the notebook.

You have now discovered that not all features are driving the Gaussian Naive Bayes algorithm in the same way.

You also now understand how to use a Gaussian Naive Bayes algorithm to perform machine learning.

Multinomial Naive Bayes

MultinomialNB implements the naive Bayes algorithm for multinomial distributed data, and it is one of the two classic naive Bayes variants used in text classification (where the data are typically represented as word vector counts, although tf-idf vectors are also known to work well in practice).

The distribution is parametrized by vectors $\theta_y = (\theta_{y1}, \ldots, \theta_{yn})$ for each class y, where n is the number of features (in text classification, the size of the vocabulary), and θ_{yi} is the probability $P(x_i | y)$ of feature i appearing in a sample belonging to class y.

The parameters θ_y are estimated by a smoothed version of maximum likelihood, that is, relative frequency counting:

$$\hat{\theta}_{yi} = \frac{N_{yi} + \alpha}{N_y + \alpha n}$$

where $N_{yi} = \sum_{x \in T} x_i$ is the number of times feature i appears in a sample of class y in the training set T, and $N_y = \sum_{i=1}^{|T|} N_{yi}$ is the total count of all features for class y.

The smoothing priors $\alpha \geq 0$ account for features not present in the learning samples and prevents zero probabilities in further computations. Setting $\alpha = 1$ is called Laplace smoothing, while $\alpha < 1$ is called Lidstone smoothing.

Open the sample example Jupyter Notebook: Chapter 004 Example 008.ipynb in your Jupyter software.

Execute the complete notebook.

The core part is C, and the model is activated by: mnb = MultinomialNB()

You should get a result that is as follows:

> Number of mislabeled Rose Feature F01, F02, F03 and F04 points
> out of a total 150 points: 7 => 4.667 %

This demonstrates that the change in model type from Gaussian Naive Bayes to Multinomial Naive Bayes for this specific data set made it worse from 4% to 4.7%.

Tip Try many other machine learning (ML) models for each of your data sets!

Understand the impact of the ML model and the data combination for each of your models.

Bernoulli Naive Bayes

BernoulliNB applies the naive Bayes training and classification algorithms for data that is distributed according to multivariate Bernoulli distributions; that is, there may be multiple features but each one is assumed to be a binary value (Bernoulli, Boolean) variable. This class requires samples to be characterized as binary-valued feature vectors; if offered any extra kind of data, a BernoulliNB instance possibly will binarize its input (depending on the binarize parameter).

The decision rule for Bernoulli naive Bayes is based on:

$$P(x_i|y) = P(i|y)x_i + (1 - P(i|y))(1 - x_i)$$

This varies from the multinomial NB's rule in that it explicitly penalizes the non-occurrence of a feature i that is an indicator for class y, where the multinomial variant would simply ignore a non-occurring feature. In the case of text classification, word occurrence vectors (rather than word count vectors) might be used to train and use this classifier.

BernoulliNB might perform better on some data sets, especially those with shorter documents. It is advisable to evaluate both models, if time permits.

Open the sample example Jupyter Notebook: Chapter 004 Example 009A.ipynb in your Jupyter software.

Run the complete notebook and look for the result.You will get following result:

```
Number of mislabeled Rose Feature F01, F02, F03 and F04 points out of a
total 600 points : 400 =>  66.667 %
```

To some degree the process went erroneously. The issue is the data!!!

Remember the algorithm needs to be binary valued.

Warning The algorithm assumes that data is correct. It does not report the non-binary-valued data.

Let's test it with another binary-valued feature set for the same outcomes or target.

Open the sample example Jupyter Notebook: Chapter 004 Example 009B.ipynb in your Jupyter software.

Run the complete notebook and look for the result.You should get three identical results:

```
GaussianNB: Number of mislabeled Rose Feature F01, F02, F03 and F04 points
out of a total 600 points : 53 => 8.833 %
MultinomialNB: Number of mislabeled Rose Feature F01, F02, F03 and F04
points out of a total 600 points : 35 =>  5.833 %
BernoulliNB: Number of mislabeled Rose Feature F01, F02, F03 and F04 points
out of a total 600 points : 200 =>  33.333 %
```

The model is not great, but it is at least more consistent across the three types of models.

The lesson here is that a change in the data's type from double to binary may not always get picked up by the model calculations but can still cause issues.

Tip Always perform an assessment of the data types as part of your pre-processing during the data engineering of your features.

Linear and Quadratic Discriminant Analysis

Linear Discriminant Analysis

This is a classifier with a linear decision boundary, generated by fitting class-conditional densities to the data and using Bayes' rule.

The model fits a Gaussian density to each class, if all classes share the same covariance matrix.

The fitted model is used to reduce the dimensionality of the input by projecting it to the most discriminative directions.

Open the sample example Jupyter Notebook: Chapter 004 Example 010A.ipynb in your Jupyter software.

The data loaded is highly specific (See Part B) format.

The result is:

```
[[-1. -1.]
 [-2. -1.]
 [-3. -2.]
 [ 1. -1.]
 [ 2.  1.]
 [ 3.  2.]]
```

Scalers

The data loaded can be transformed using a range of data transformers.

These scalers enable the machine learning to handle characteristics of the data that will impact the machine learning capability.

I will now take you through a series of transformers to introduce them to you and explain how and when you use them in our data engineering.

Standard Scaler

In Part C/1 - Transform Data - Standard Scaler

The Standard Scaler transformer creates features by removing the mean and scaling to unit variance from the data set.

The example shows how to use:

```
transformer = StandardScaler(copy=True, with_mean=True, with_std=True).
fit(X_raw)
```

The scaler converts:

```
Raw Data:
[[-1. -1.]
 [-2. -1.]
 [-3. -2.]
 [ 1.  1.]
 [ 2.  1.]
 [ 3.  2.]]
Transform Data:
[[-0.46291005 -0.70710678]
 [-0.9258201  -0.70710678]
 [-1.38873015 -1.41421356]
 [ 0.46291005  0.70710678]
 [ 0.9258201   0.70710678]
 [ 1.38873015  1.41421356]]
```

Robust Scaler

In Part C/2 - Transform Data - Robust Scaler:

This scaler removes the median and scales the data according to the quantile range of the data set.

The example shows how to use:

```
transformer = RobustScaler(copy=True, with_centering=True, with_
scaling=True, quantile_range=(25.0, 75.0)).fit(X_raw)
```

The scaler converts:

```
Raw Data:
[[-1. -1.]
 [-2. -1.]
 [-3. -2.]
 [ 1.  1.]
 [ 2.  1.]
 [ 3.  2.]]
```

```
Transform Data:
[[-0.28571429 -0.5        ]
 [-0.57142857 -0.5        ]
 [-0.85714286 -1.         ]
 [ 0.28571429  0.5        ]
 [ 0.57142857  0.5        ]
 [ 0.85714286  1.         ]]
```

Tip This scaler features uses statistics that are robust to outliers.

Binarize

In Part C/3 - Transform Data – Binarize:

This transformer performs Boolean thresholding of the data set.

The example shows how to use:

```
X3a=binarize(X_raw, copy=True,  threshold=1.75)
```

The scaler converts:

```
Raw Data:
[[-1. -1.]
 [-2. -1.]
 [-3. -2.]
 [ 1.  1.]
 [ 2.  1.]
 [ 3.  2.]]
Transform Data:
[[0. 0.]
 [0. 0.]
 [0. 0.]
 [0. 0.]
 [1. 0.]
 [1. 1.]]
```

Normalize

In Part C/4 - Transform Data – Normalize:

This scale input vectors individually to a unit norm of a data set.

The example shows how to use:

```
X4a=normalize(X_raw, copy=True, norm='l1', axis=1, return_norm=False)
```

The scaler converts:

```
Raw Data:
[[-1. -1.]
 [-2. -1.]
 [-3. -2.]
 [ 1.  1.]
 [ 2.  1.]
 [ 3.  2.]]
Transform Data:
[[-0.5        -0.5       ]
 [-0.66666667 -0.33333333]
 [-0.6        -0.4       ]
 [ 0.5         0.5       ]
 [ 0.66666667  0.33333333]
 [ 0.6         0.4       ]]
```

Quantile Transformer

Part C/5, C/6, C/7, and C/8 - Transform Data - Quantile Transformer:

Transform features using quantiles information. The process transforms the features to track a uniform or a normal distribution.

The example shows how to use:

```
transformer = QuantileTransformer(output_distribution='normal',n_
quantiles=4, random_state=0).fit(X_raw)
```

The scaler converts:

```
Raw Data:
[[-1. -1.]
 [-2. -1.]
 [-3. -2.]
 [ 1.  1.]
 [ 2.  1.]
 [ 3.  2.]]
Transform Data:
[[-0.70710678 -0.70710678]
 [-0.89442719 -0.4472136 ]
 [-0.83205029 -0.5547002 ]
 [ 0.70710678  0.70710678]
 [ 0.89442719  0.4472136 ]
 [ 0.83205029  0.5547002 ]]
```

Now that you can perform transformation, you see that in Part D the example enables you to select one of the transformations.

For the main example, I suggest we use the data set as is by selecting `X=X0`.

Tip Try the different transformers and learn how they influence the machine learning's performance and effectiveness.

Now we can get back to the machine learning.

In Part E you run the model by using `clf = LinearDiscriminantAnalysis(solver='eigen', shrinkage='auto')` and `clf.fit(X, y)`.

Linear Discriminant Analysis is a classifier with a linear decision boundary, produced by fitting class-conditional densities to the data sets and with Bayes' rule fits a Gaussian density to each class, assuming that classes have been entirely shared by the identical covariance matrix. The fitted model can be used to reduce the dimensionality of the input by projecting it to the most discriminative directions.

You can investigate different solvers:

The Solver in use can be any of these possible values:

- 'svd': Singular value decomposition (default). This solver does not compute a covariance matrix for the data set. This solver is endorsed for data with a larger number of features.
- 'lsqr': The Least squares solution can be integrated with shrinkage to solve the problem.
- 'eigen': The Eigenvalue decomposition can be integrated with shrinkage to solve the problem.

Part F prepares you the plot for the results of your machine learning.

Your result is shown in Figure 4-18.

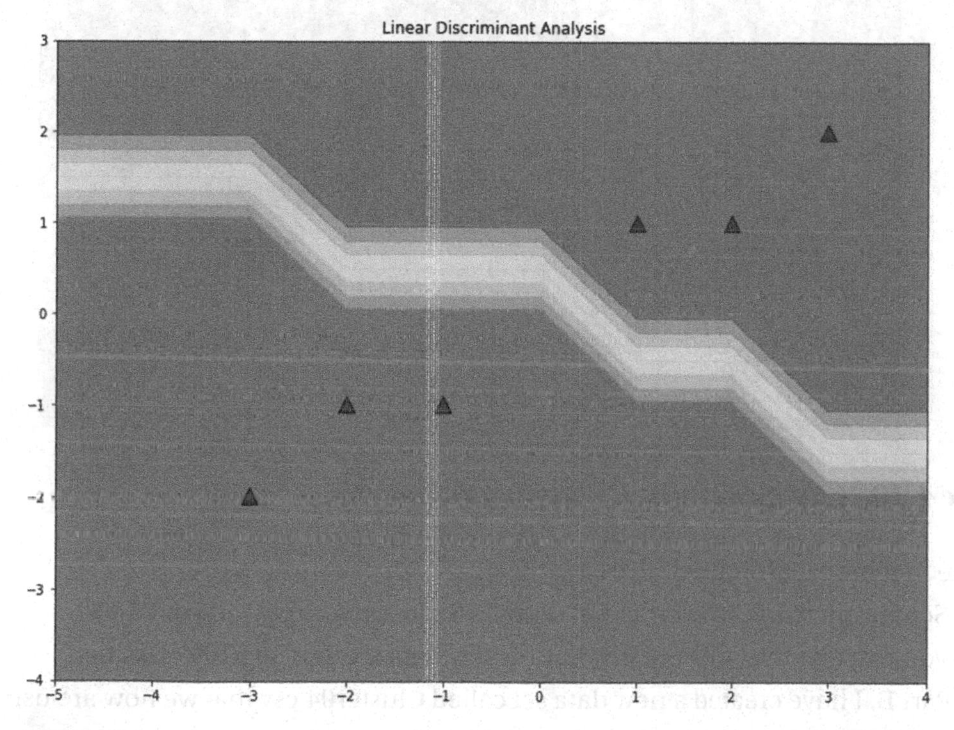

Figure 4-18. *Simple Linear Discriminant Analysis*

The model predicts a clear line between the two classes. This is a great outcome for the modeling process.

You can now close this example.

You have now successfully performed a series of transformations and a Linear Discriminant Analysis on a simple data set.

Next, we will investigate a more complex data set.

Open the sample example Jupyter Notebook: Chapter 004 Example 010B.ipynb in your Jupyter software.

Run the complete notebook.

Your result is shown in Figure 4-19.

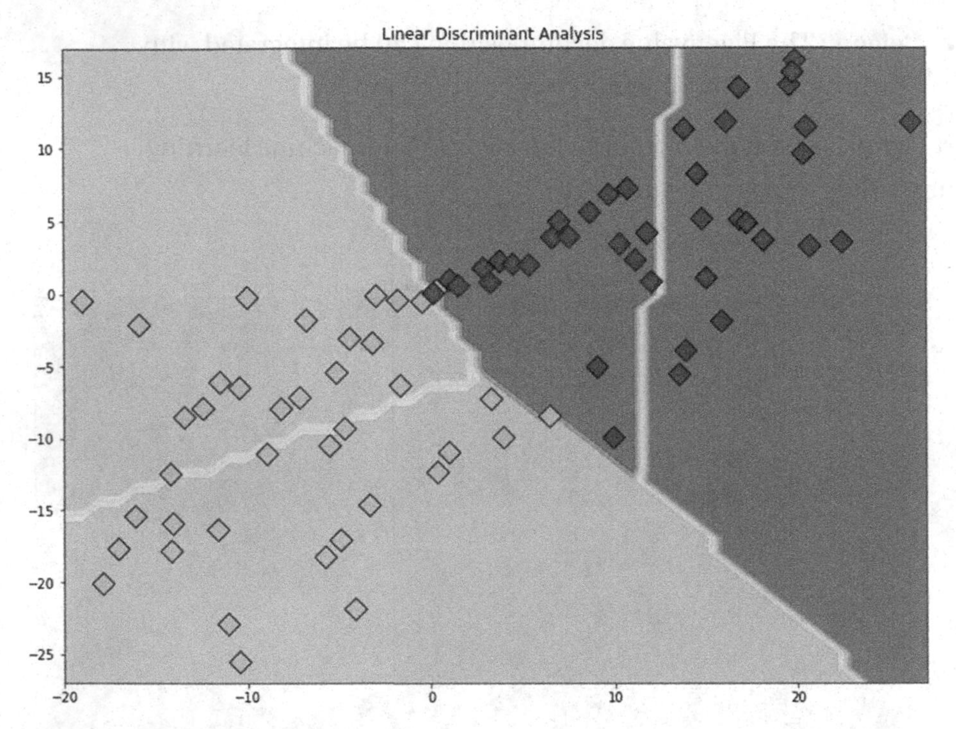

Figure 4-19. *Complex Linear Discriminant Analysis*

The machine learning achieves a score of:

F1 Score (micro): 0.7468 or 74.68% and F1 Score (macro): 0.7505 or 75.05%.

As you can observe, this means that 1 in 4 points are erroneously classified.

In Part B, I have created a new data set called Cluster04.csv that we now are using to demonstrate how the machine learning could evolve if you use it many times to predict the outcomes you need.

You can now save and close the current notebook.

Next you open the sample example Jupyter Notebook: Chapter 004 Example 010C. ipynb in your Jupyter software.

Run the notebook completely.

Your result is shown in Figure 4-20.

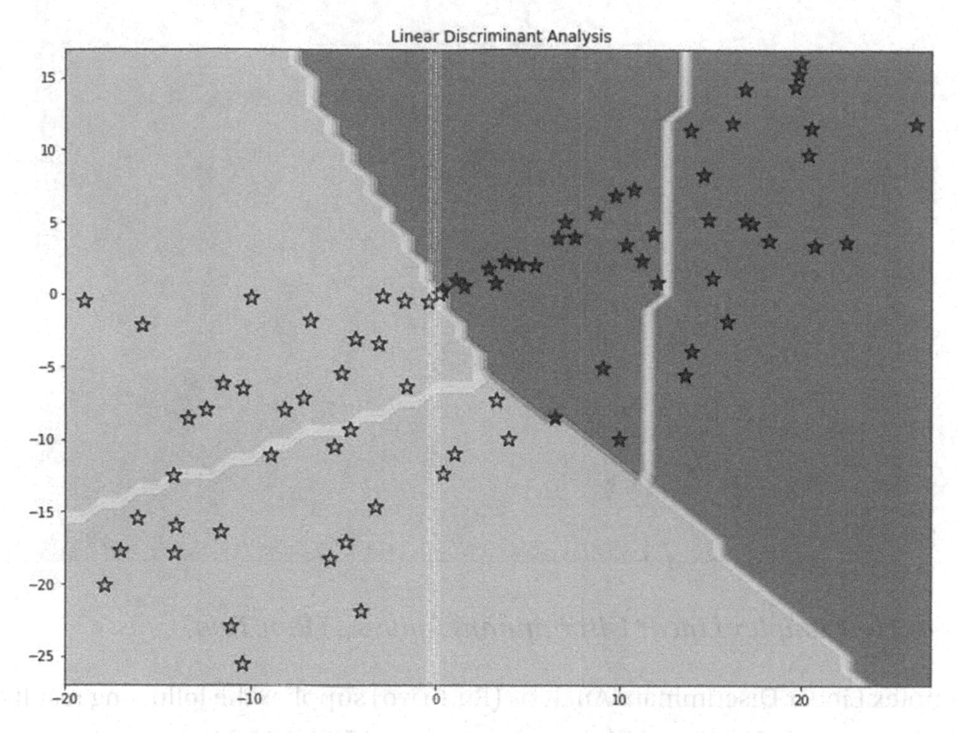

***Figure 4-20.** Complex Linear Discriminant Analysis (Run One)*

Your F1 Score (micro) is 0.87341772 and F1 Score (macro) is 0.87565352. That is an improvement due to the prediction now accepted by the model as training data.

Warning This is only a simulation. Performing the machine learning in this manner is not a common practice.

It does show what the true impact would be if the business follows the predictions by 100%. I have used this line of investigation several times to demonstrate what is possible if the customer gets a 100% take-up of the suggested predictions.

Run the notebook again.

Your result is shown in Figure 4-21.

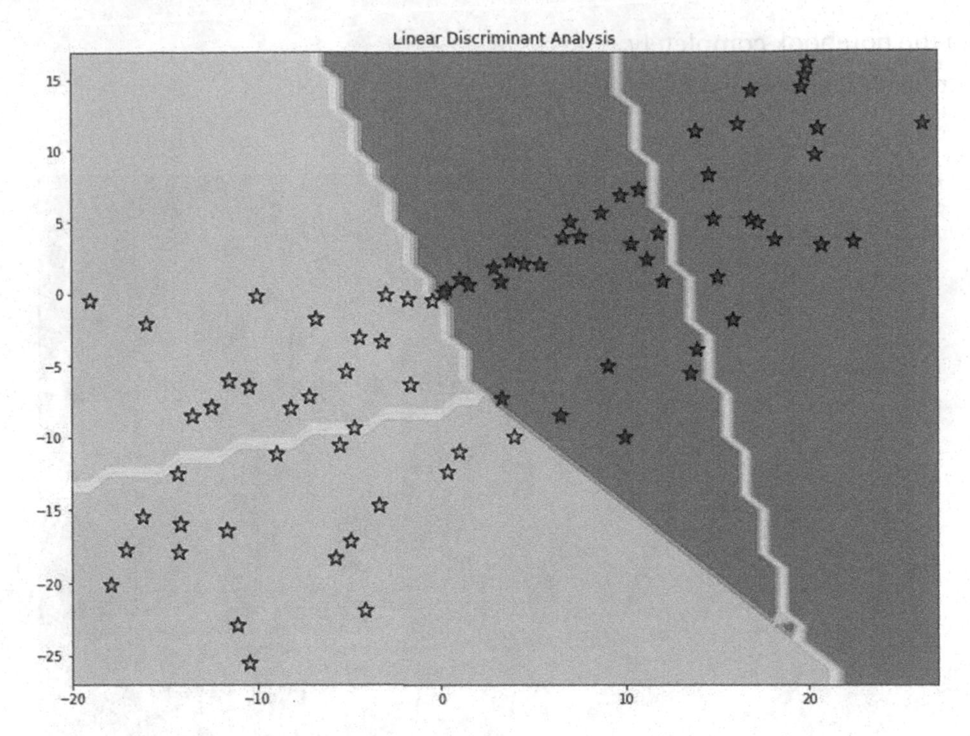

Figure 4-21. *Complex Linear Discriminant Analysis (Run Two)*

Complex Linear Discriminant Analysis (Run Two) supplies the following results: F1 Score (micro) is 0.91139241 and F1 Score (macro) is 0.91392064. More improvements!

Complex Linear Discriminant Analysis (Run Three) supplies F1 Score (micro) is 0.93670886 and F1 Score (macro) is 0.93809072.

Complex Linear Discriminant Analysis (Run Four) supplies F1 Score (micro) is 0.95727848 and F1 Score (macro) is 0.95884775.

Complex Linear Discriminant Analysis (Run Nine) supplies that after five more runs, F1 Score (micro) is 0.99040744 and F1 Score (macro) is 0.98882348.

This a highly statistical significance result; at more than 99%, this model can predict most of the data set correctly.

You can close this notebook.

You should now understand that the combination of data set, transformation, and good machine learning model results in good predictions.

Quadratic Discriminant Analysis

Open the sample example Jupyter Notebook: Chapter 004 Example 011.ipynb in your Jupyter software.

The machine learning for the Quadratic Discriminant Analysis follows the same process than the Linear one but uses a quadratic model to generate a quadratic decision boundary.

Your result is shown in Figure 4-22.

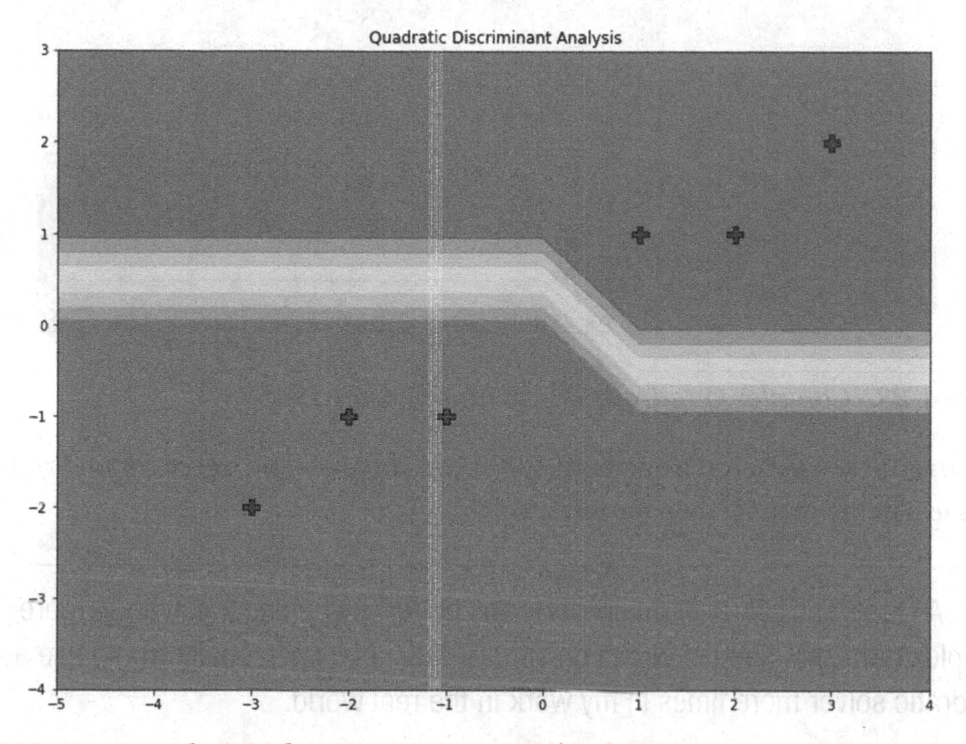

***Figure 4-22.** Simple Quadratic Discriminant Analysis*

As the data is simple in structure, the differences are not that clear between the linear and quadratic analysis for the data set.

You can close the notebook now.

Let evaluate the same analysis with a more complex data set.

Open the sample example Jupyter Notebook: Chapter 004 Example 011B.ipynb in your Jupyter software.

Run the complete notebook.

Your result is shown in Figure 4-23.

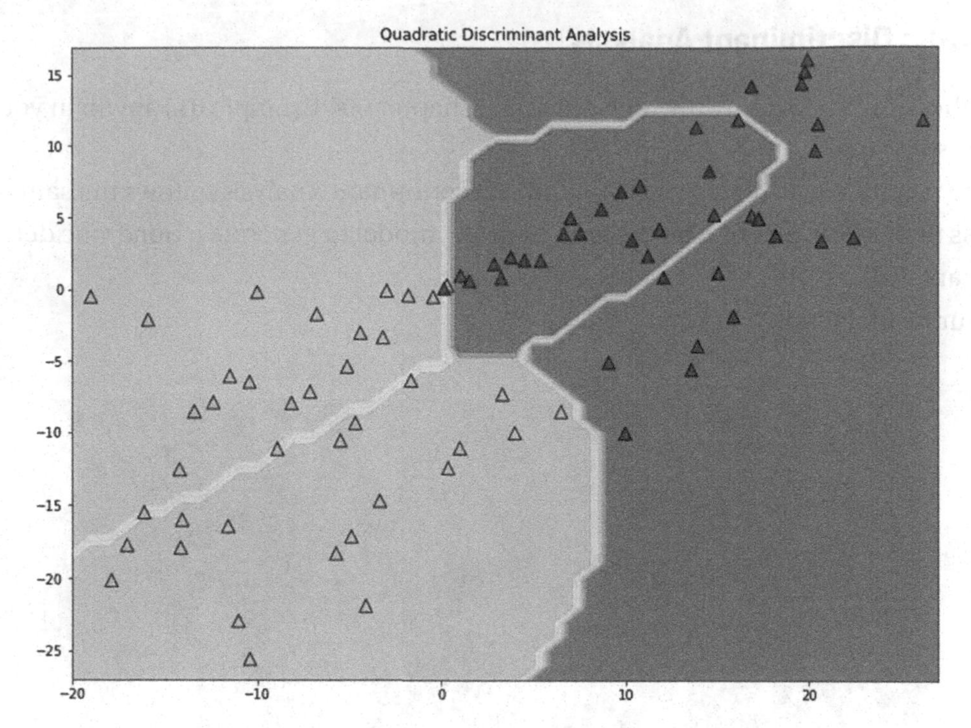

Figure 4-23. *Complex Quadratic Discriminant Analysis*

The result shows how the quadratic solver can curve around the common overlap of values in a better manner than the linear solver.

Tip A Quadratic solver normally performs better than a linear solver on more complex data sets and not worse on simple data sets. I personally would use a Quadratic solver more times in my work in the real world.

You can close the notebook now.

You have now completed both the Linear Discriminant Analysis and Quadratic Discriminant Analysis.

This example shows the difference between the two methods (Figure 4-24).

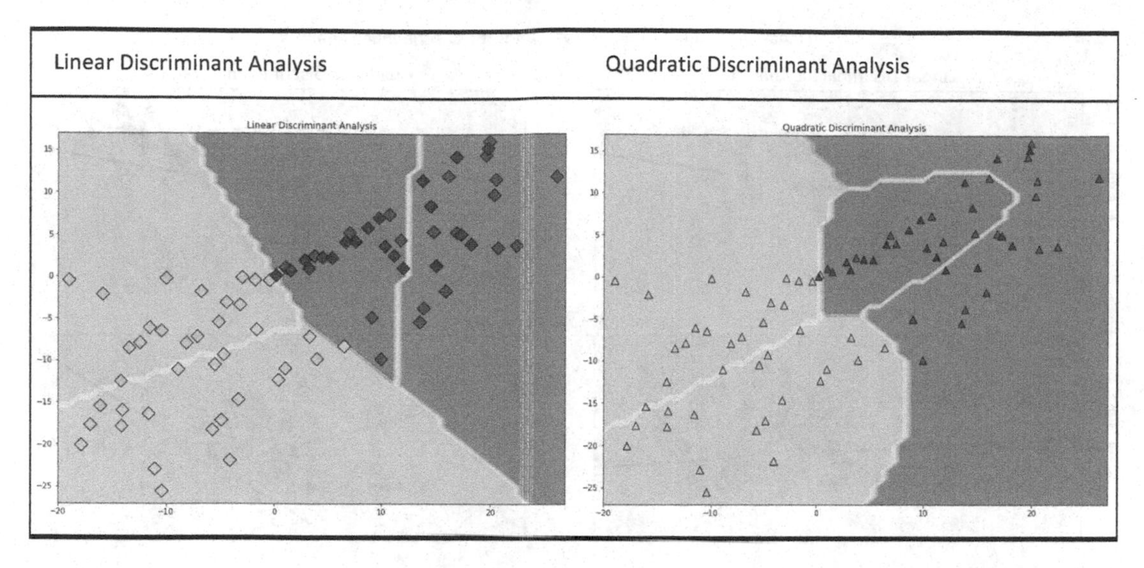

***Figure 4-24.** Compare Linear and Quadratic Discriminant Analysis Results*

Can you see how the quadratic boundary is not a straight line but a quadratic curve?

Linear and Quadratic Discriminant Analysis with Covariance Ellipsoid

This example plots the covariance ellipsoids of each class and the decision boundary learned by Linear Discriminant Analysis (LDA) and Quadratic Discriminant Analysis (QDA). The ellipsoids display the double standard deviation for each class. With LDA, the standard deviation is identical for an entire set of classes, while each class has its individual standard deviation with QDA.

Open the sample example Jupyter Notebook: Chapter 004 Example 012.ipynb in your Jupyter software.

This example demonstrates how the covariance ellipsoid clusters the data for the two Linear and Quadratic Discriminant Analysis models differ in the results they deliver using the same data sets but different models and hyperparameters.

Your result is shown in Figure 4-25.

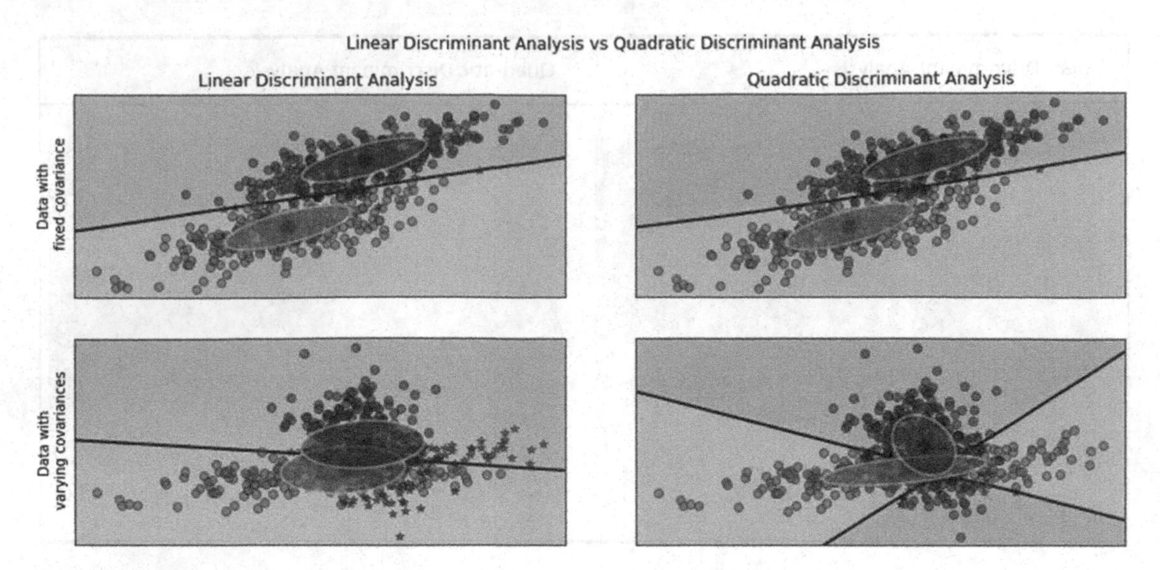

Figure 4-25. *Linear vs Quadratic Discriminant Analysis*

The comparison unmistakably demonstrates how these linear vs. quadratic discriminant analyses results indicate the quadratic solver achieving better results on this data set.

You can close the notebook now.

Graphics and Images

You should now be able to manipulate matrixes with numbers and labels with ease.

Next, we are trying a new type of data source, that is, graphics or images.

Open the sample example Jupyter Notebook: Chapter 004 Example 013.ipynb in your Jupyter software.

I will show you how to process the pictures of handwriting to determine the number the person wrote.

Run the notebook to process the machine learning. We will collect all the projections once we have discussed them.

Random 2D Projection

Reduce dimensionality through sparse random projection while giving a guarantee of similar embedding quality: more memory efficient and faster computation of the projected data.

```
rp = random_projection.SparseRandomProjection(n_components=2, random_state=321)
```

(See result: Table 4-1.)

Computing SVD Projection

This transformer performs linear dimensionality reduction by means of truncated singular value decomposition (SVD):

```
X_pca = decomposition.TruncatedSVD(n_components=2).fit_transform(X)
```

(See result: Table 4-1.)

Linear Discriminant Analysis Projection

LDA uses a classifier with a linear decision boundary and the Bayes' rule.

```
X_lda = discriminant_analysis.LinearDiscriminantAnalysis(n_components=2).
fit_transform(X2, y)
```

(See result: Table 4-1.)

We just use the same machine learning solver you used before on a graphic; that is the power of these ML solvers.

Isomap Embedding

Isomap Embedding uses nonlinear dimensionality reduction through isometric mapping of the data set.

```
X_iso = manifold.Isomap(n_neighbors, n_components=2).fit_transform(X)
```

(See result: Table 4-1.)

Locally Linear Embedding (LLE)

Locally linear embedding (LLE) pursues a lower-dimensional projection of the data, which preserves distances within local neighborhoods.

LLE Embedding

```
clf = manifold.LocallyLinearEmbedding(n_neighbors, n_components=2,
method='standard', eigen_solver='auto')
```

(See result: Table 4-1.)

Modified LLE Embedding

One of the recognized issues with LLE is the regularization problem. This happens when the number of neighbors is greater than the number of input dimensions; the matrix defining each local neighborhood is rank deficient. The modified LLE fixes this issue successfully.

```
clf = manifold.LocallyLinearEmbedding(n_neighbors, n_components=2,
method='modified', eigen_solver='auto')
```

(See result: Table 4-1.)

Hessian LLE Embedding

Hessian Eigen mapping is an alternative method of solving the same regularization problem of LLE.

```
clf = manifold.LocallyLinearEmbedding(n_neighbors, n_components=2,
method='hessian', eigen_solver='dense')
```

(See result: Table 4-1.)

Local Tangent Space Alignment (LTSA) Embedding

Local tangent space alignment (LTSA) is algorithmically similar to LLE but rather than focusing on preserving neighborhood distances as in LLE, LTSA seeks to characterize the local geometry at each neighborhood via its tangent space, and it performs a global optimization to align these local tangent spaces to learn the embedding.

```
clf = manifold.LocallyLinearEmbedding(n_neighbors, n_components=2,
method='ltsa', eigen_solver='dense')
```

(See result: Table 4-1.)

Multidimensional Scaling (MDS) Embedding

Multidimensional Scaling (MDS) pursues a low-dimensional format of the data in which the distances map indirectly to the distances in the original high-dimensional space presented to the model.

```
clf = manifold.MDS(n_components=2, n_init=1, max_iter=100)
```

(See result: Table 4-1.)

This is the last of the LLE style embedding models.

Totally Random Trees Embedding

An ensemble of totally random trees is an unsupervised transformation of a data set to a high-dimensional sparse representation.

```
hasher = ensemble.RandomTreesEmbedding(n_estimators=200, random_state=0,
max_depth=5)
pca = decomposition.TruncatedSVD(n_components=2)
```

(See result: Table 4-1.)

Spectral Embedding

Spectral embedding for nonlinear dimensionality reduction creates an affinity matrix specified by the indicated function and relates spectral decomposition to the corresponding graph using laplacian eigenmaps. The consequential transformation is set by the value of the eigenvectors for each data point.

```
embedder = manifold.SpectralEmbedding(n_components=2, random_state=0,
eigen_solver="arpack")
```

(See result: Table 4-1.)

t-distributed Stochastic Neighbor Embedding (t-SNE)

t-SNE is an implementation to visualize high-dimensional data. It adapts similarities between data points to joint probabilities and attempts to minimize the Kullback-Leibler divergence between the joint probabilities of the low-dimensional embedding and the high-dimensional data.

```
tsne = manifold.TSNE(n_components=2, init='pca', random_state=0)
```

(See result: Table 4-1.)

Tip Use PCA for dense data or TruncatedSVD for sparse data to reduce the dimensions, if the dimensions are over fifty dimensions, as this will subdue the noise and increase the speed of the computation of pair-wise distances between data samples.

The notebook produces the following output results (Table 4-1).

Table 4-1. *Graphics and Images*

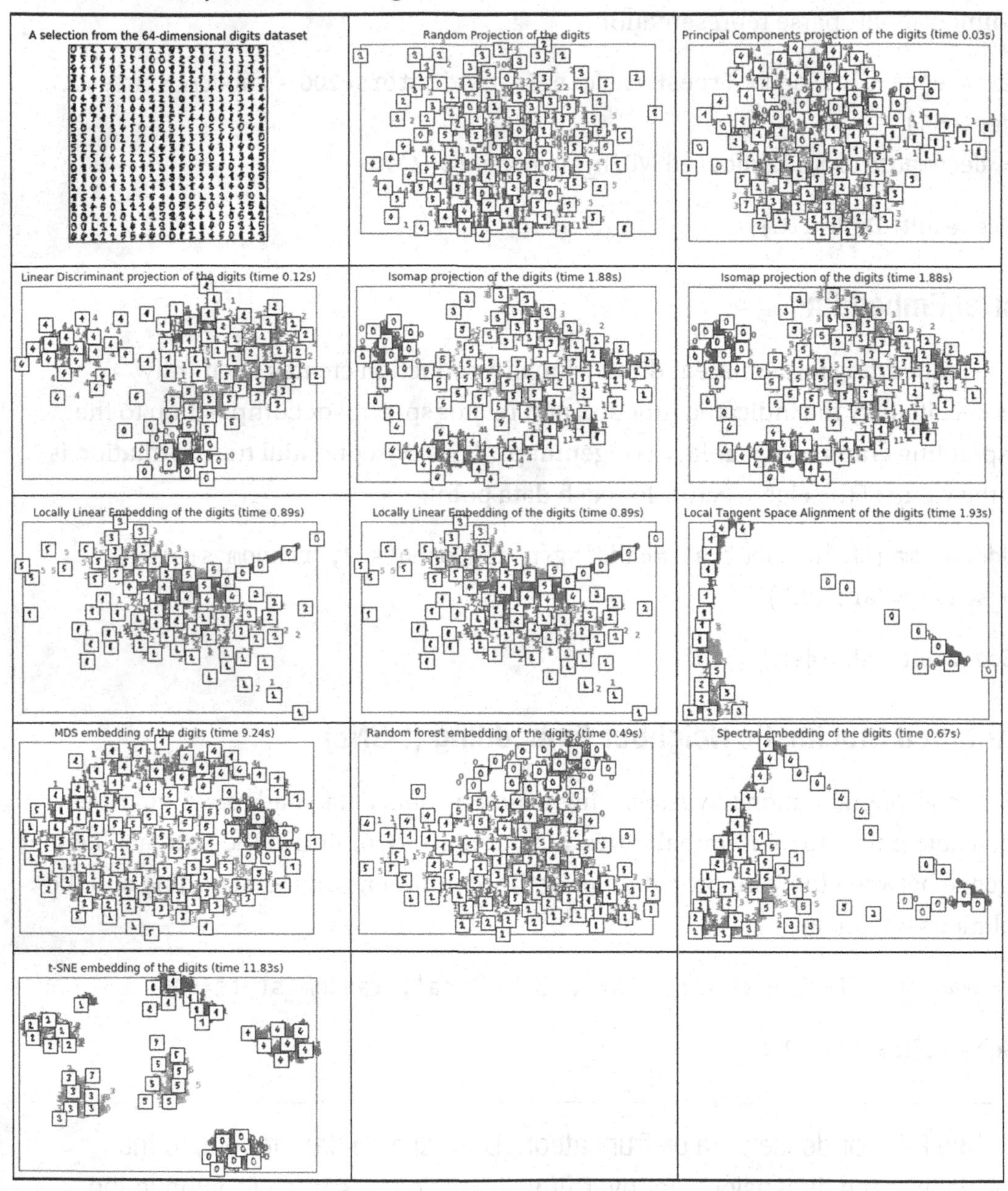

You can now accomplish machine learning on images to categorize associated images together.

Decision Boundaries

To understand the true nature of how dissimilar techniques perform on diverse problems is to visualize the prediction space by comparing plots of decision boundaries. The decision boundaries are created by looping through a sample or the complete population and plotting the prediction of every point onto a graph. This shows precisely where the boundaries are between the data categorizations.

Open the sample example Jupyter Notebook: Chapter 004 Example 014.ipynb in your Jupyter software.

I will now take you through a Classifier comparison of several classifiers in scikit-learn on synthetic data sets. The point of this example is to demonstrate the nature of decision boundaries of dissimilar classifiers. The plots show training points in solid colors and testing points are semi-transparent. The lower right displays the classification accuracy.

Your results are shown in Figure 4-26.

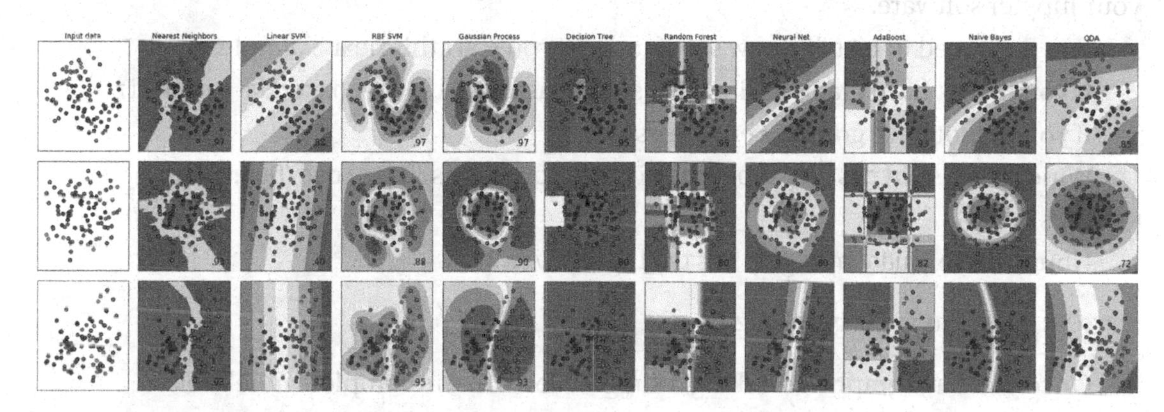

***Figure 4-26.** Decision boundaries*

You can clearly compare the diverse results of the models on the same data set.

Warning Can you see that the different algorithms will result in varied classifications for each type?

This is a major concern if you choose a mismatched algorithm.

Tip I personally always try a minimum of three different algorithms on each data training set.

Now that you have seen the decision boundaries and the impact of the algorithms on these decision boundaries, I suggest we investigate other different algorithms.

This would be a logically good point to take a break. We have covered a fair amount of the tools needed to generate machine learning models that I have seen used to train solutions in the real world.

So, I suggest ... Have a coffee or a tea and then we'll start again when you are done.

Decision Trees

Decision Trees (DTs) are a non-parametric supervised learning method used for classification and regression. The goal is to create a model that predicts the value of a target variable by learning simple decision rules inferred from the data features in the training data set.

Open the sample example Jupyter Notebook: Chapter 004 Example 015A.ipynb in your Jupyter software.

You may need to install a new set of tools as you start this next example.

If you do not have graphviz library, use `conda install -c anaconda graphviz`

If you do not have graphviz engine use:

Windows: Use `https://graphviz.gitlab.io/_pages/Download/Download_windows.html`

Ubuntu: use `sudo apt-get install graphviz`

```
CentOS: yum install graphviz
```

Once you have graphviz working, I suggest you run the complete notebook.

The machine learning model generates in Part C:

```
clf = tree.DecisionTreeClassifier(
max_features=None,
presort=True,
criterion='gini',
class_weight='balanced',
random_state=321
)
```

The important selection is the criterion='gini' as it selects the "gini" for the Gini impurity selection method and class_weight='balanced' as the 'balanced' mode usages the values of y to automatically adjust weights inversely proportional to class frequencies in the input data as n_samples / (n_classes ∗ np.bincount(y)).

The model scores a Score: 0.9067

It produces a 93-node tree.

Your notebook ecosystem generates a decision tree and displays it as an image (Figure 4-27) in Part F.

Your results are shown in Figure 4-27.

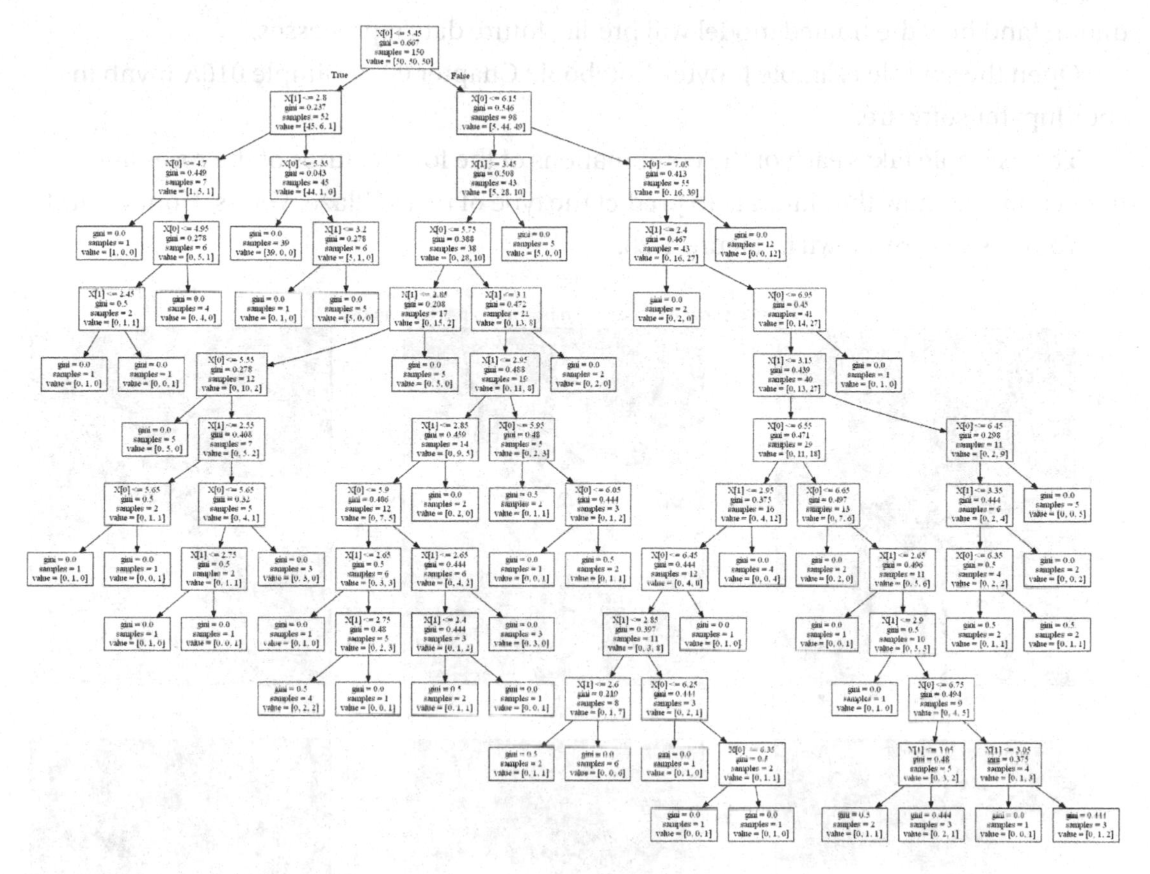

Figure 4-27. *Decision Trees*

Good progress; you have successfully generated your decision tree.

I have included an example Jupyter Notebook:

- Chapter 004 Example 015B.ipynb - Use more than one target.
- Chapter 004 Example 015C.ipynb - Use more classes.
- Chapter 004 Example 015D.ipynb - Use more features and classes.
- Chapter 004 Example 015E.ipynb - Change to criterion='entropy'.
- Chapter 004 Example 015F.ipynb - Change to ExtraTreeClassifier engine.

You should now have a comprehensive knowledge of how trees act in response to different classification hyperparameters.

Decision Tree Classifier (Roses)

I suggest we now investigate the visualization of the solutions decision surface to understand how the trained model will predict future data it processes.

Open the sample example Jupyter Notebook: Chapter 004 Example 016A.ipynb in your Jupyter software.

The example takes each of the combinations of the four features of the roses and then compares how they interact to predict the type of roses (Blaze, Venus, Honey milk).

Your results are shown in Figure 4-28.

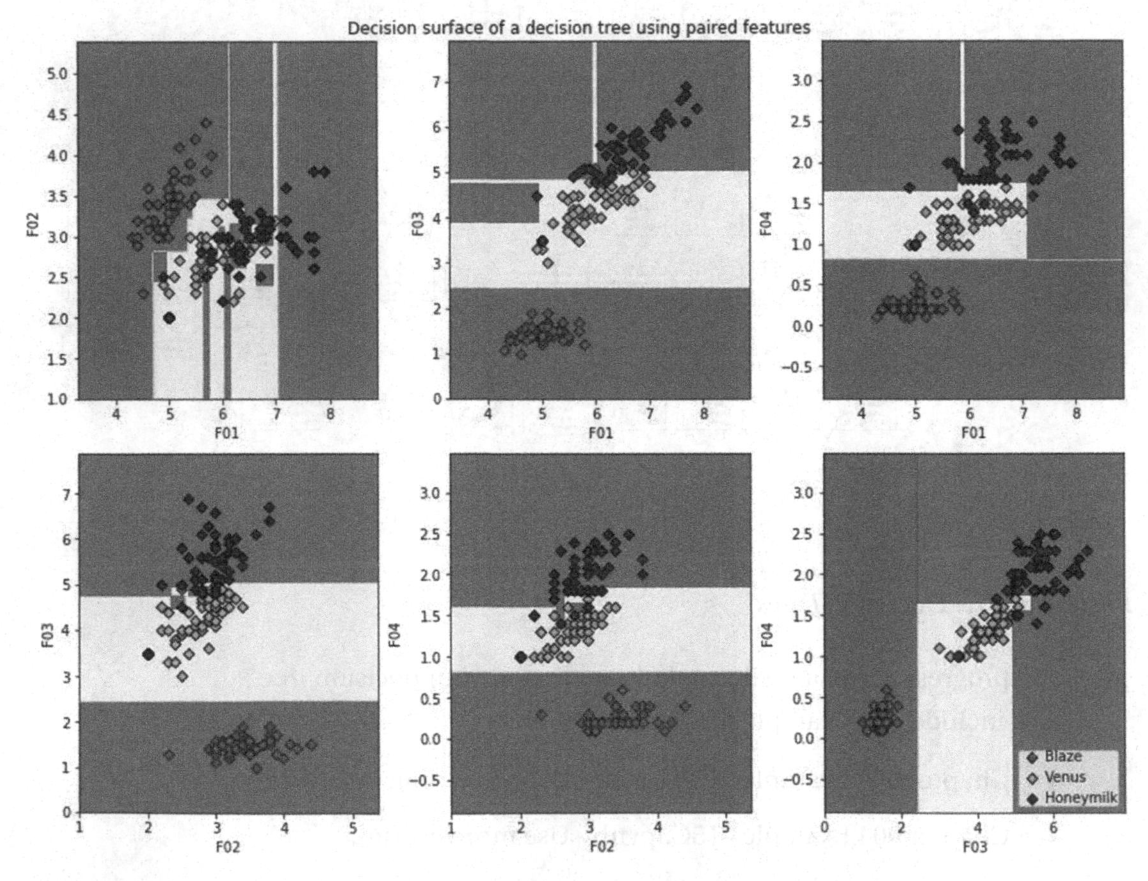

Figure 4-28. *Decision Tree (Roses)*

You can clearly observe that the combinations of features have distinct dissimilar decision surfaces.

I normally plot these results, as it shows my customers what predict values would deliver.

Tip Finding the decision surfaces is only 50% of the real-world effort; the other 50% is explaining the results to the business to explain the reasons for these insights.

Decision Tree Regressor

Decision trees apply to regression problems with similar ease by using the DecisionTreeRegressor class.

You can now investigate how the decision trees achieve a regression capability.

Open the sample example Jupyter Notebook: Chapter 004 Example 017A.ipynb in your Jupyter software.

Run the complete notebook.

Your Results are shown in Figure 4-29.

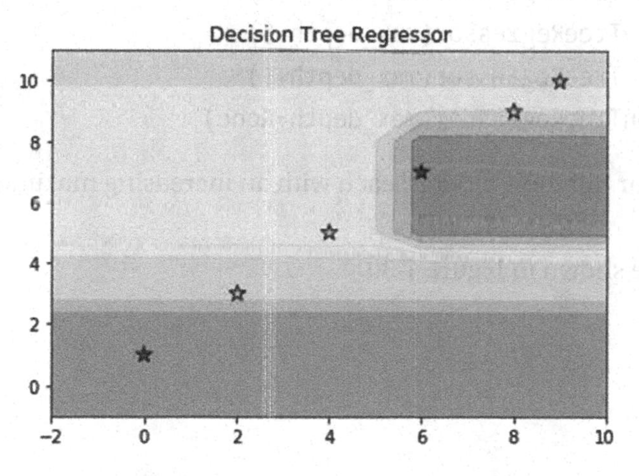

Figure 4-29. *Decision Tree Regressor*

If you look at Part C, you will see a familiar model:

```
clf = tree.DecisionTreeRegressor(max_features=None, presort=True,
criterion='mse', random_state=321, splitter='best')
```

The important hyperparameters are:

```
criterion='mse'
```

The mean squared error (MSE) is equal to variance reduction as a feature selection criterion and minimizes the L2 loss using the mean of each terminal node.

```
splitter='best'
```

The 'best' to choose the best split every time, as this pays in processing cycles for a more accurate tree model.

Decision Tree Regressor (max_depth)

The other hyperparameter you can tune on a tree is the maximum depth of the tree.

The following example will assist you in understanding the impact of the max depth on the effectiveness of the tree model.

Open the sample example Jupyter Notebook: Chapter 004 Example 018A.ipynb in your Jupyter software.

Run the complete notebook.

The core process is:

```
regr_1 = DecisionTreeRegressor(max_depth=2)
regr_2 = DecisionTreeRegressor(max_depth=5)
regr_3 = DecisionTreeRegressor(max_depth=7)
regr_4 = DecisionTreeRegressor(max_depth=None)
```

This creates four different models, each with an increasing maximum depth for the tree.

Your results are shown in Figure 4-30.

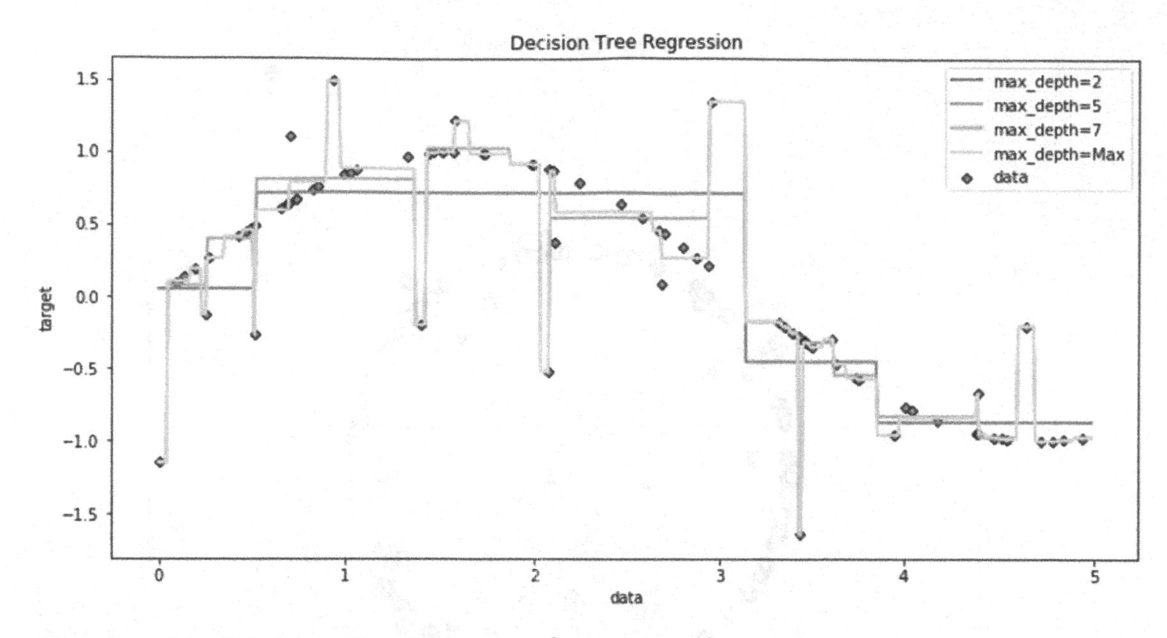

Figure 4-30. *Decision Tree Regressor (max_depth)*

The tree's score increases as the max depth increases.

```
Score max_depth=2  :  0.7630
Score max_depth=5  :  0.9539
Score max_depth=7  :  0.9923
Score max_depth=Max:  1.0000
```

You can also test the same with ExtraTreeRegressor using example this: Chapter 004 Example 018A.ipynb.

Multi-output Decision Tree Regression

The next example will show you how to deal with multi-output trees.

Open the sample example Jupyter Notebook: Chapter 004 Example 019.ipynb in your Jupyter software.

Your results are shown in Figure 4-31.

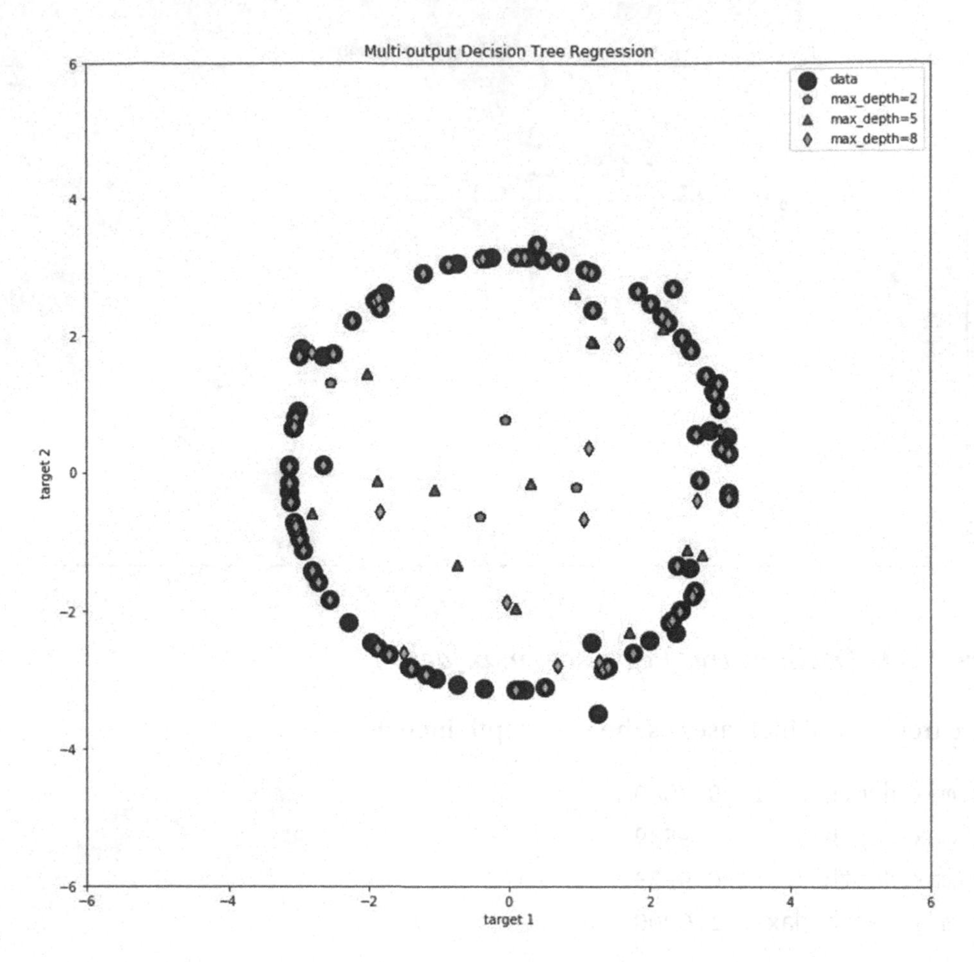

Figure 4-31. *Multi-output Decision Tree Regression*

The change in the max depth results in the following results on the scores:

```
Score max_depth=2         :  0.1081 with    7 nodes or    3.5176 % of tree
Score max_depth=5         :  0.5295 with   53 nodes or   26.6332 % of tree
Score max_depth=8         :  0.8797 with  161 nodes or   80.9045 % of tree
Score max_depth=None      :  1.0000 with  199 nodes or  100.0000 % of tree
```

This is the last example for the machine learning trees.

You should now be able to design the trees and tune the hyperparameters. This point is a good time to close out any previous notebooks and enjoy a drink.

k-Nearest Neighbors 3-Class Classification

The classifier we are implementing is the k-nearest neighbors vote algorithm.

Neighbors-based classification is an instance-based learning or non-generalizing learning algorithm that performs classification by computing a simple majority vote of the nearest neighbors for each point in the data set.

Open the sample example Jupyter Notebook: Chapter 004 Example 020.ipynb in your Jupyter software.

```
for algorithmitem in ['auto', 'ball_tree', 'kd_tree', 'brute']:
    for weightitem in ['uniform', 'distance']:
        for n_neighborsitem in range(3,7,1):
            clf = neighbors.KNeighborsClassifier(n_neighbors=n_
            neighborsitem, weights=weightitem, algorithm=algorithmitem)
```

The two core hyperparameters are:

n_neighbors = range of 3, 4, 5 and 6, this is a constant set to determine the closes neighbors.

weights=weights from list [‘uniform,’ ‘distance’]:

- ‘uniform’: uniform weights. All points in each neighborhood are weighted equally.
- ‘distance’: weight points by the inverse of their distance. So the closer neighbors to a query point will have a bigger influence than neighbors that are farther away.

The results are shown in Figure 4-32.

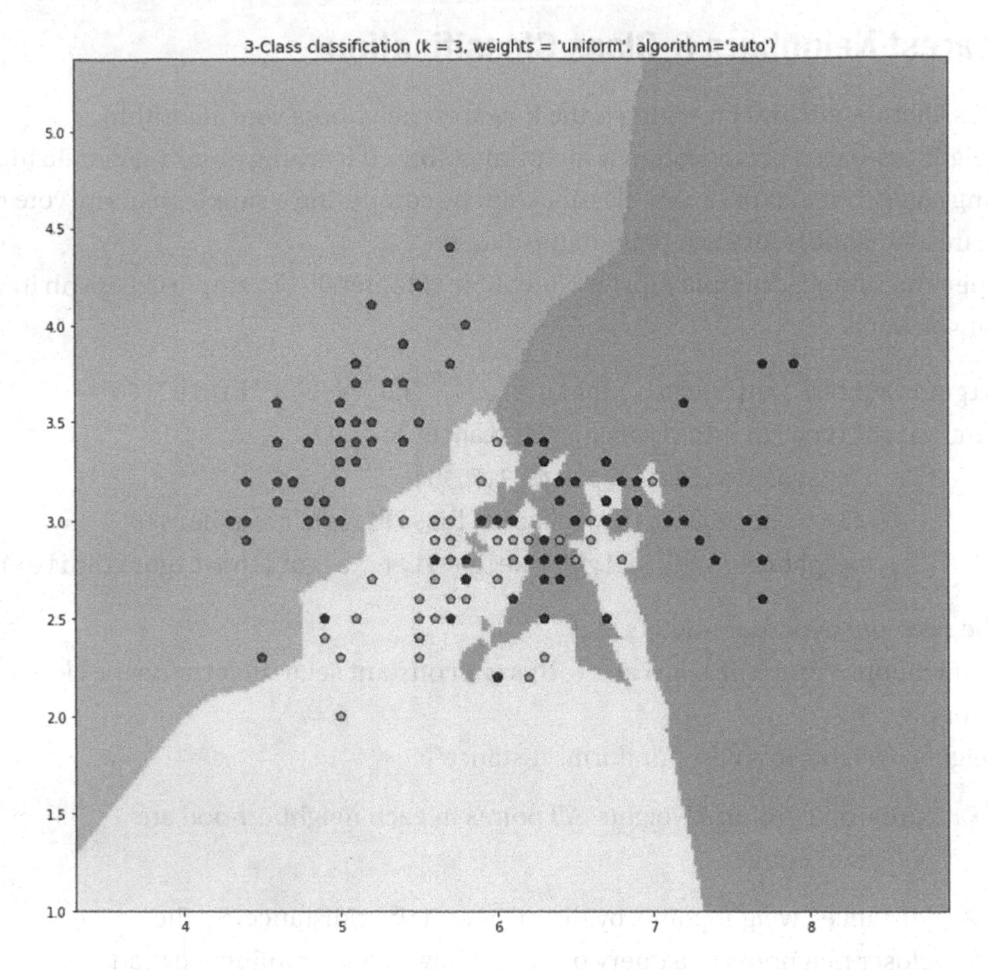

Figure 4-32. *3-Class classification (k = 3) - uniform weights - auto algorithm*

Figure 4-33 shows how the k = 6 changes the resulting decision areas by causing decision islands.

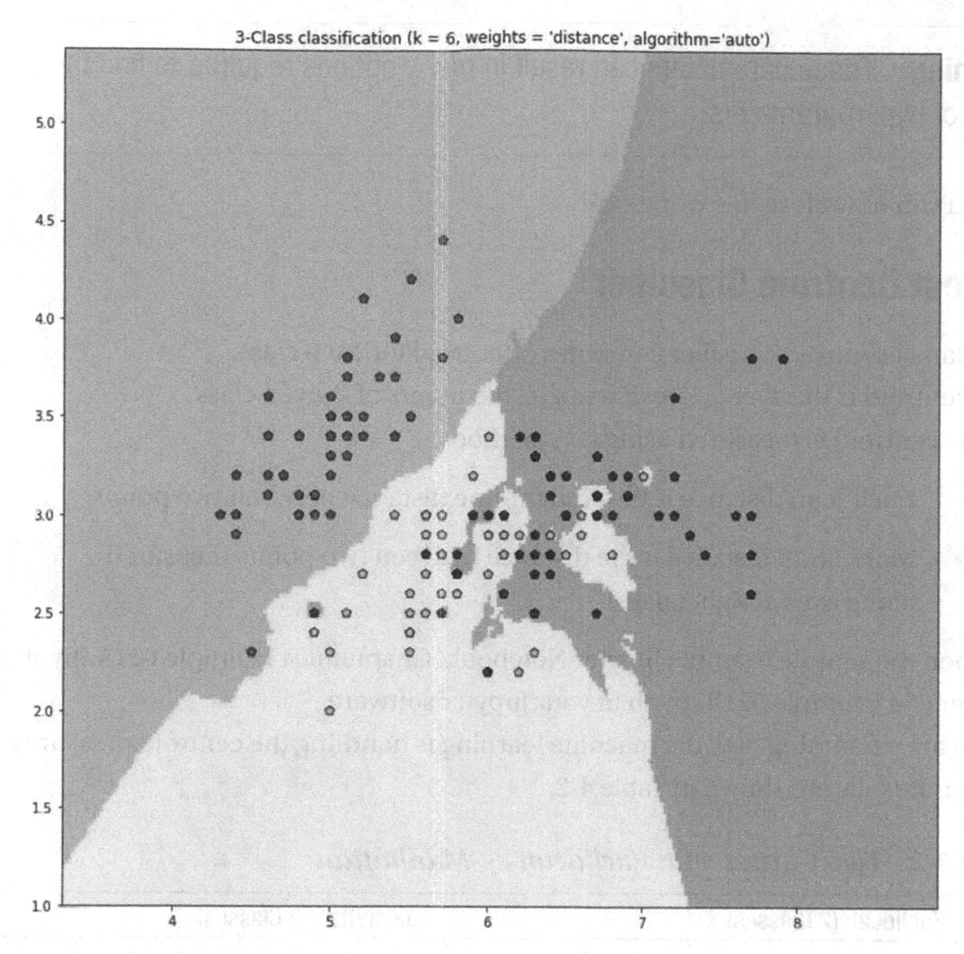

***Figure 4-33.** 3-Class classification (k = 6) distance weights - auto algorithm*

Tip Can you observe the attempt at classifying a blue point at (5, 2.5) that results in an island cluster forming? This is a sign of overfitting.

If you investigate the results, you can observe that a range of cluster effects that reveal how the data set is clustered with different hyperparameters.

The main behavior you should investigate is the islands of classification that are forming as in the real world; this behavior is common in the geospatial, geodemographic segmentation, and route planning.

Warning These parameters can result in many options required to find the correct hyperparameters.

You can now close this workbook.

Nearest Centroid Classifier

The nearest centroid classifier computes a centroid for each class.

A centroid is the average position of all the points of a given class.

The centroid is measured using two methods:

- Euclidean distance is the straight-line distance between two points.
- Manhattan distance is the distance between two points measured along axes at right angles.

Open the sample example Jupyter Notebook: Chapter 004 Example 021A.ipynb and Chapter 004 Example 021B.ipynb in your Jupyter software.

We are examining how the machine learning is handling the centroid measures.

Your Results are shown in Table 4-2.

Table 4-2. *Two Classes with Euclidean vs. Manhattan*

Euclidean (2 Classes)	Manhattan (2 Classes)

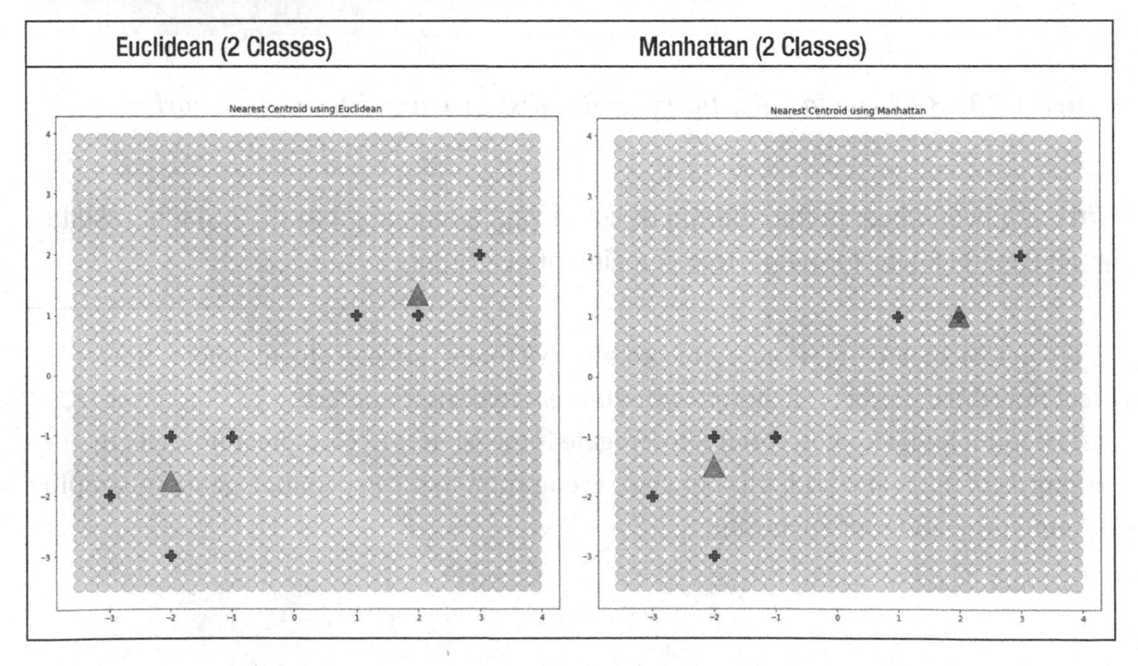

You can clearly observe how the simple, different measuring technique impacts the centroid calculations.

These changes can have major impacts on the solution space.

Warning Beware the 'manhattan' setting as it has unexpected outcomes that most people in the real-world solution space do not immediately understand. Most people use straight-line distances as manhattan uses right-angle travel.

I will now show you another small issue with this process.

Let's investigate what three classes would generate (Table 4-3).

Table 4-3. *Three Classes with Euclidean vs. Manhattan*

Euclidean (3 Classes)	Manhattan (3 Classes)
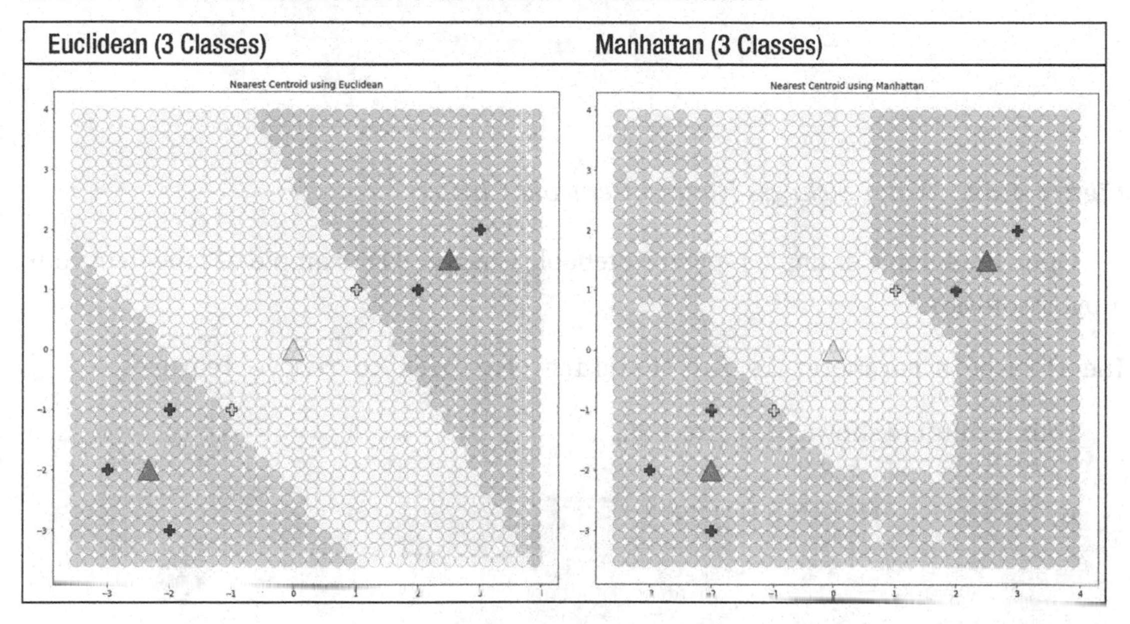	

Wow, look at the right-hand 'manhattan'-based plot. There are several decision areas that appear as nonlogical white areas. The reason is a rounding issue between the training and prediction for this data set.

The other more insignificant error is that the one white point is predicted as white but is within the blue domain.

This is because it is on a diagonal path to the two neighboring blue areas.Open the sample example Jupyter Notebook: Chapter 004 Example 021E.ipynb to investigate a real-world water sickness example (Figure 4-34).

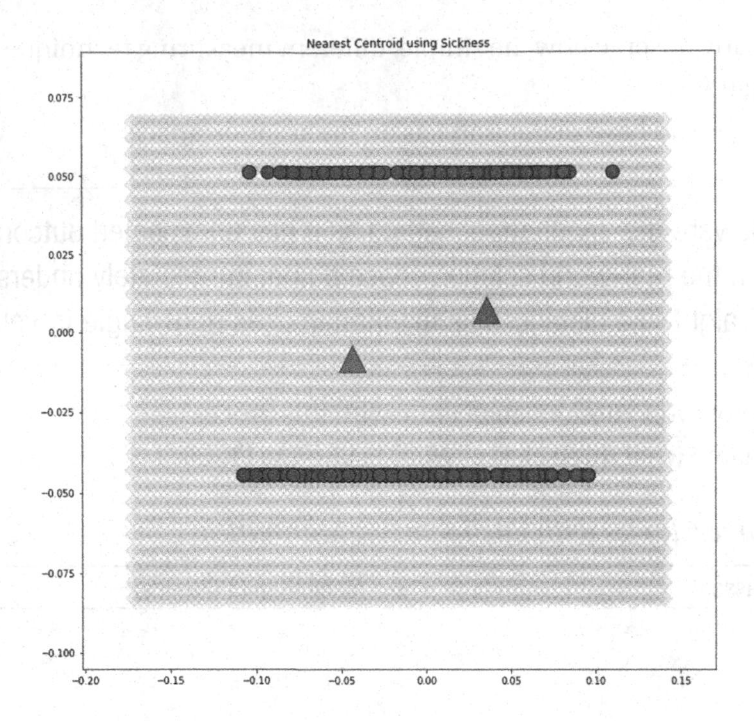

Figure 4-34. *Water Sickness Nearest Centroid Cluster*

Open the sample example Jupyter Notebook: Chapter 004 Example 022.ipynb in your Jupyter software.

`The threshold parameter is for shrinking centroids to remove features.`

Your results are shown in Figure 4-35.

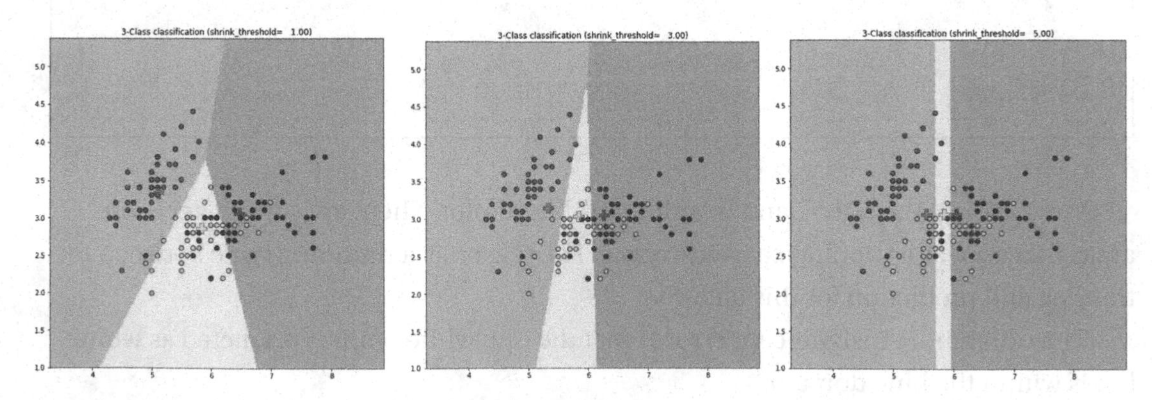

Figure 4-35. *Shrink threshold parameters*

Warning Experiment with your shrink threshold parameters because an inappropriate parameter can change the complete outcome (Table 4-4).

Table 4-4. *Lost Features through Shrink Threshold*

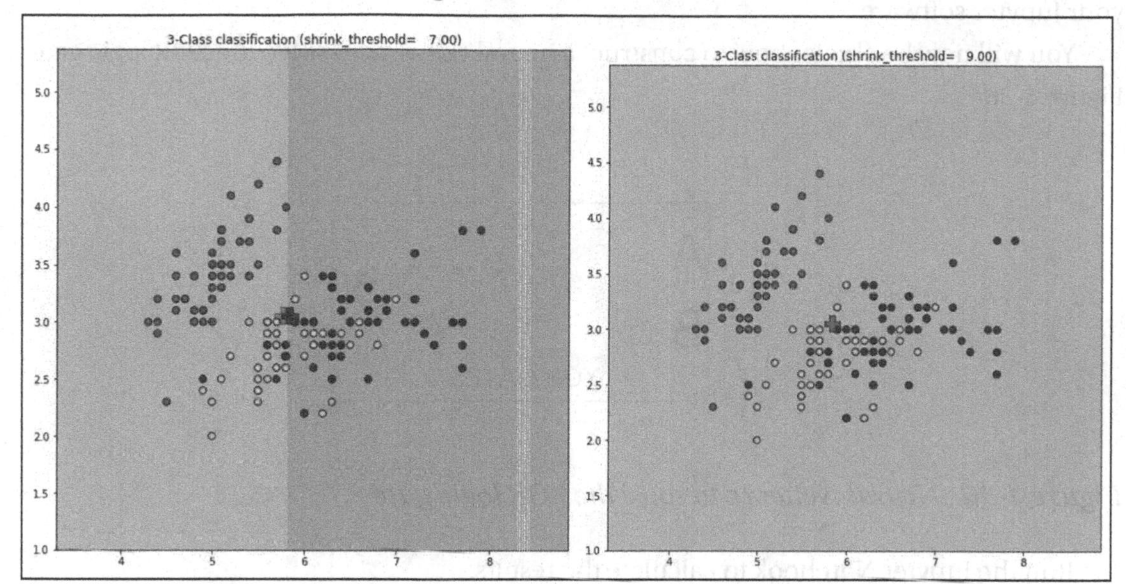

You can close the notebook.

By this point in the chapter, you should be able to tune the hyperparameters on various ML models.

Neural Networks

A neural network is an interconnected group of nodes, similar to the enormous network of neurons in a human brain. In the artificial neural network, each node characterizes an artificial neuron, and an arrow signifies a connection from the output of one artificial neuron to the input of another.

Neural networks enable the simulations of the decision process of a human. At the moment, the capabilities of the network are mostly a reduced or simplified version of the decision process. Currently neural networks are trained to perform a narrow projection of real-world decision processing.

Neural networks can, for example, be trained to identify different pictures and then used to identify similar images in the frames of a CCTV feed.

I will use the following decision process. You have two doors on an entrance to the building; the rule for a safe state is if either door is open. So, if both doors are closed, you cannot enter; or if both doors are open, there is an error that means a security guard has to be dispatched.

Open the sample example Jupyter Notebook: Chapter 004 Example 023A.ipynb in your Jupyter software.

You will now be shown how to construct a neural network to solve the XOR logic gate Figure 4-36.

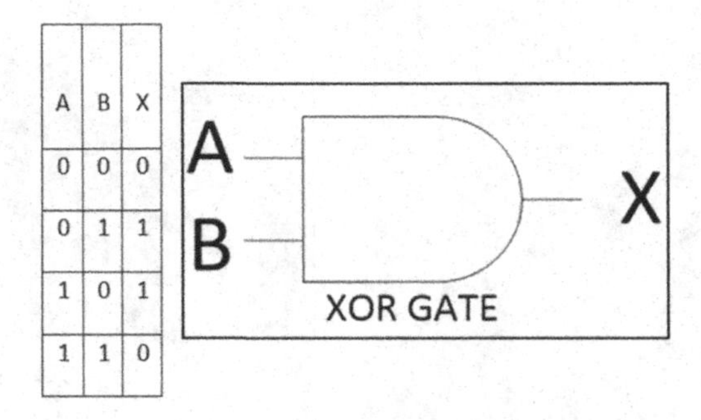

Figure 4-36. *Neural Network to solve the XOR logic gate*

Run the Jupyter Notebook to calculate the results.

Your Results:

```
------------------------------------
Run XOR Gate Network: [2, 4, 1]
------------------------------------
epochs: 000000
epochs: 010000
epochs: 020000
epochs: 030000
epochs: 040000
epochs: 050000
epochs: 060000
epochs: 070000
epochs: 080000
epochs: 090000
```

```
------------------------------------
Results:
------------------------------------
[0,0] p=0.000285 y(p)=0 y(t)=0
[0,1] p=0.981030 y(p)=1 y(t)=1
[1,0] p=0.982048 y(p)=1 y(t)=1
[1,1] p=0.000511 y(p)=0 y(t)=0
```

As you can see, the network gives an adequate suggestion of the forecast outcome of the network.

This example is a simple binary decision flow, but it can be industrialized to take over the predictions on extremely complex and highly dimensional problem with thousands or even millions of variables influencing the outcome of the process.

The field of neural networks is an area of the machine learning field that will evolve numerous times over the coming years to assist data science experts to model complex problems that currently seem outside our current capabilities.

Prediction The neural network will be available as a plugin hardware module to most mobile devices by 2025. The edge computing will use the neural network by 2020 as a prediction model.

You can run the following Jupyter Notebooks to test other logical gates:

- Chapter 004 Example 023B.ipynb - OR Gate
- Chapter 004 Example 023C.ipynb - AND Gate
- Chapter 004 Example 023D.ipynb - NXOR Gate

The prediction of logical gates output is used in robotics to predict the output from an IoT device against known expected outputs.

Multilayer Perceptron

A multilayer perceptron (MLP) is a class of feedforward artificial neural network. An MLP consists of, at least, three layers of nodes: an input layer, a hidden layer, and an output layer. Except for the input nodes, each node is a neuron that uses a nonlinear activation function. MLP utilizes a supervised learning technique called backpropagation for training.

Regularization Parameter 'alpha'

This varying regularization in a multilayer perceptron is a comparison of different values for the regularization parameter 'alpha' on your synthetic data sets. The plot shows that different alphas yield different decision functions.

- Alpha is a parameter for a regularization term, aka penalty term, which avoids overfitting by constraining the size of the weights. Increasing alpha may fix high variance (a sign of overfitting) by encouraging smaller weights, resulting in a decision boundary plot that appears with lesser curvatures.
- Likewise, decreasing alpha may fix high bias (a sign of underfitting) by encouraging larger weights, potentially resulting in a complicated decision boundary.

Example:

Open the sample example Jupyter Notebook: Chapter 004 Example 024.ipynb in your Jupyter software.

You can run the complete Jupyter Notebook to observe the impact of the alpha hyperparameter.

Your results are shown in Figure 4-37.

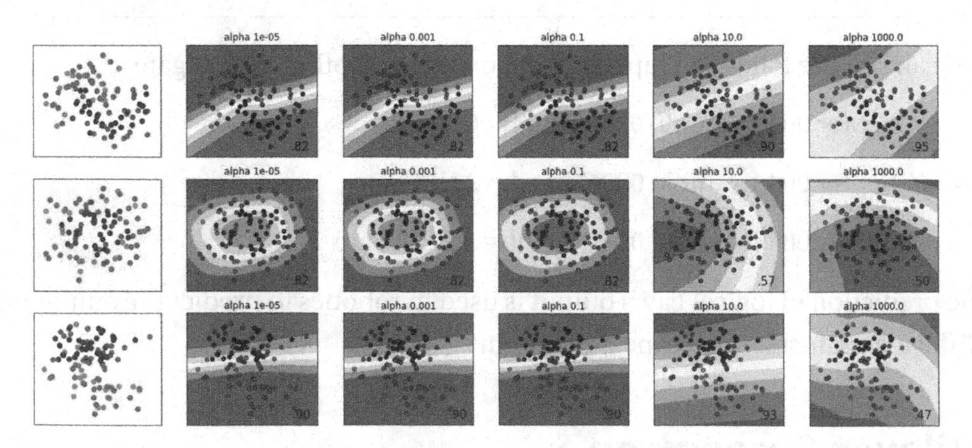

Figure 4-37. *Data Impact Study*

Stochastic Learning Strategies

You will at this point compare stochastic learning strategies for MLPClassifier by visualizing some training loss curves for different stochastic learning strategies, including SGD and Adam. Because of time constraints, I suggest that you use several small data sets, for which L-BFGS might be more suitable. But this algorithm can process larger data sets also with similar results with the only penalty that the process will run slower as the volume increases.

Warning Results can be highly dependent on the value of the learning rate parameter.

Open the sample example Jupyter Notebook: Chapter 004 Example 025.ipynb in your Jupyter software.

A Multilayer Perceptron Classifier is a model that optimizes the log-loss function using LBFGS or stochastic gradient descent.

I have constructed an example that loops through all the possible hyperparameters and creates a universal plot to demonstrate to different outcomes.

Your results are shown in Figure 4-38.

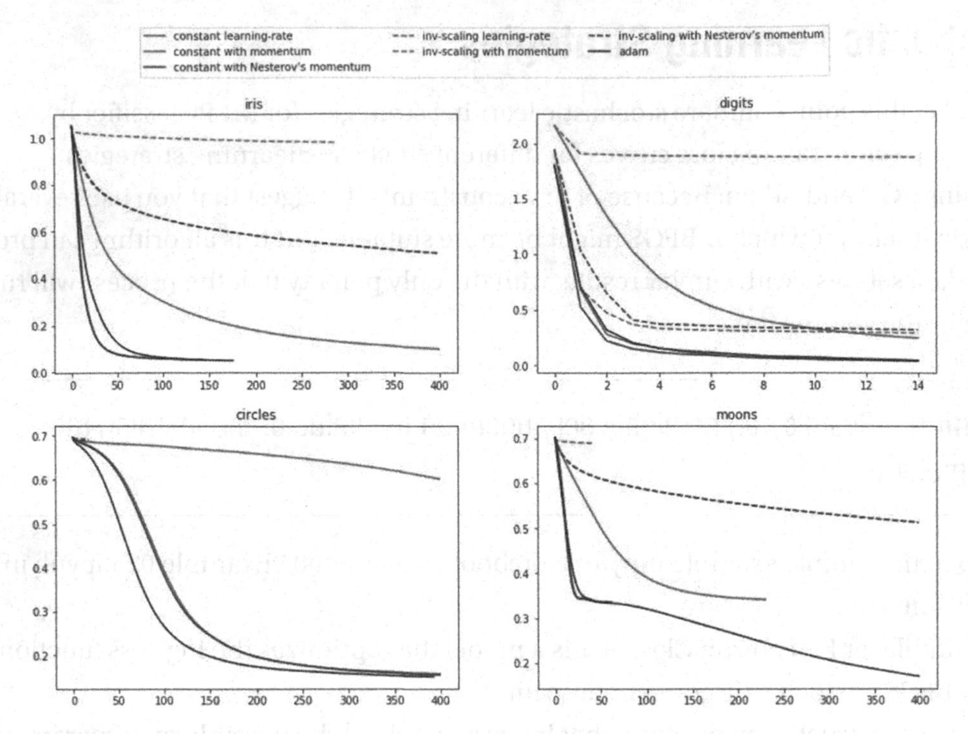

***Figure 4-38.** Stochastic Learning Strategies*

This example shows you that different learning parameters yield significant diverse results against various data sets.

What's Next?

You have successfully completed Chapter 4 and the first part of supervised learning.

Next, I will cover additional supervised machine learning techniques in Chapter 5 and the second part of supervised learning.

First, I suggest you go and make a cup of coffee or tea. Enjoy this time to rethink your fresh knowledge and get organized for the next chapter of *Industrialized Machine Learning.*

CHAPTER 5

Supervised Learning: Advanced Algorithms

This is Part 2 of the Supervised Learning process. You are now studying the more advanced machine learning algorithms in the supervised learning ecosystem.

Tip These algorithms are commonly used in the industrialized machine learning ecosystem as they solve many of the common problems that you will discover during the examples and their solutions.

Boosting (Meta-algorithm)

The machine learning term "Boosting" refers to a family of algorithms that adapt weak learners to strong learners. These are useful when the features are not yielding strong-enough learners for the algorithms. Boosting algorithms use a set of the low accuracy classifiers to create a highly accurate classifier.

The basic concept (Figure 5-1) is to chain the machine learning into a chain of processing to achieve an improved outcome.

A. F. Vermeulen, *Industrial Machine Learning*, https://doi.org/10.1007/978-1-4842-5316-8_5

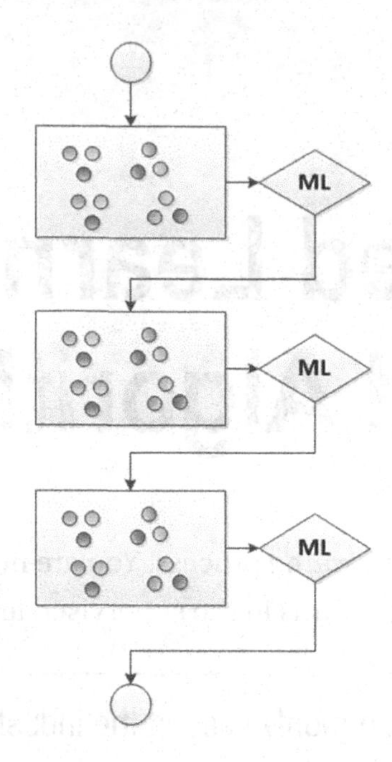

***Figure 5-1.** ML Boosting Concept*

That is a complex introduction. I will quickly explain the concept of weak learners.

A "weak learner" is an ML algorithm (for regression/classification) that provides accuracy marginally better than random guessing.

AdaBoost (Adaptive Boosting)

The AdaBoost or Adaptive boosting is a method that enables the data scientist to enhance the outcomes of the data processing to get better results.

An AdaBoost classifier is a meta-estimator that begins by fitting a classifier on the original data set and then fits additional copies of the classifier on the same data set but where the weights of incorrectly classified instances are adjusted such that subsequent classifiers focus more on difficult cases.

Example:

Open the sample example Jupyter Notebook: Chapter 005 Example 001A.ipynb in your Jupyter software.

Using Part A, you activate the core libraries you will require.

The following three imports are important for the work you need to perform.

```
from sklearn.ensemble import AdaBoostClassifier
```

The AdaBoost classifier uses AdaBoost-SAMME set of algorithms.

This specific example is using the SAMME discrete boosting algorithm.

The next structure you need is the machine learning algorithm that you want to boost.

You are using a Support Vector Classification (SVC) algorithm.

```
from sklearn.svm import SVC
```

The data engineering uses a standard scaler to prepare the features.

```
from sklearn.preprocessing import StandardScaler
```

The transformer standardizes features by removing the mean and scaling to unit variance.

Now that you have the basic tools, I will show you how to build a boosting solution.

Part B and Part C supply the data you will need. I suggest we use the data on my roses.

In Part D you use the transformer to perform the scaling of the features you require.

```
transformer = StandardScaler(copy=True, with_mean=True, with_std=False)
```

This transforms the data to support a better training process.

The transformer results are:

```
Features: 3
Samples: 420
Scale: None
Mean: [5.82341667 3.07140952 1.22077619]
variance: None
```

In Part E you create the SVC ready for the AdaBoost to use.

```
svc=SVC(max_iter=5000,
        gamma='auto',
        class_weight='balanced',
        probability=True,
        kernel='linear',
        random_state=0,
        verbose=False)
```

This SVC will be used by the AdaBoost.

In Part F you will perform a single cycle of the boost:

```
clf1 = AdaBoostClassifier(algorithm='SAMME',
                          n_estimators=1,
                          base_estimator=svc,
                          learning_rate=1,
                          random_state=0)
clf1.fit(X_train_scale, y_train)
score1 = clf1.score(X_test_scale,y_test)
```

The score result is: 67.2222%

In Part G you perform 5 cycles, and the score is: 67.7778% - Improvement!

In Part H you perform 10 cycles, and the score is: 73.33333% - Improvement!

In Part I you perform 20 cycles, and the score is: 73.3333% - No Improvement?

Warning The boost will only improve to a point, and then you are simply just wasting clock cycles for no effective gain.

You have successfully boosted a 67.2222% to 73.3333% that is a 9.09% improvement.

Note The machine learning can be any of the machine learning algorithms in this book.

You can close the notebook you have successfully completed, your first boost example.

Open the sample example Jupyter Notebook: Chapter 05 Example 001B.ipynb in your Jupyter software.

We are going to complete another example, but this example is using a different base machine learning algorithm and scaler.

See Part D for the scaler called the RobustScaler. This scaler transforms features using statistics that are robust to outliers as it removes the median and scales the data according to the quantile range.

Part E is the point where you get to use the ExtraTreeClassifier as the base ML process.

In Part I you will use a 1,000-cycle boost.

Tip To enable the Jupyter Notebook to show the 1,000 cycle, you need to perform this modification:

```
%%javascript

IPython.OutputArea.prototype._should_scroll = function(lines) {

return false;

}
```

As you can observe in Part J, you achieve a score improvement from 70.270% to 97.200%, so a 38.324% improvement! That is the power of boosting.

You can close this notebook.

Next, we will investigate the value domain results of a boost solution.

Open the sample example Jupyter Notebook: Chapter 05 Example 001C.ipynb

Execute the complete notebook.

You achieve a 69.440% to 82.220% improvement with the solution's boosting.

I want you to look at the lack of improvement between 1 and 5 boosts.

If we plot the prediction value domain, you will observe that there is a major value shift in the value domain.

You will get following four results (Figure 5-2 and Figure 5-3).

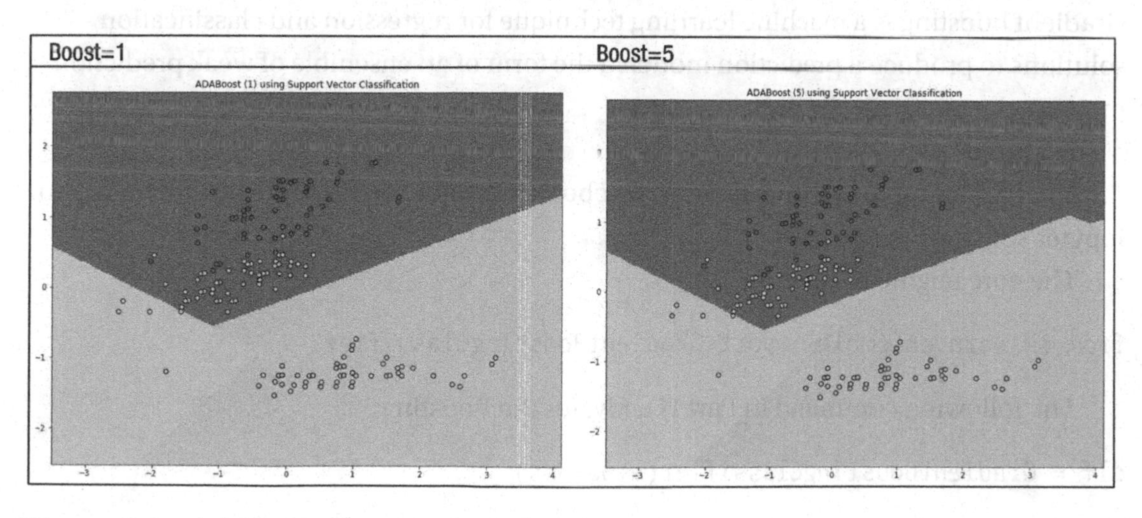

***Figure 5-2.** AdaBoost changes the value domain*

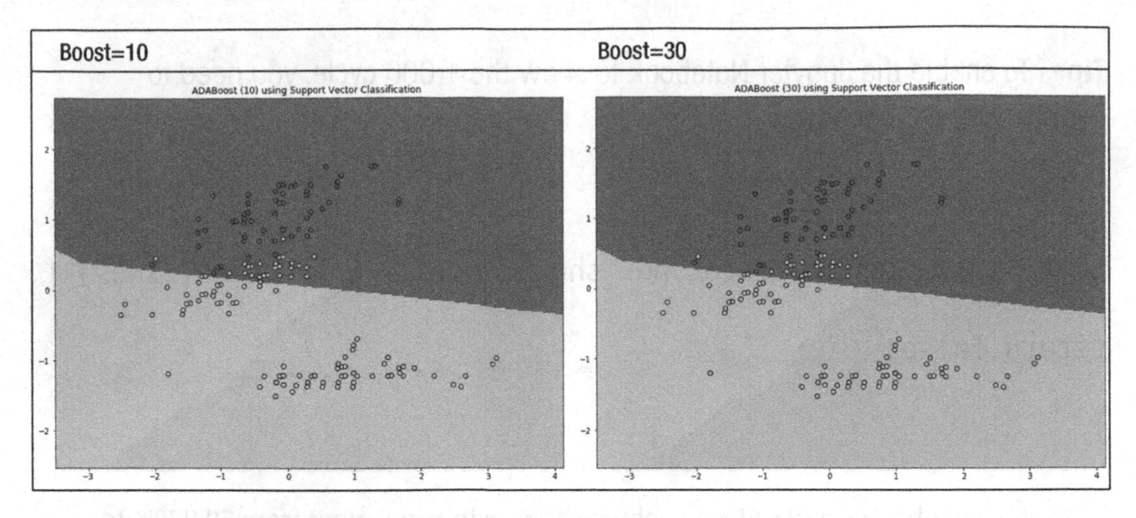

***Figure 5-3.** AdaBoost does not change the value domain*

Warning Due to the lack of test data on the domain boundaries, the shift does not show up in the score as you are not testing in those edge boundaries.

Always check if your data covers the whole value domain!

You can close the notebook. Well done; you can now perform boosting.

Gradient Tree Boosting

Gradient boosting is a machine learning technique for regression and classification solutions to produce a prediction model in the form of an ensemble of weak prediction models, characteristically decision trees.

Example:

Open the sample example Jupyter Notebook: Chapter 005 Example 002.ipynb in your Jupyter software.

The core engine is:

```
from sklearn.ensemble import GradientBoostingClassifier
```

The following command in Part H activates the Boosting:

```
clf = GradientBoostingClassifier(**params)
```

Your results are shown in Figure 5-4.

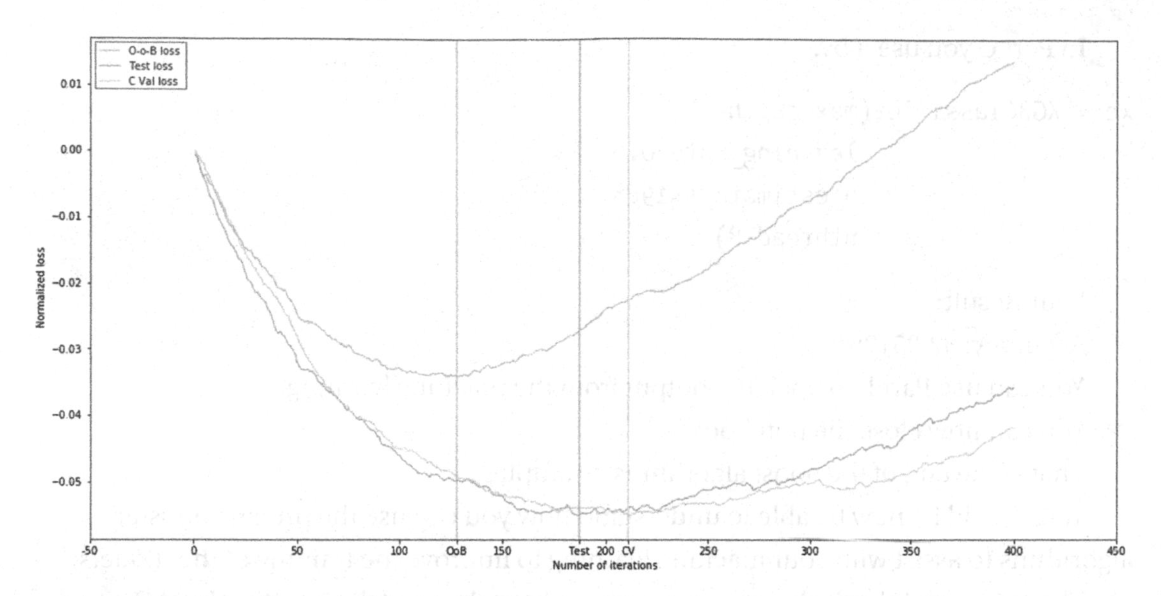

***Figure 5-4.** Gradient Tree Boosting*

You can close the notebook.

XGBoost

XGBoost provides a gradient boosting framework. It is extremely successful as a machine learning library.

The process supports the distributed processing frameworks Apache Hadoop, Apache Spark, and Apache Flink. This makes this framework highly scalable and is one of the most used libraries for that reason.

The Python library is installed using:

```
conda install -c conda-forge xgboost
or
conda install -c anaconda py-xgboost (This one was my personal choice)
```

Example:

Open the sample example Jupyter Notebook: Chapter 005 Example 003.ipynb in your Jupyter software.

In Part A, you load the xgboost engine:

```
from xgboost import XGBClassifier
```

In Part C you use it by:

```
xc = XGBClassifier(max_depth=12,
                   learning_rate=0.05,
                   n_estimators=1968,
                   nthread=8)
```

Your Result:

Accuracy: 82.05128%

You can use Part E to open the output from the machine learning.

You can now close the notebook.

That is the end of the boost algorithms' examples.

You should by now be able to understand how you can use the process booster algorithms to assist with your machine learning to improve the training of the models.

The model training is the most expensive part of the modeling part of the IML.

Next, we are going to investigate one of the up-and-coming algorithms, that is, TensorFlow.

TensorFlow

TensorFlow is an open source software library for dataflow programming across a range of tasks. It is a symbolic math library and is also used for machine learning applications such as neural networks.

TensorFlow was developed by the Google Brain team in November 2015. TensorFlow can run on multiple CPUs and GPUs.

Tip The Machine Learning Crash Course (MLCC) training course from Google will enable you to gain better basic preparation for TensorFlow.

If you want to read more about the background, I suggest you visit: `https://www.tensorflow.org/learn`

Tip The TensorFlow community is a strong and well-developed group.

`https://www.tensorflow.org/tutorials`

Tip As TensorFlow was developed by the Google Brain team. I suggest using: `https://colab.research.google.com/notebooks/welcome.ipynb`

You simply upload the examples and you can use them.

Note I have used my own TensorFlow installations for this book.

You install the TensorFlow library into Python:

```
conda install -c conda-forge tensorflow
```

I will show you the basic building blocks of using TensorFlow for your machine learning.

TensorFlow has a massive capability in processing data into insightful information.

Open the sample example Jupyter Notebook: Chapter 005 Example 004.ipynb in your Jupyter software.

Note You are directly working with TensorFlow.

Later in this chapter, I will show you how to use Keras as an interface.

In Part A, you will observe that the TensorFlow interface is loaded via import tensorflow as tf.

In Part B, you will perform some basic mathematics to introduce you to the basic concepts.

You declare constants as follows: `a = tf.constant(5)`. This is to achieve a=5.

So let's perform these calculations.

```
a=5.000, b=10.000, c=20.000, d=12.000, e=89.000
x=5.000 + 10.000 + 20.000
Addition with constants: x = 35
y=5.000 x 10.000 x 20.000
Multiplication with constants: y = 1000
z=2.000 ^ 12.000
Power with constants: z = 4096
s=sqrt(89.000)
Square root with constants: s = 9
```

I will show you how to use TensorFlow to perform Matrix.

Create a Constant op that produces a 1x3 matrix. The op is added as a node to the default graph. The value returned by the constructor represents the output of the Constant op.

```
Create Constant that produces a 1x4 matrix.
matrix1 = tf.constant([[10., 11., 12.,13.]])
Create another Constant that produces a 4x1 matrix.
matrix2 = tf.constant([[14.],[15.],[16.],[17.]])
```

Create a Matmul op that takes 'matrix1' and 'matrix2' as inputs. The returned value, 'product', represents the result of the matrix multiplication.

```
product = tf.matmul(matrix1, matrix2)
```

To run the matmul op, we call the session 'run()' method, passing 'product'.

The output of the op is returned in 'result' as a numpy `ndarray` object.

```
with tf.Session() as sess:
    result = sess.run(product)
    print(result)
```

Your results:

[[718.]]

Investigate the Part G calculations to test the capability of the system: Change `i=1000` and see the power!

Congratulations you can perform basic TensorFlow processing.

Prediction TensorFlow is currently a major workhorse in the machine learning ecosystem. I want to predict that this library with its accompanying ecosystem will become a leading system by 2025.

The following terminology is important before you start running the next examples:

Epoch

An epoch is a full iteration over the given data samples. The number of epochs is the amount of times the algorithm is going to cycle or iterate. The number of epochs directly affects the result of the training step. Too few epochs could cause your algorithms to only reach a local minimum, but the more epochs you use, this will result in a better result when you reach a global minimum or at least a better local minimum.

Layers

Machine learning models are assembled as the composition and stacking of logical chains known as layers. The separation of layers with precise functional purposes results in more simple and reusable layers.

TensorFlow provides a good set of preconfigured useful common layers that support an effective and effiecient processing environment for your machine learning.

Your models will consist of several layers of useful operations created by these layers.

Estimators

Estimators are TensorFlow's most scalable and production-oriented model type that enables you to use premade Estimators that enable you to take you processing work to a higher conceptual level than the base TensorFlow supports. This is a feature of TensorFlow that made it highly successful.

Checkpoints

Checkpoints are versions of the model created during training and then saved for later use. This is highly useful to determine the success of new models against older models.

You can now validate how to use TensorFlow in other data processing examples.

Open the sample notebooks:

- Jupyter Notebook: Chapter 005 Example 005A.ipynb

  See results in Chapter-005-Example-005A-01.txt
- Jupyter Notebook: Chapter 005 Example 005B.ipynb

  See: Accuracy of 0.9333
- Jupyter Notebook: Chapter 005 Example 005C.ipynb

  See results in Chapter-005-Example-005C-01.txt

After you complete these workbooks, you are done with TensorFlow.

Bayesian Statistics

Bayesian statistics is a theory in the field of statistics based on the Bayesian interpretation of probability where probability expresses a degree of belief in an event, which can change as new information is gathered, rather than a fixed value based upon frequency or propensity.

Bernoulli Trial

A Bernoulli trial is a random experiment with only two outcomes, usually labeled as "success" or "failure," in which the probability of success is exactly the same every time the trial is carried out. The probability of success is given by θ, which is a number between 0 and 1. Thus $\theta \in [0,1]$.

Open the sample example Jupyter Notebook: Chapter 005 Example 006A.ipynb in your Jupyter software.

In this example we are going to use a game called "Heads-or-Tails" to demonstrate how you can predict how to handle a "success" or "failure" solution.

This could be:

- You won the customer or lost them.
- Your product works or fails.
- The robot picks the strawberry or leaves it on the land.

You can apply this to any bi-outcome solution.

Now run the complete notebook as you seek success by achieving a coin on heads!

Your outcomes are:

```
Perform Test (001):     0 trials,     0 heads,     0 tails ->  0.0000 %
Perform Test (002):     2 trials,     1 heads,     1 tails -> 50.0000 %
Perform Test (003):    10 trials,     5 heads,     5 tails -> 50.0000 %
Perform Test (004):    20 trials,     9 heads,    11 tails -> 45.0000 %
Perform Test (005):    50 trials,    23 heads,    27 tails -> 46.0000 %
Perform Test (006):   500 trials,   246 heads,   254 tails -> 49.2000 %
Perform Test (007):  1000 trials,   488 heads,   512 tails -> 48.8000 %
Perform Test (008): 10000 trials,  4957 heads,  5043 tails -> 49.5700 %
```

Your graphical results are shown in Figure 5-5.

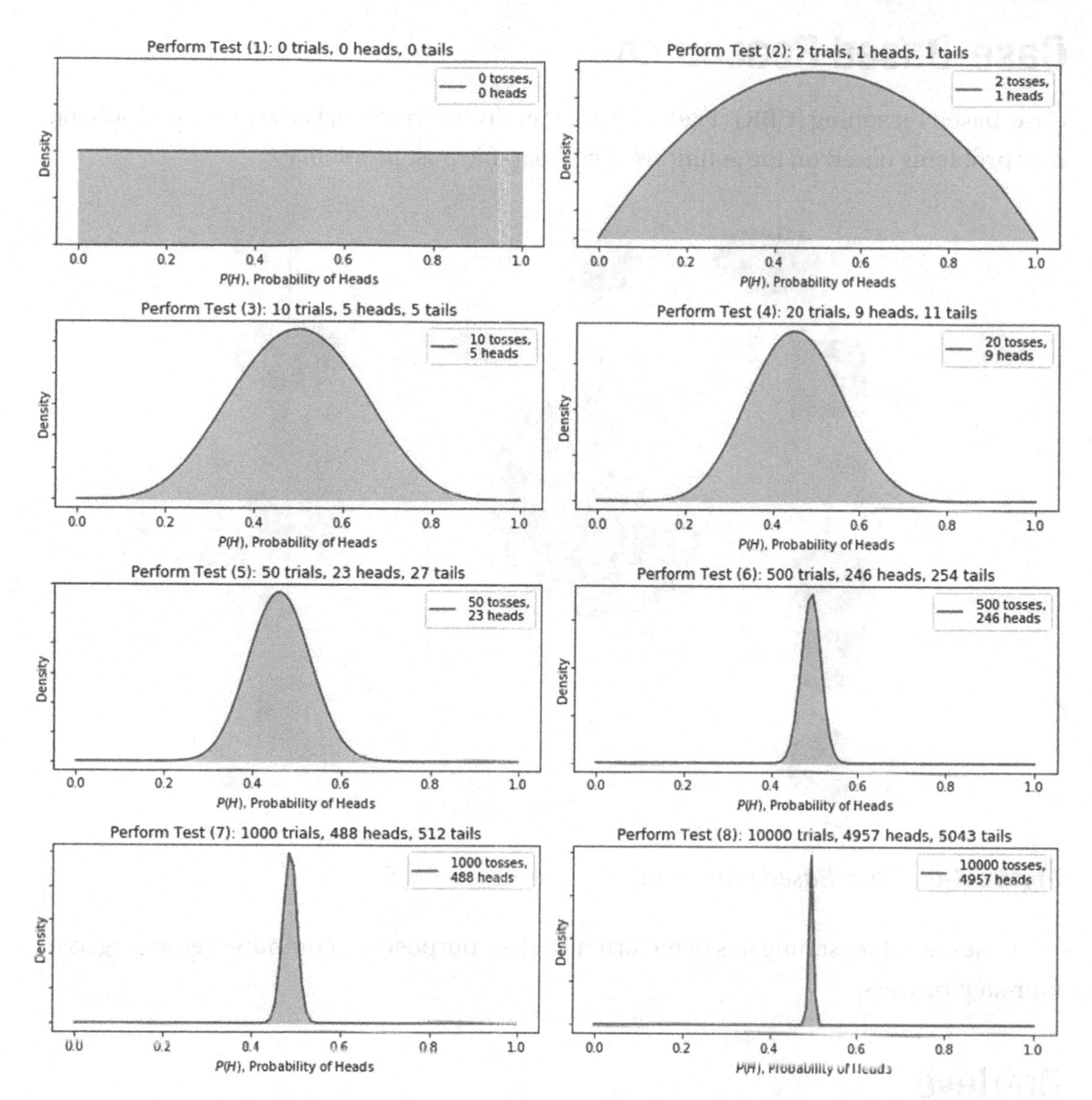

Figure 5-5. *Bernoulli trial*

You can observe that as the amount of trials increases the "Probability for Heads" to increase closer to the fair 50% you would expect.

You can save your notebook or you can experiment with: `number_of_trials = [0, 2, 10, 20, 50, 500, 1000, 10000]`

See: Chapter 005 Example 006A.ipynb for a Fibonacci with 20 trails.

Next, we will discuss common knowledge that you need to understand for performing IML in general.

Case-Based Reasoning

Case-based reasoning (CBR) (Figure 5-6), generally interpreted, is the process of solving new problems based on the solutions of comparable past problems.

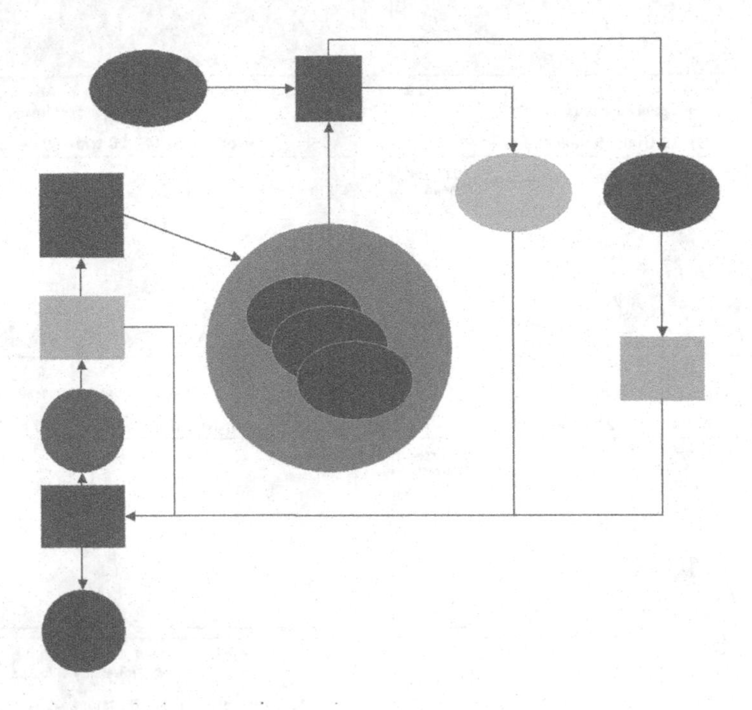

Figure 5-6. *Case-Based reasoning*

Case-based reasoning has been formalized for purposes of computer reasoning as a four-step process.

Retrieve

Given a target problem, retrieve from the memory cases pertinent to solving it. A case consists of a problem, its solution, and, characteristically, explanations about how the solution was derived. For example, suppose Angus wants to prepare cinnamon coffee. Being a novice barista, the most relevant experience he can recall is one in which he successfully made plain coffee. The procedure he followed for making the plain coffee, together with explanations for decisions made along the way, constitute Angus's reference case.

Reuse

Map the solution from the previous case to the target problem. This may involve adapting the solution as needed to fit the new situation. In the coffee example, Angus must adapt his reference solution to include the addition of cinnamon.

Revise

Having mapped the previous solution to the target situation, test the new solution in the real world (or a model) and, if required, revise. Suppose Angus adapted his coffee solution by adding cinnamon sticks to the hot water. After mixing, he discovers that the coffee has pieces of cinnamon stick not just the flavor - an undesired effect. This suggests the following revision: delay the addition of water to coffee until after the water is filtered and then use it to make the coffee.

Retain

After the solution has been successfully adapted to the target problem, store the resulting experience as a new case in memory. Angus, consequently, records his newfound procedure for making cinnamon coffee, thereby enriching his set of stored experiences, and better preparing him for future coffee-making demands.

Reinforcement Learning

Reinforcement Learning (RL) (Figure 5-7) is a subscience of Machine Learning. It's considered a Hybrid of supervised and unsupervised Learning. It simulates human learning based on trial and error in the same way Angus now can now make cinnamon coffee. We will return to this in detail in Chapter 9.

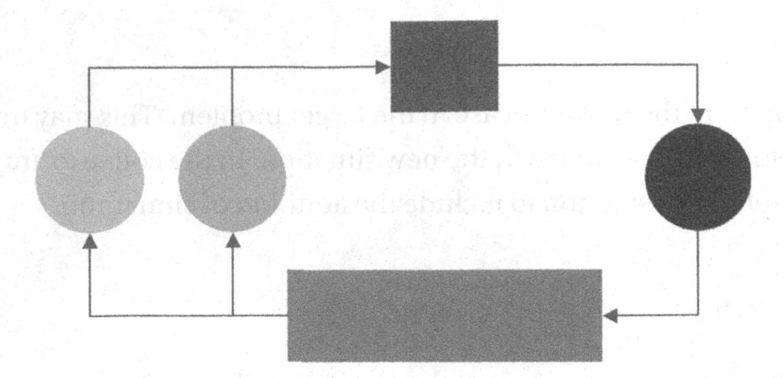

Figure 5-7. *Reinforcement Learning*

Inductive Logic Programming

Inductive Logic Programming (ILP) is a research area formed at the intersection of Machine Learning and Logic Programming.

Gaussian Process Regression

In probability theory and statistics, a Gaussian process is a stochastic process, so that every finite collection of random variables has a multivariate normal distribution, that is, every finite linear combination of values is normally distributed. The distribution of a Gaussian process is the joint distribution of random variables and a distribution over functions within a continuous domain, for example, time or space.

A machine learning algorithm that involves a Gaussian process uses lazy learning and a measure of the similarity between points (the kernel function) to predict the value for an unseen point from training data.

Example:

Open the sample example Jupyter Notebook: Chapter 005 Example 007.ipynb in your Jupyter software.

I have used the sklearn.gaussian_process.GaussianProcessRegressor engine.

I am using this machine learning process to also demonstrate how adding noise to the observations of the function y = 1.968 ∗ (x ∗ np.cos(x)) improves the probability of predicting the values appropriately.

For Part A, you simply use the observations as is, that is, noiseless.

```
gp = GaussianProcessRegressor(kernel=kernel,
                              optimizer='fmin_l_bfgs_b',
                              alpha=1e-10,
                              n_restarts_optimizer=10,
                              random_state=0)
```

In Part B, you add noise via Tikhonov regularization into the alpha hyperparameter of the GaussianProcessRegressor engine:

```
y = f(X).ravel(); dy = 0.5 + 1.0 * np.random.random(y.shape);
noise = np.random.normal(0, dy); y += noise
gp = GaussianProcessRegressor(kernel=kernel,
                              optimizer='fmin_l_bfgs_b',
                              alpha=dy ** 2,
                              n_restarts_optimizer=10,
                              random_state=0)
```

Your results are shown in Figure 5-8 and Figure 5-9.

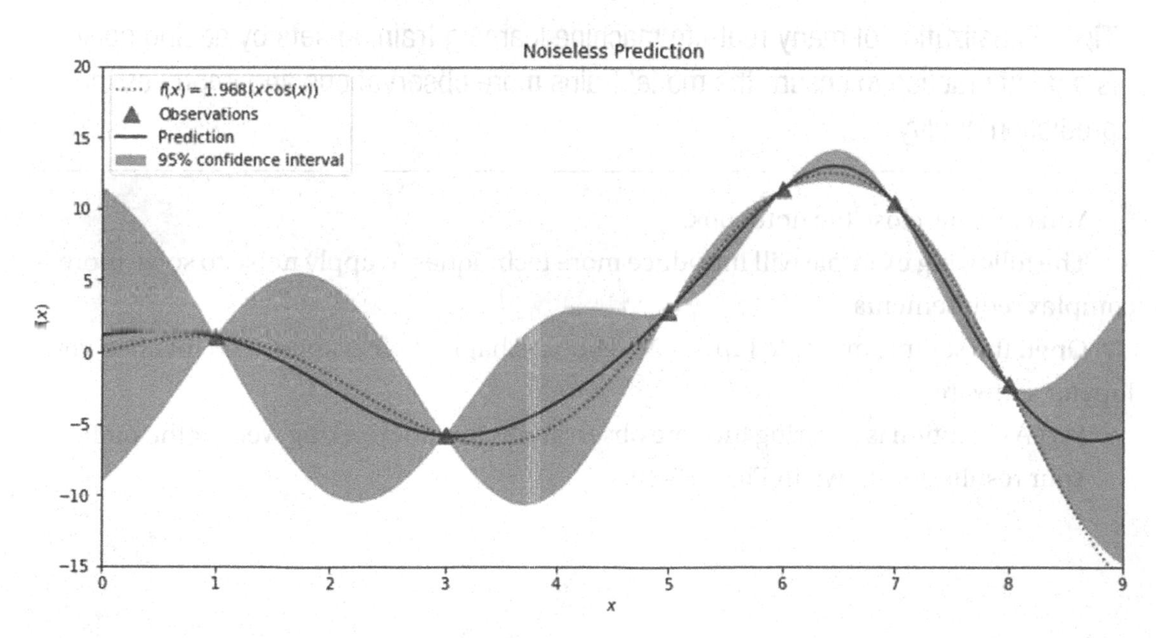

Figure 5-8. *Gaussian process regression (1)*

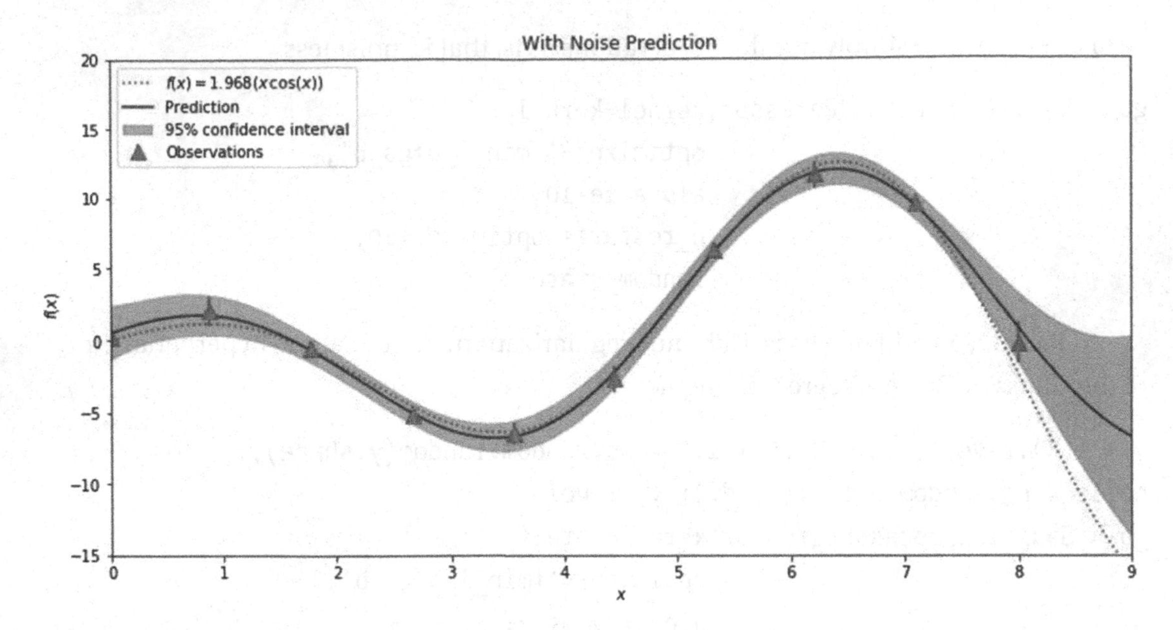

***Figure 5-9.** Gaussian process regression (2)*

You can clearly see that the improvement with added noise is significant.

Tip Penalization of many real-life machine learning training sets by adding noise is a good practice to ensure the model trains more observations and improves the prediction quality.

You can now close the notebook.

The following example will introduce more techniques to apply noise to solve more complex requirements.

Open the sample example Jupyter Notebook: Chapter 005 Example 008.ipynb in your Jupyter software.

Part A's solution is covering the core observations but not working well for the outliers. Your results are shown in Figure 5-10.

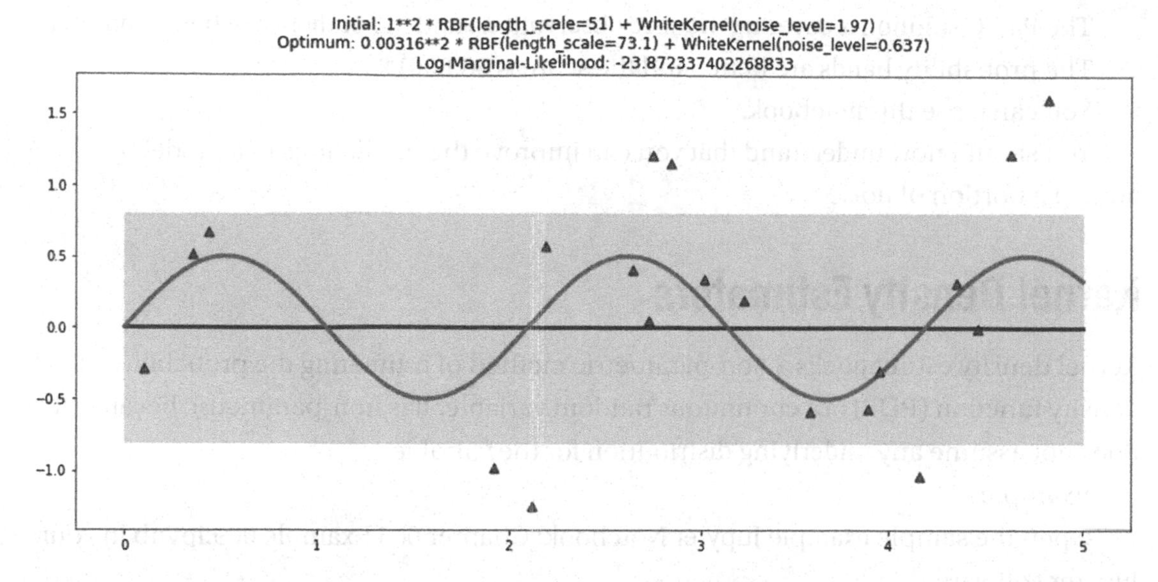

***Figure 5-10.** Gaussian process regression using Tikhonov regularization (1)*

The probability bands are also causing issues with the predictions and have been inaccurate with a score of -4.3556%.

You need a better addition of the noise via the kernel functions.

The Part B solution is covering the core observations and the outliers in a better manner. The probability bands are better and the score is 66.6001%.

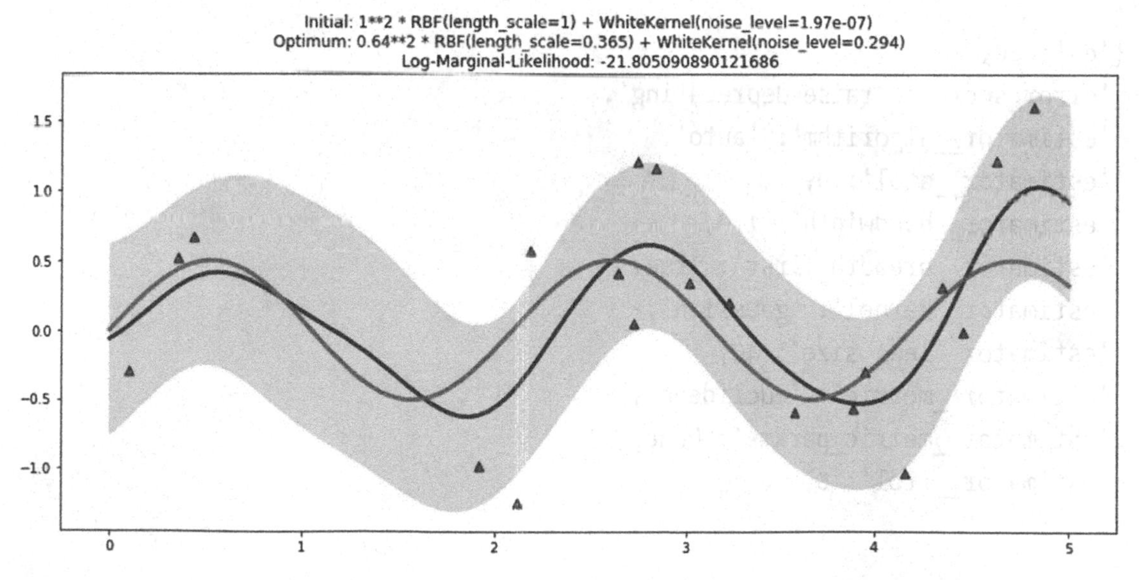

***Figure 5-11.** Gaussian process regression using Tikhonov regularization (1)*

The Part C solution is covering the core observations and the outliers in a better manner.

The probability bands are better and the score is 66.6001%.

You can close this notebook.

You should now understand that you can improve the predictions of a model by adding a portion of noise.

Kernel Density Estimators

Kernel density estimation is a non-parametric method of estimating the probability density function (PDF) of a continuous random variable. It is non-parametric because it does not assume any underlying distribution for the variable.

Example:

Open the sample example Jupyter Notebook: Chapter 005 Example 009.ipynb in your Jupyter software.

In this example we are investigating a set of known values, that is, handwritten digits; and then we infer a set of digits we could expect.

You will use a Principal Component Analysis (PCA) process to perform linear dimensionality reduction using Singular Value Decomposition of the data to project into a lower-dimensional space.

After that you will perform a Kernel density estimation using a grid search cross-validation for five cycles.

```
{'cv': 10,
 'error_score': 'raise-deprecating',
 'estimator__algorithm': 'auto',
 'estimator__atol': 0,
 'estimator__bandwidth': 1.0,
 'estimator__breadth_first': True,
 'estimator__kernel': 'gaussian',
 'estimator__leaf_size': 40,
 'estimator__metric': 'euclidean',
 'estimator__metric_params': None,
 'estimator__rtol': 0,
```

```
'estimator': KernelDensity(algorithm='auto', atol=0, bandwidth=1.0,
breadth_first=True,
       kernel='gaussian', leaf_size=40, metric='euclidean',
       metric_params=None, rtol=0),
'fit_params': None,
'iid': True,
'n_jobs': -1,
'param_grid': {'bandwidth': array([ 0.1       ,  0.1274275 ,  0.16237767,
0.20691381,  0.26366509,
        0.33598183,  0.42813324,  0.54555948,  0.6951928 ,  0.88586679,
        1.12883789,  1.43844989,  1.83298071,  2.33572147,  2.97635144,
        3.79269019,  4.83293024,  6.15848211,  7.8475997 , 10.        ]),
 'leaf_size': array([35, 36, 37, 38, 39, 40, 41, 42, 43, 44, 45, 46, 47,
 48, 49, 50, 51, 52, 53, 54]),
 'kernel': array(['gaussian', 'tophat'], dtype='<U8'),
 'algorithm': array(['kd_tree', 'ball_tree'], dtype='<U9')},
'pre_dispatch': '2*n_jobs',
'refit': True,
'return_train_score': 'warn',
'scoring': None,
'verbose': 1}
```

The best parameters are:

```
{'algorithm': 'kd_tree',
 'bandwidth': 3.79269019073225,
 'kernel': 'gaussian',
 'leaf_size': 51}
```

The grid search has found the optimum parameter set for the solutions training data. Results are shown in Figure 5-12.

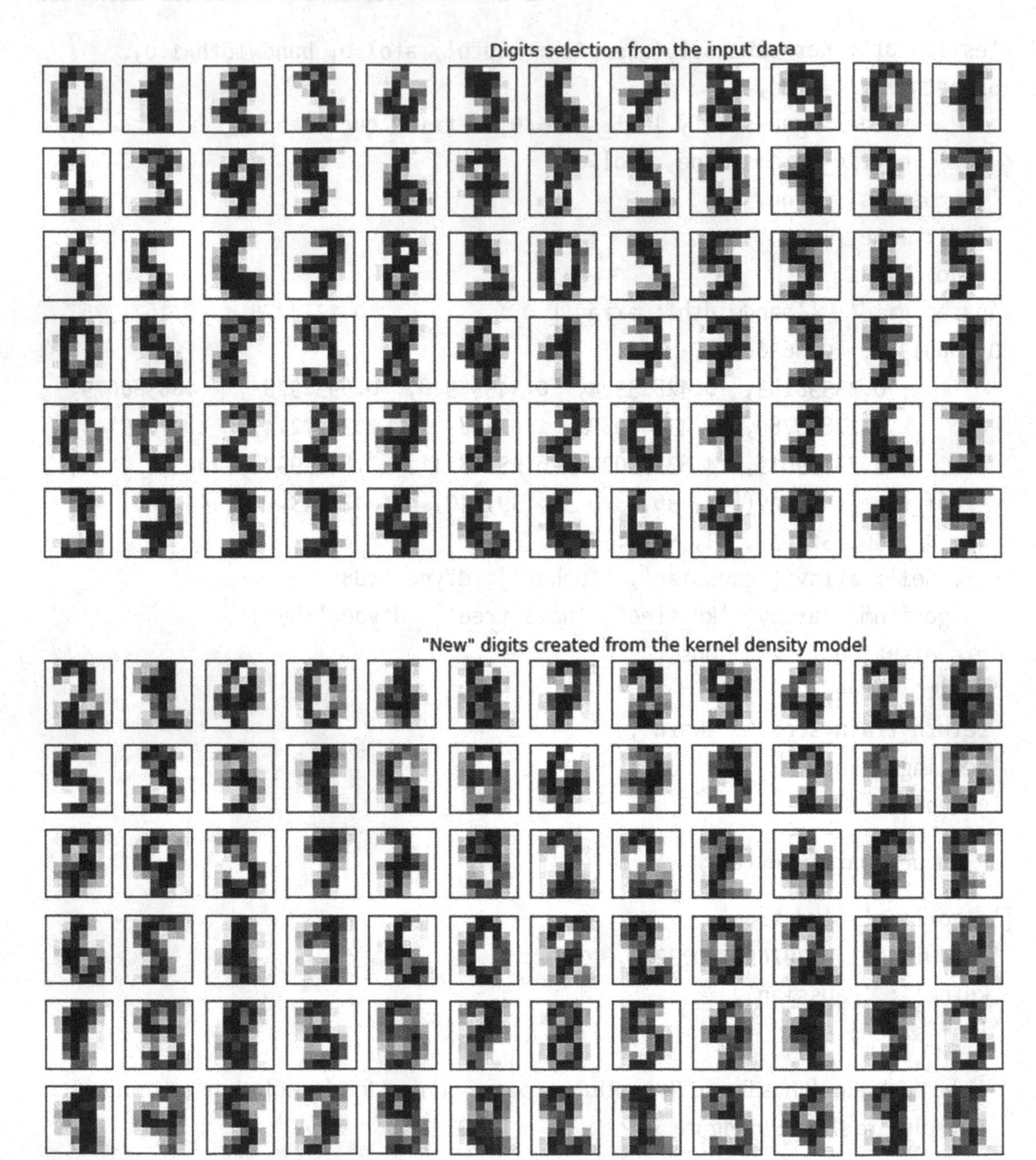

Figure 5-12. *Kernel density estimation using grid search cross-validation*

The second block of digits generated is clearly possible matches for the model for the handwritten digit in the range 0, 1, 2, 3, 4, 5, 6, 7, 8, or 9.

You can now successfully use the Kernel density estimation.

Please close the notebook.

Mayavi 3-Dimensional Visualizers

Next I want to introduce you to the Mayavi scientific data 3-dimensional visualizers that I use for my data science and machine learning projects.

See: https://docs.enthought.com/mayavi/mayavi/

To use the Mayavi 3D scientific data visualization and plotting in Python, you will need:

```
conda install -c anaconda mayavi
```

Open the sample example Jupyter Notebook: Chapter 02 Example 010A.ipynb in your Jupyter software.

This visualizer toolkit enables you to create really detailed 3D visualizations.

If you execute the notebook, you will get following 3D image (Figure 5-13) in your results directory.

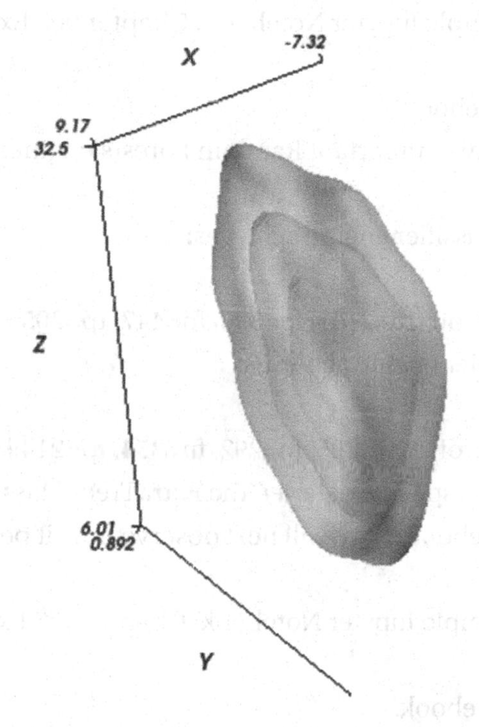

***Figure 5-13.** Kernel density estimation 3D with Mayavi 3D Engine*

Note You can interact with these images by zooming in and out … Or rotating them.

If you explore: Chapter 02 Example 010B.ipynb and Chapter 02 Example 010C.ipynb, you will see the complexity of visualization that is possible.

This is a great visualization toolkit. Please devote time to understand Mayavi's abilities.

Random Forests

Random forests are a combination of tree predictors such that each tree depends on the values of a random vector sampled independently and with the same distribution for all trees in the forest. The generalization error for forests converges to a limit as the number of trees in the forest becomes large.

The generalization error of a forest of tree classifiers depends on the strength of the individual trees in the forest and the correlation between them.

Example:

Open the sample example Jupyter Notebook: Chapter 005 Example 011.ipynb in your Jupyter software.

Run the complete notebook.

You are investigating two important Random Forests engines.

Your results will be:

The RandomForestClassifier engine achieves:

Score 83.140%

With confusion matrix of: (tn=2101, fp=396, fn=447, tp=2056)

The ExtraTreesClassifier engine achieves:

Score 87.080%

With confusion matrix of: (tn=2205, fp=292, fn=354, tp=2149)

This shows that for this specific data set, the ExtraTreesClassifier is better.

You can close the notebook as we will next observe how it performs against some real data.

Open the sample example Jupyter Notebook: Chapter 005 Example 012.ipynb in your Jupyter software.

Run the complete notebook.

Your results will classify the data for three colors of Roses.

Predicted Species	Floribunda (White)	Rosa kordesii (Red)	Rosaceae (Blue)
Actual Species			
Rosaceae (Blue)	0	0	41
Floribunda (White)	53	4	0
Rosa kordesii (Red)	0	45	0

The four Features impact the outcome with the following impact of %:

```
Leaf Length (mm)    :  9.3581%
Leaf Width (mm)     :  3.2167%
Stem length (mm)  : 43.2882%
Stem width (mm)    : 44.1370%
```

You can now close your notebook.

You have completed random forests.

I will next show you how to handle data that is not so perfect and needs some preprocessing to be useful.

Handling Imbalanced Data Sets

Imbalanced classes put "accuracy" out of business. This is a surprisingly common problem in machine learning (specifically in classification), occurring in data sets with a disproportionate ratio of observations in each class.

Standard accuracy no longer reliably measures performance, which makes model training much trickier.

I will show you how to handle these inaccurate data sets through showing you how to preprocess the data sets to fix the imbalance in the data structures.

For our first example, we will explain how these target variables have three classes.

R for right-heavy, that is, when var3 $*$ var4 > var1 $*$ var2

L for left-heavy, that is, when var3 ∗ var4 < var1 ∗ var2

B for balanced, that is, when var3 ∗ var4 = var1 ∗ var2

Using the imbalanced-learn library

```
conda install -c conda-forge imbalanced-learn
```

The imbalanced-learn library we will be using was started in August 2014 by Fernando Nogueira.

Open the sample example Jupyter Notebook: Chapter 005 Example 013.ipynb in your Jupyter software.

For this example, you will use several standard data sets that will enable you to compare the machine learning engines and their behavior against these data sets.

Your results are shown in Figure 5-14.

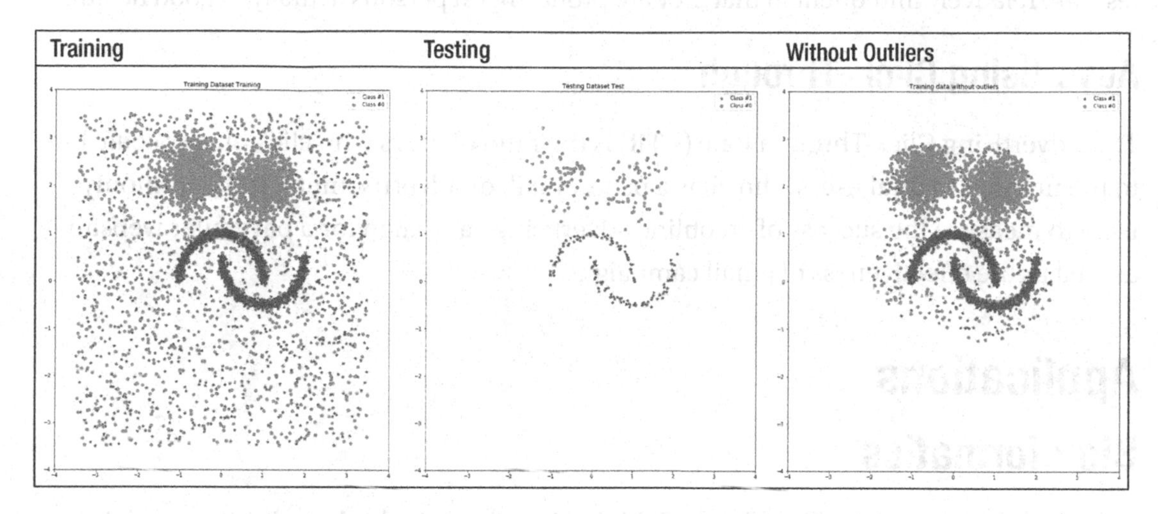

***Figure 5-14.** Handle imbalanced Data*

Imbalanced classes appear in many domains, including the following ones.

Fraud Detection

Fraud detection is a stimulating problem. The fact is that fraudulent transactions are rare in organizational data as is; in reality, they represent a very small fraction of activity within an organization. The challenge is that a small percentage of activity can quickly turn into money losses without the right tools and systems in place.

Criminals are cunning. As old-fashioned fraud schemes fail to pay off, fraudsters have learned to change their tactics. The great news is that with improvements in machine learning, systems can learn, adapt, and uncover emerging patterns for preventing fraud.

Spam Filtering

Email filtering is the processing of email to organize it according to specified criteria. Most advanced states are the automatic processing of incoming messages.

Disease Screening

Screening, in medicine, is a strategy used in a population to identify the possible presence of an as-yet-undiagnosed disease in individuals without signs or symptoms. These normally include individuals with presymptomatic or unrecognized symptomatic disease. Screening tests are relatively infrequent in that they are proficient on persons actually in good health.

Advertising Click-Through

The Advertising Click-Through Rate (CTR) is the ratio of users who click on a specific link to the number of total users who view a page, email, or advertisement. It is commonly used to measure the success of an online advertising campaign for a particular website as well as the effectiveness of email campaigns.

Applications

Bioinformatics

Bioinformatics is an interdisciplinary field that improves methods and software tools for investigating biological data. It is an interdisciplinary field of information engineering, biology, computer science, mathematics, and statistics to examine and understand biological data. Bioinformatics has been used via machine learning techniques that apply advanced mathematical and statistical techniques. Bioinformatics is collectively set up for the identification of candidates' genes and single nucleotide polymorphisms (SNPs) in DNA research. The extreme speed and volume proficiency of the ML processing formulates this as an exceptional tool in the hands of a good data scientist and machine learning engineer.

The application of industrialized machine learning to the field has already outclassed the effective discoveries of several former techniques by years; we are now performing levels of research that would have not been possible if it was not for the advances in machine learning.

Database Marketing

Database marketing is a form of direct marketing that uses databases of customers to generate targeted lists for direct marketing communications. Databases include customers' names and addresses, phone numbers, emails, information requests, purchase histories, and any other related data that can be legally and accurately collected Information for these databases is collected while the customer is performing basic business activities within their daily lives. These significant databases are in this day and age treasured as highly exclusive company assets and highly sought-after resources in the world of supervised learning.

Human-in-the-Loop

However, due to the dependence of properly and verified data to train the supervised learning system, the arrival of machine learning has also generated new jobs for humans.

These processes are called "Human-in-the-Loop," and this technique enables the ML to perform supervised learning for the data that is clearly classified; but for the items that are ambiguous or is a clear outlier, the decision is reversed to a human operator. This new decision is then added to the training data and the model is retrained.

The tagging of data is become big business worldwide. Many legitimate data firms collect personal information and are selling customer data gathered from many public sources. Public records are just one option, but information that people voluntarily shares through social media and business networking sites over and over again makes for a considerable richer harvest for a customer database.

This creates a scenario in which data that is not illegal to possess but that was obtained through illicit means can pass between data source owners without the identification of how illegitimate databases are now blended with legitimate ones that focus on public information. The personal data then becomes entrenched in its public availability even though it should not have been made public in the first place.

The final result can be a practically complete collection of an individual's personal information and life history available for sale, and at affordable prices like never before.

Warning Before you simply accept data into your machine learning environment, ensure you have the comprehensive data lineage and legal permission to process that data.

I suggest you look at regulations like General Data Protection Regulation 2016/679. It is a regulation in European Union (EU) law on data protection and privacy for all individuals within the EU and the European Economic Area and Data Protection Act 2018.

In the United States, look at Federal Trade Commission Act (15 U.S.C. §§41–58) (FTC Act) for data protection rules and principles, including the obligations on the data controller and the consent of data subjects; rights to access personal data or object to its collection; and security requirements. It furthermore covers cookies and spam, data processing by third parties, and the international transfer of data. These rules explain the fine points of the powers data regulator, its enforcement powers. and sanctions and remedies.

Warning Getting this wrong will expose you and your company to fines and other legal actions.

Machine Learning Methodology

I want to introduce you to a cross-industry standard process for data mining, commonly recognized by its acronym CRISP-DM. This is a data mining process model that defines a universal approach for data scientists and is used to tackle machine learning projects.

Seven steps of Machine Learning are shown in Figure 5-15.

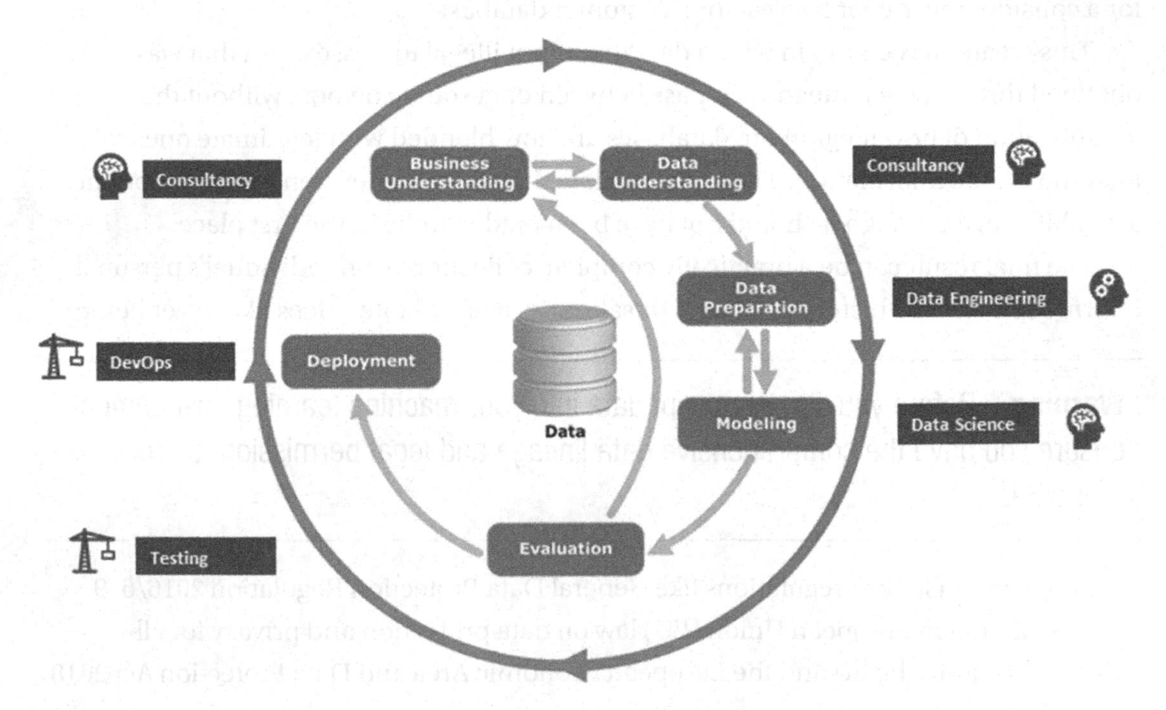

***Figure 5-15.** Cross-industry standard process for data mining (CRISP-DM)*

Who Does What in CRISP-DM?

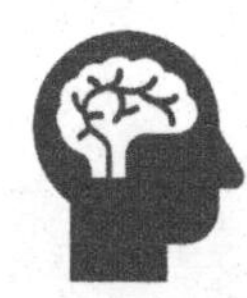

Data Scientists

The Data Scientists enable the team to select and implement the most effective and efficient algorithms and machine learning. Data scientists convert data into actionable insights.

Data Engineer

The Data Engineer enables the team to extract, transform, and load the required data into the data science structures as well as develop data features for use in modeling.

DevOps ensures controlled, managed release into production for ongoing effective and efficient Return-on-Investment through automation, continuous testing, and integration.

CRISP-DM Cycle

See Figure 5-15.

Business Understanding

The first section in CRISP-DM requires that you understand in detail what the business has as requirements. It is highly likely that you will receive competing or conflicting requirements on what the true requirements are from the business. If you do not get the phase to work 100% correctly, you will end up with producing at best the right answers to the wrong questions. Make sure you collect all relevant functional and nonfunctional requirements to ensure you have a complete view of the business's expectations.

What Are the Desired Outputs of the Project?

The major items to check at this point are to identify the priority of the business requirements and also cluster the related desired outputs together to improve the throughput of the data science and data engineering capability of your team.

Note I use an Agile scrum-style project management methodology where you simply place all desired outputs in the backlog first. Then we, as a team, process the backlog into logical sprints of work.

At this point in the process, your list of work is finalized.

Set Objectives

In this phase you must find the primary objective from a business perspective. This core or primary objective is the one answer you must always answer.

Example: If your main objective is: "How to stop high-value customers from leaving?" All your objectives must support this one question.

Secondary objectives could be:

- Who are my high-value customers?
- Why are they leaving?
- How do I stop them?

The system objectives must support these main questions.

- What data is needed?
- What format is the data in?

Produce Project Plan

The ML solution requires the same project plans any other project requires. At this point, your standard project management rules apply. I normally use an Agile project process as it works well for my customers.

Business Success Criteria

You must determine what is the minimum criteria from the business to achieve success with the project.

I use SMART criteria:

- Specific - Be precise in what you achieve. So, 1,000 less customers leaving is good.
- Measurable - Is the required outcome quantifiable? Can you determine what is the status?
- Achievable - It the target possible? Preventing 100,000 customers per month is not a reasonable target.
- Realistic - Make sure the target is possible.
- Time bound - There has to be a time by which targets are met.

Assess the Current Situation

Undertake a detailed fact-finding about all aspects of the current business situation. Make sure everybody accept that your findings are the starting baseline.

Make sure everybody involved understands the criteria.

Determine Data Goals

Determine the detailed goals of the machine learning process. Always relate back to the business requirements.

Business Success Criteria

Describe the intended outputs of the project that enable the achievement of the business objectives.

Data Science, Data Engineering, and Machine Learning success criteria

Define the criteria for a successful outcome to the project in technical terms To do this, explain how and what you will measure as success criteria.

Produce Project Plan

Produce an intended plan for all the data processing goals and how these are helping to achieve the business goals. Your plan should have specified the steps to perform during the rest of the project, including the initial selection of machine learning tools and development of new techniques.

Data Understanding

The second phase of the CRISP-DM process is to determine the true state of the data and the metadata. Understand the providence (Where data comes from?) and lineage (What processing was done to the data before you use it?).

Explore Data

This consists of what querying, data visualization, and reporting you will use. These may include:

Distribution of key attributes (for example, the target attribute of a prediction task)
Relationships between pairs or small numbers of attributes
Results of simple aggregations
Properties of significant subpopulations
Simple statistical analyses

These analyses must contribute to or refine the data description and quality reports, and feed into the transformation and other data preparation steps needed for further analysis.

Data Exploration Report

Perform structured data exploration and examination of all data subsets.

Verify Data Quality

Examine the quality of the data, addressing questions such as:

- Is the data complete (does it cover all the items required)?
- Is the base data correct, or does it contain errors and, if there are errors, how common are they?
- Are there missing values in the data? If so, how are they represented, where do they occur, and how common are they?

Data Quality Report

Produce data quality verification. Solutions to data quality problems, as a rule, depend heavily on both data and business knowledge.

Data Preparation

Select Your Data

Decide what data that you're using for analysis.

Rationale for Inclusion/Exclusion

Determine data to be included or excluded and the details for these decisions.

Clean Your Data

By cleaning the data quality, you will save on downstream processing of the analysis techniques for the estimation of missing data by modeling.

Data Cleaning Report

Describe what decisions and actions you took to resolve data quality problems.

Feature Engineering

Plan `Feature Engineering` to extract new attributes that are constructed from one or more existing attributes

Modeling

Select Modeling Technique

As the first step in modeling, you'll select the actual modeling technique that you'll be using. Although you may have already selected a tool during the business understanding phase, at this stage you'll be selecting the specific modeling technique, for example, decision-tree building with C5.0, or neural network generation with back propagation. If multiple techniques are applied, perform this task separately for each technique.

Test Design

Plan for training, testing, and evaluating the models. Plan the actions required for processing sample data sets into training, test, and validation data sets.

Build Model

Run the model on the prepared data set to create one or more models.

Parameter Settings

Any model requires specific parameters that can be adjusted. List the parameters and their chosen values, along with the rationale for the choice of parameter settings.

Models

These are the actual models produced by the modeling tool.

Model Assessment

Summarize the results of the produced models (e.g., in terms of accuracy), and rank their quality in relation to each other.

Revised Parameter Settings

Plan improvements within the model assessment, revise parameter settings, and tune them for the next modeling run. Iterate model building and assessment until you strongly believe that you have found the best model(s).

Evaluation

Assessment of Data Science and Machine Learning Results

Summarize assessment results in terms of business success criteria, including a final statement regarding whether the project already meets the initial business objectives.

Approved Models

After evaluating models with respect to business success criteria, the produced models that meet a selection of criteria become the approved models.

Review of Process

Summarize the process review and highlight activities that have been missed and those that should be repeated.

Determine Next Steps

Plan the assessment and the process review; you now decide how to advance.

Deployment

Summarize your deployment strategy, including the necessary steps and how to perform them.

Monitoring and Maintenance Plan

Summarize the monitoring and maintenance strategy, including the necessary steps and how to perform them.

Experience Documentation

Summarize important experience gained during the project.

Well done; you have reached the end of the CRISP-DM methodology.

How Do You Use This New Knowledge?

Now that you have the basic standard methodology of CRISP-DM, I want to introduce you to a processing system that I personally have been using for the last 10+ years. It has been the core subject of research for two MSc projects, and my PhD uses the basic principles for the new research processes.

It is also used by my data science teams.

I also teach and advise my clients from my position of Head of Data Science into how to use this process.

It is covered in the following books:

- *Practical Hive: A Guide to Hadoop's Data Warehouse System* by Scott Shaw, Andreas François Vermeulen, Ankur Gupta, and David Kjerrumgaard (Apress, 2016).
- *Practical Data Science: A Guide to Building the Technology Stack for Turning Data Lakes into Business Assets* by Andreas François Vermeulen (Apress, 2018).

Rapid Information Factory Ecosystem

The Rapid Information Factory ecosystem is a convention of techniques I used for my individual processing developments. The processing route of the book will be formulated from this basis, but you are not bound to exclusively use it. The tools I discuss in this chapter are available to you without constraints. The tools can be used in any configuration or permutation that is suitable for your specific ecosystem.

I recommend you start an implementation of formulating an ecosystem of your own or simply adopting mine. You have the prerequisites to become accustomed with a set of tools you know well and can deploy proficiently.

Remember that your data lake will have its own properties and features. So, adopt your tools to those particular characteristics on the ecosystem in which you work.

R-A-P-T-O-R Data Science Process Using Data Lake

What Is R-A-P-T-O-R?

This stands for Retrieve-Assess-Process-Transform-Organize-Report. See Figure 5-16.

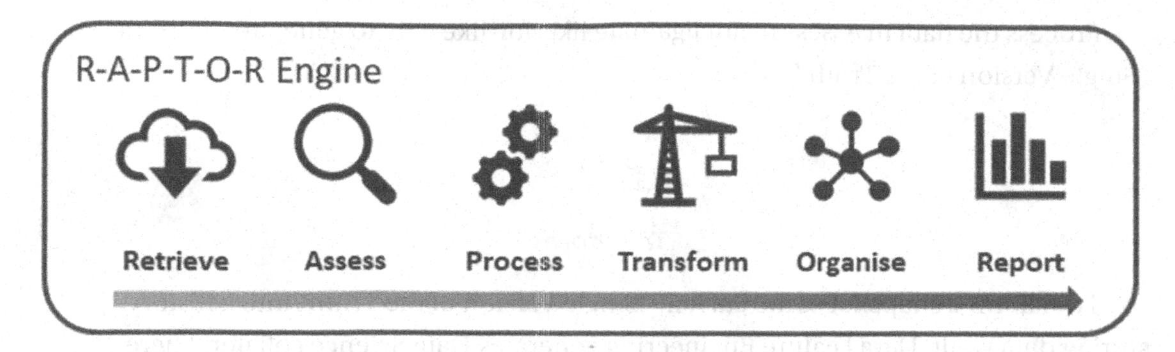

Figure 5-16. *R-A-P-T-O-R Engine*

The Rapid Information Factory (RIF)'s R-A-P-T-O-R engine guides the data from the data lake via the data pipe stages of retrieve (import external data sources), assess (data quality), process (amalgamate), transform (Single Version of the Truth), organize (feature engineering), and report (business insight).

The R-A-P-T-O-R engine (Figure 5-16) transforms the data as it is distributed between the different zones in the data lake.

Retrieve the data from the source system into the data engineering environment. This is the data as is in the source systems.

Assess the data in Retrieve to identify any data quality issues, and remediate the data to ensure a better-quality data science data source.

Process the data in Assess to amalgamate like-for-like data to generate a current "Single Version of the Truth."

Transform

Transform a snapshot of the current "Single Version of the Truth" into a history-storing data vault. Data Feature Engineering generates Data Science collateral here.

Organise

Organize the "Data Vault" into sets of data collections ready for the data science models to process. Generate Data Warehouse, Data Marts, Training, Test, and Result Data sets.

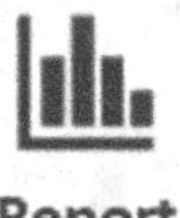

Report

Report the results of the Data Science as preapproved deliverables for each of the Data Science Cases ready for the standard reporting tools to consume.

What Is a Data Lake?

A data lake is a storage repository for a massive amount of raw data. It stores data in native format in anticipation of future requirements. You will acquire insights during this book on why this is extremely important for practical data science and engineering solutions. While a schema-on-write data warehouse stores data in predefined databases, tables, and records structures, a data lake uses a less-restricted schema-on-read-based architecture to store data. Each data element in the data lake is assigned a distinctive identifier and tagged with a set of comprehensive metadata tags.

A data lake is typically deployed using distributed data object storage to enable the schema-on-read structure. This means that business analytics and data mining tools access the data without a complex schema. Using a schema-on-read methodology supports you loading your data as is and starts to get value from it instantaneously.

I will discuss and provide more details on the reasons for using a schema-on-read storage methodology in Chapters 6, 7, 8, and 9.

Our deployment onto a cloud is a cost-effective solution to use Amazon Simple Storage Service (Amazon S3) to store the base data for the data lake.

I will demonstrate how you feasibly use cloud technologies to provision your data science work. It is, however, not needed to go to the cloud for the examples in this book as they will easily be processed on a laptop.

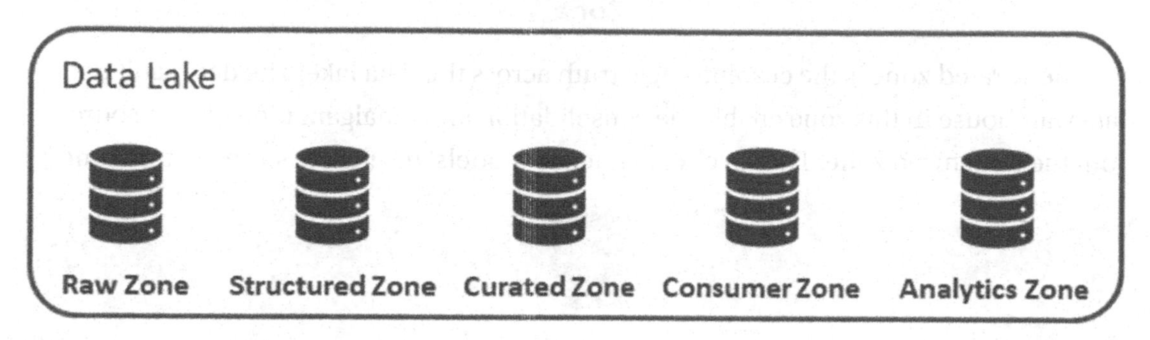

Figure 5-17. *Data Lake Six Zones*

This approach can leverage the existing technology stack, thereby reducing risk and cost.

Data Lake Zones

See Figure 5-17.

The raw zone is the entry point of all data outside the data lake data. It is the endpoint for several data extracting processing solutions.

The structured zone is used to convert the raw data into enhanced data sources. This zone's data is in harmonized formats to assist the next zone's processing capability. Any data quality issues are resolved in this zone.

The curated zone is the current single truth across the data lake. The data vault and data warehouse in this zone enable the consolidation and amalgamation of data sources from the structured zone. The "live" data science models' results are stored in this zone.

The consumer zone is the area where different data marts for the business insights are stored, ready for the end users to visualize their business insights. This is the primary zone for the majority of the businesspeople's questions.

Tip The consumer zone is the only area that is exposed to the active customer base in agreement to the approved Service-Level Agreements (SLA) to ensure the rest of the system is capable of proper maintenance and evolvement of capability.

The analytics zone is the "sandbox" of the data lake. This zone is used to design, develop, and train new and novel data science and machine learning solutions before DevOps transfers it to the curated zone.

What Is a Data Vault?

Data vault modeling, designed by Dan Linstedt, is a database modeling method that is intentionally structured to be in control of long-term historical storage of data from multiple operational systems.

The data vaulting processes transforms the schema-on-read data lake into a schema-on-write data vault.

The data vault is designed into the schema-on-read query request and then executed against the data lake.

I have also seen the results being stored as a schema-on-write format to persist the results for future queries.

The techniques for both of these techniques are discussed in Chapter 8. At this point I only expect you to understand the rudimentary structures required to formulate a data vault.

The structure is built from three basic data structures: Hubs, Links, and Satellites. Let's examine the specific data structures to clarify why they are compulsory.

Hubs

Hubs contain a list of unique business keys with a low propensity to change. Hubs contain a surrogate key for each hub item and metadata classification of the origin of the business key.

The hub is the core backbone of your data vault, and I will discuss in more detail how and why you use this structure in Chapter 9.

Links

Associations or transactions between business keys are modeled using link tables. These tables are essentially many-to-many join tables, with specific additional metadata.

The link is a singular relationship between hubs to ensure the business relationships are accurately recorded to complete the data model for the real-life business.

In Chapter 9, I will explain how and why you would need specific relationships.

Satellites

The hubs and links form the structure of the model, but store no chronological characteristics or descriptive characteristics of the data. These characteristics are stored in appropriated tables identified as satellites.

The satellites are the structures that store comprehensive levels of the information on business characteristics and are normally the largest volume of the complete data vault data structure. In Chapter 9, I will explain how and why these structures work so well to model the real-life business characteristics.

The appropriate combination of hubs, links, and satellites supports the data scientist to construct and store prerequisite business relationships. This is a highly in-demand skill as a data modeler.

The transformation to this schema-on-write data structure is discussed in detail in Chapter 9 to point out why a particular structure supports the processing methodology.

I will explain in Chapter 9 ... Why you would need particular hubs, links. and satellites.

What Next?

You have successfully completed Chapter 3.

You now have achieved a good working knowledge of the supervised learning machine learning process.

You are familiar with the CRISP-DM methodology, and this ensures your machine learning transcends from experiments and hypotheses to IML.

I shared my own knowledge and processes I use via the RIF in my daily work as a machine learning consultant.

This process will be covered in Chapter 15 during our full-scale Industrialization project in which we will cover the practical use of the techniques and the framework of the RIF.

Next, I will cover unsupervised machine learning in Chapter 6.

CHAPTER 6

Unsupervised Learning: Using Unlabeled Data

Unsupervised Learning is a type of machine learning that acquires insight by inferring a function to describe hidden structures from unlabeled, uncategorized data. The classification or categorization is not included in the training observations. There is consequently no right or wrong evaluation of the learner and no evaluation of the accuracy of the learned insights that is output by the related algorithm used.

On one hand, unsupervised learning algorithms can perform more complex processing tasks than supervised learning systems. On the other hand, unsupervised learning can be more unpredictable than the other machine learning models.

Algorithms

I will discuss a few of the core and commonly used algorithms in data processing using unsupervised learning. I will start with one of the most common algorithms.

K-Nearest Neighbor Algorithm

BallTree is for fast generalized N-point problems.

Open the sample example Jupyter Notebook: Chapter-006A-001- k-Nearest-Neighbor.ipynb in your Jupyter software.

You are investigating the closest four and eight neighbors.

The NearestNeighbors with ball_tree engine performs the processing.

```
nbrs1 = NearestNeighbors(n_neighbors=ncnt1,  algorithm='ball_tree' ).fit(X)
```

Your results are shown in Figure 6-1 (four neighbors) and Figure 6-2 (eight neighbors).

A. F. Vermeulen, *Industrial Machine Learning*, https://doi.org/10.1007/978-1-4842-5316-8_6

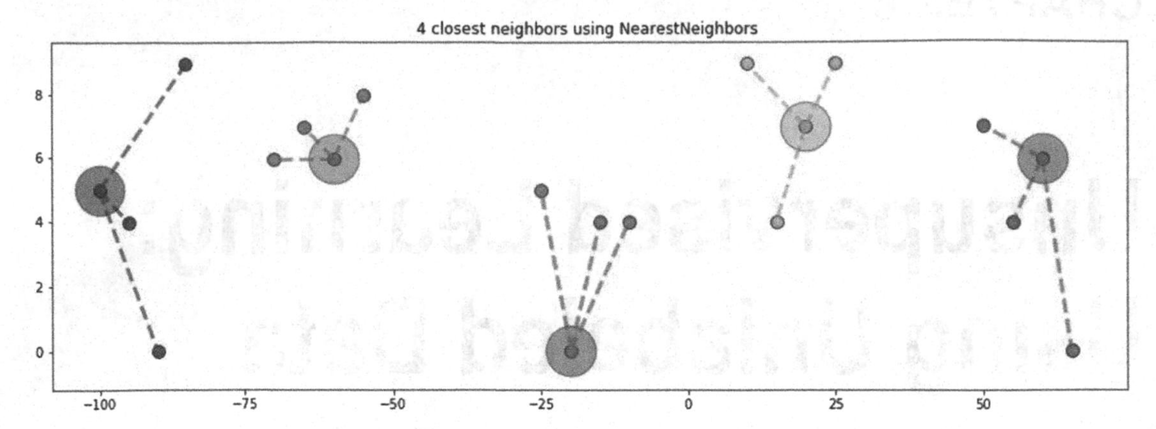

Figure 6-1. *Four nearest neighbors*

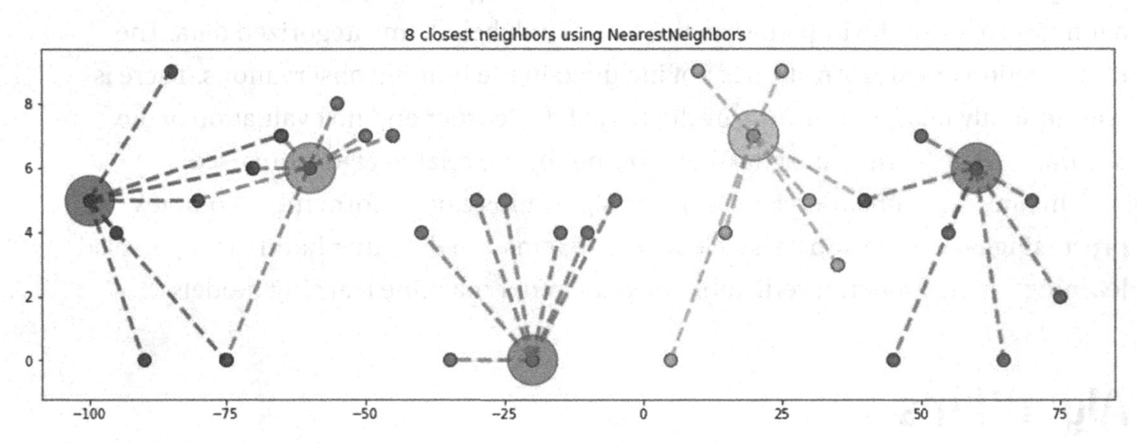

Figure 6-2. *Eight nearest neighbors*

You can observe how the closest four neighbors have no joint neighbors, but when expanding to the closest eight neighbors, you start observing joint neighbors. This type of joint neighbors' investigation is used in e-commerce to decide who you will recommend or have interaction with in your area or in health care to conclude who you may possibly have passed that nasty flu to for the duration of the winter.

Please examine Jupyter Notebook: Chapter-006B- 001-k-Nearest-Neighbor.ipynb and Jupyter Notebook: Chapter-006C- 001-k-Nearest-Neighbor.ipynb to observe how to apply the same algorithm to a postcode/zip code solution for the United Kingdom and the United States of America.

There is also Jupyter Notebook: Chapter-006D- 001-k-Nearest-Neighbor.ipynb and Jupyter Notebook: Chapter-006E- 001-k-Nearest-Neighbor.ipynb. Can you determine where these locations are in the world?

The ability to determine the closest points of interest (Nearest Neighbor) is an ability that is use in IML for mechatronics to determine the shortest or closest source of a specific resource or workspace.

In Chapter 15, I will demonstrate how this is used to determine a mining location.

Clustering K-Means

Voronoi Cells

K-means clustering is a method of vector quantization, originally from signal processing, which is popular for cluster analysis in data mining. K-means clustering aims to partition n observations into k clusters in which each observation belongs to the cluster with the nearest mean, serving as a prototype of the cluster. This results in a partitioning of the data space into Voronoi cells.

Open the Jupyter notebook called: Chapter-006-002-Clustering-K-Means-01-02.ipynb

I will now show you the true data distribution for five clusters (see Figure 6-3).

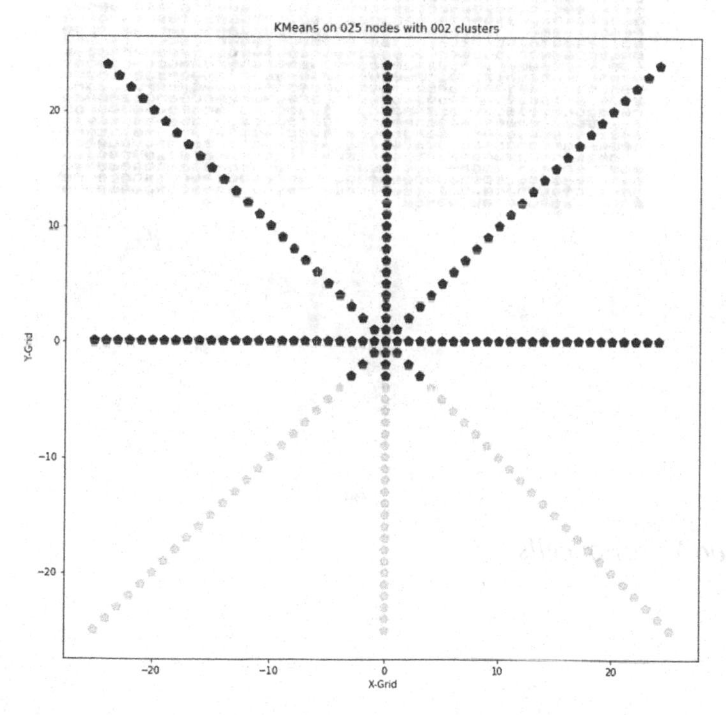

Figure 6-3. *Cross-Grid for 2 Clusters*

The K-Means have generated a two-cluster split of the data.

Warning The K-Means on this dataset is not 100% splitting into two clusters. You get the following results:

(k-means++) Cluster 1: 1450 (58.0000%) vs. Cluster 2: 1050 (42.0000%)

(random) Cluster 1: 1050 (42.0000%) vs. Cluster 2: 1450 (58.0000%)

This indicates that the two algorithms cause a dissimilar outcome due to the different starting states of the machine learning algorithms.

Let's investigate the Voronoi cells that these two clusters form (see Figure 6-4).

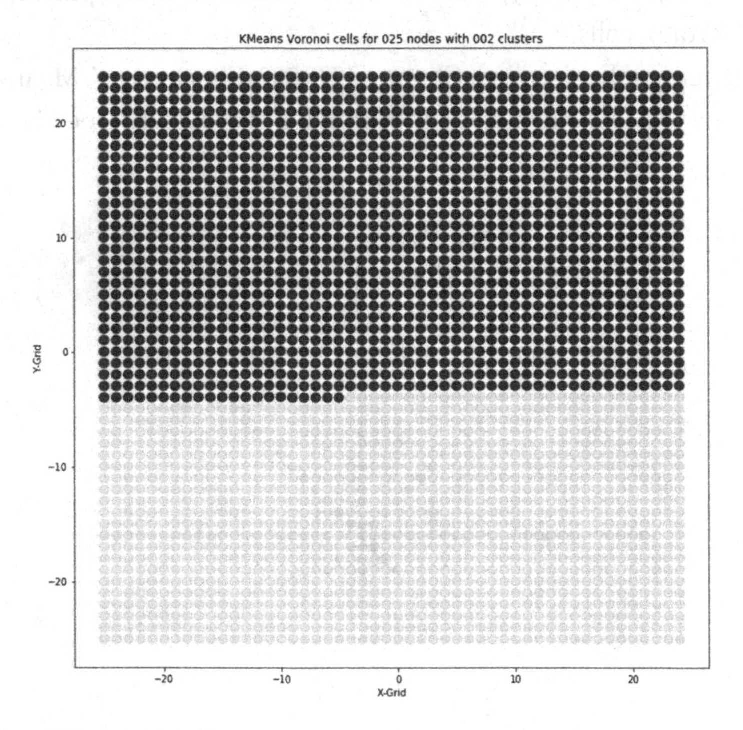

Figure 6-4. *Two Voronoi cells*

These results show you clearly that the two clusters form 2 Voronoi cells.

If you want to experiment with the cluster, you can modify the cluster count in this piece of code:

N=int(2) is Clusters required.

M=int(25) is Nodes in grid.

If you open the Jupyter notebook called: Chapter-006-002-Clustering-K-Means-01-13.ipynb

The base of this investigation is:

N=int(13)

M=int(25)

So, we would expect 13 clusters. Your results are show in Figures 6-5 and 6-6.

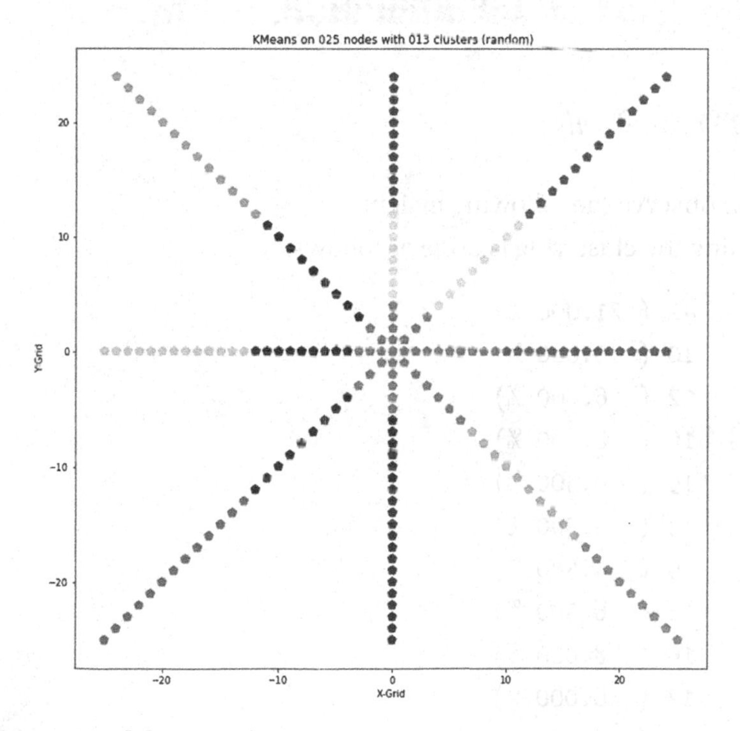

Figure 6-5. *Cross-Grid for 13 Clusters*

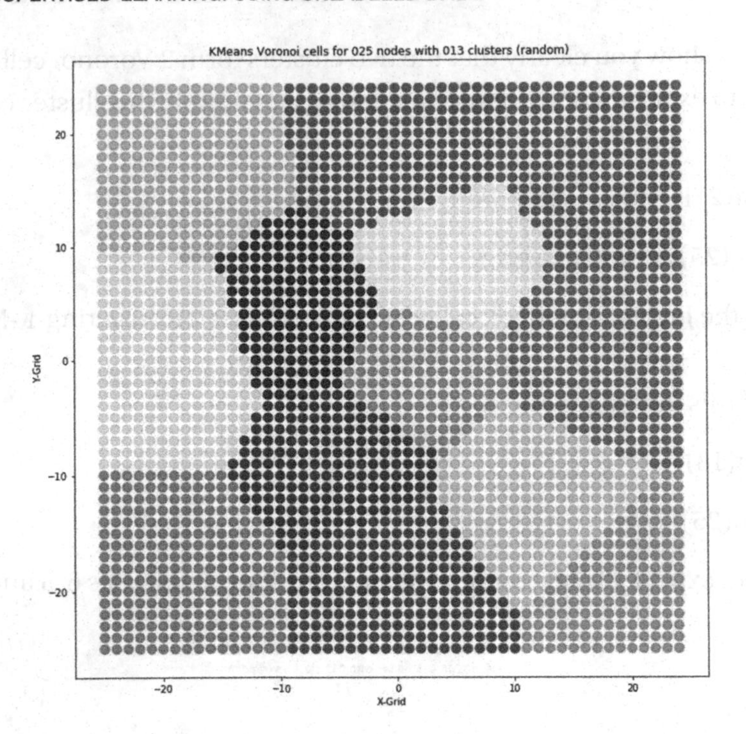

Figure 6-6. *13 Voronoi cells*

I suggest you observe the following insight:

During training the clustering is done as follows:

```
Cluster   1 :   42 ( 21.000 %)
Cluster   2 :   10 (  5.000 %)
Cluster   3 :   12 (  6.000 %)
Cluster   4 :   16 (  8.000 %)
Cluster   5 :   19 (  9.500 %)
Cluster   6 :   13 (  6.500 %)
Cluster   7 :    9 (  4.500 %)
Cluster   8 :   13 (  6.500 %)
Cluster   9 :   16 (  8.000 %)
Cluster  10 :   12 (  6.000 %)
Cluster  11 :   10 (  5.000 %)
Cluster  12 :   10 (  5.000 %)
Cluster  13 :   18 (  9.000 %)
----------------------
Cluster     :  200
```

But when the Voronoi cells are tested, you find a different distribution:

```
Cluster   1 :  142 (  5.680 %)
Cluster   2 :  173 (  6.920 %)
Cluster   3 :  209 (  8.360 %)
Cluster   4 :  227 (  9.080 %)
Cluster   5 :  166 (  6.640 %)
Cluster   6 :  226 (  9.040 %)
Cluster   7 :  149 (  5.960 %)
Cluster   8 :  240 (  9.600 %)
Cluster   9 :  234 (  9.360 %)
Cluster  10 :  204 (  8.160 %)
Cluster  11 :  187 (  7.480 %)
Cluster  12 :  186 (  7.440 %)
Cluster  13 :  157 (  6.280 %)
----------------------
Cluster     : 2500
```

The Voronoi cells are the true prediction spread of the model.

I have generated a small number of additional combinations, so please investigate:

- Chapter-006-002-Clustering-K-Means-01-04.ipynb
- Chapter-006-002-Clustering-K-Means-01-07.ipynb
- Chapter-006-002-Clustering-K-Means-01-09.ipynb

These all show how the spread varies with the increase in clusters.

The next step is a change to observe a bigger cluster process.

```
Chapter-006-002-Clustering-K-Means-01-20.ipynb
N=int(20)
M=int(100)
```

You will observe how the two algorithms give altered results:

```
Cluster   1 : (k-means++) 1653 (  4.133 %) vs (random) 1356 (  3.390 %) (Net:   297)
Cluster   2 : (k-means++) 1342 (  3.355 %) vs (random) 2663 (  6.657 %) (Net: -1321)
Cluster   3 : (k-means++) 2929 (  7.322 %) vs (random) 2649 (  6.623 %) (Net:   280)
Cluster   4 : (k-means++) 2667 (  6.667 %) vs (random) 2679 (  6.698 %) (Net:   -12)
```

```
Cluster   5 : (k-means++) 1238 (  3.095 %) vs (random) 2996 (  7.490 %) (Net: -1758)
Cluster   6 : (k-means++) 1770 (  4.425 %) vs (random) 1478 (  3.695 %) (Net:   292)
Cluster   7 : (k-means++) 2657 (  6.643 %) vs (random) 2556 (  6.390 %) (Net:   101)
Cluster   8 : (k-means++) 2704 (  6.760 %) vs (random) 1378 (  3.445 %) (Net:  1326)
Cluster   9 : (k-means++) 2759 (  6.897 %) vs (random) 1329 (  3.322 %) (Net:  1430)
Cluster  10 : (k-means++) 2602 (  6.505 %) vs (random) 2364 (  5.910 %) (Net:   238)
Cluster  11 : (k-means++) 1449 (  3.623 %) vs (random) 1077 (  2.692 %) (Net:   372)
Cluster  12 : (k-means++) 2668 (  6.670 %) vs (random) 1711 (  4.277 %) (Net:   957)
Cluster  13 : (k-means++) 2153 (  5.382 %) vs (random) 1123 (  2.808 %) (Net:  1030)
Cluster  14 : (k-means++) 2668 (  6.670 %) vs (random) 2400 (  6.000 %) (Net:   268)
Cluster  15 : (k-means++) 1569 (  3.923 %) vs (random) 1514 (  3.785 %) (Net:    55)
Cluster  16 : (k-means++) 1528 (  3.820 %) vs (random) 2701 (  6.753 %) (Net: -1173)
Cluster  17 : (k-means++) 1304 (  3.260 %) vs (random) 2360 (  5.900 %) (Net: -1056)
Cluster  18 : (k-means++) 1711 (  4.277 %) vs (random) 1588 (  3.970 %) (Net:   123)
Cluster  19 : (k-means++) 1286 (  3.215 %) vs (random) 2672 (  6.680 %) (Net: -1386)
Cluster  20 : (k-means++) 1343 (  3.357 %) vs (random) 1406 (  3.515 %) (Net:   -63)
----------------------
Cluster     : 40000
```

Tip Always test the Voronoi cells to ensure you report the models' true outcomes

---

We will now test our new knowledge against a real solution.

Next, I will guide you through an example with data from a real-life clustering problem.

The data set has the following structure:

The rose data set is a characteristic and stress-free multiclass classification data set.

Classes	**3**
Samples per class	50
Samples total	150
Dimensionality	4
Features	real, positive

Open the Jupyter notebook called: Chapter-06-003-Clustering-K-Means-02.ipynb

If you execute the notebook you will observe that the process creates a series of K-Means engines by running them as a collection of estimators (See Part C).

Your results are produced by Part D (Figures 6-7 to 6-10).

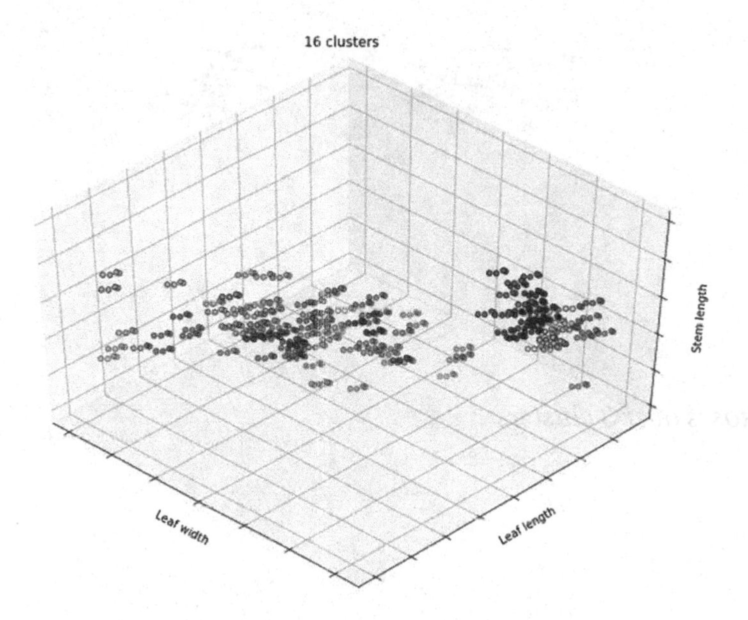

Figure 6-7. *Roses over 16 clusters*

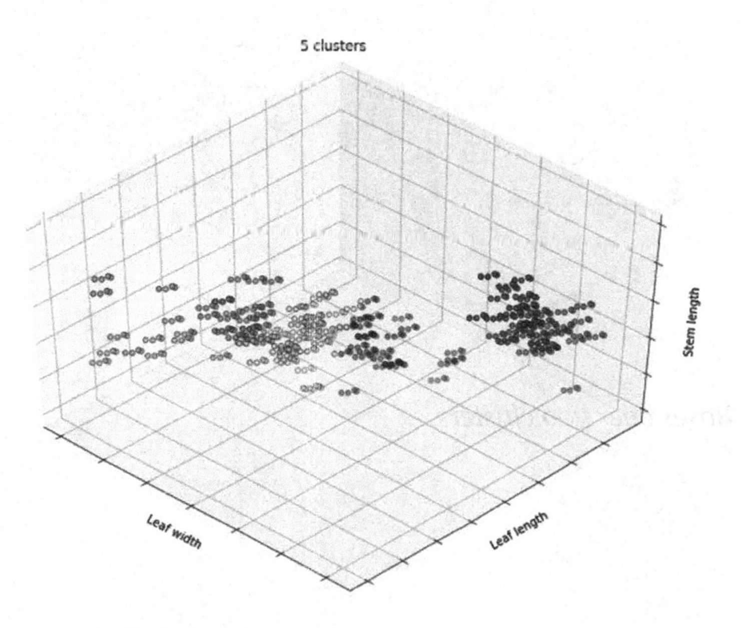

Figure 6-8. *Roses over 5 clusters*

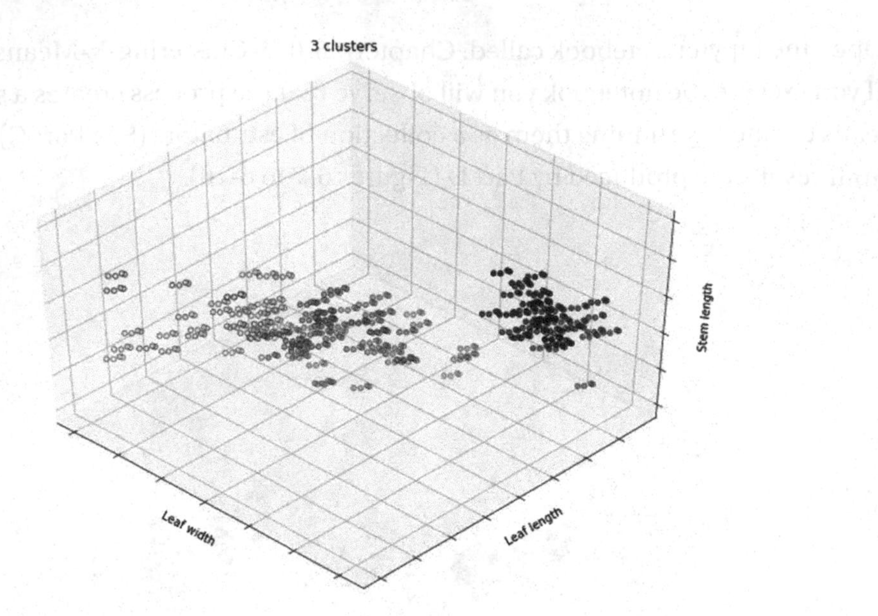

Figure 6-9. *Roses over 3 clusters*

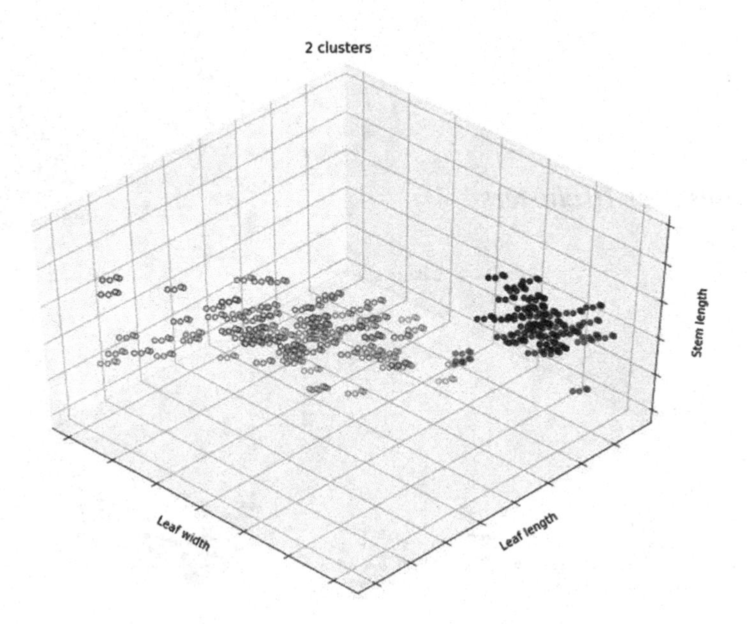

Figure 6-10. *Roses over two clusters*

The real data plots show that the three clusters are the closest to the real-life distribution of the data.

You can now close the notebook as we are now moving on to another common use for K-Means engines.

I will now show you how you can use this new knowledge about the K-Means algorithm to manipulate an example image called "China.jpg".

Open the Jupyter notebook called: Chapter-006-004A-Clustering-K-Means-China.ipynb

Run the notebook to observe how the K-Means engine is used to reduce the colors on a picture.

For more insights you should investigate Chapter-006-004B-Clustering-K-Means-Flower.ipynb

Can you see how the use of K-Means can assist in processing images?

Now that you have the basic process, I suggest we try this technique against a famous landmark, that is, Edinburgh Castle.

Open the Jupyter notebook called: Chapter-006-004C-Clustering-K-Means-Edinburgh.ipynb

The photo of Edinburgh gives you a clear demonstration of how a K-Means engine interacts with the image.

You should get the following results as shown in Figures 6-11 to 6-13.

Figure 6-11. *Original Picture*

Figure 6-12. *Picture with 64 Colors*

Figure 6-13. *Quantized Picture*

This type of analysis is useful when working with CCTV images or robotics vision.

To enable you to hone your skills, investigate Chapter-006-004D-Clustering-K-Means-Horses.ipynb

You will observe that you will be first performing several minor image processing tasks to find the horse you want to process. This is mostly required in real-world applications.

Optimum K Value

The next critical insight you need is this: How do you determine the optimum K value?

Open the Jupyter notebook called Chapter-006-005A-Clustering-K-Means-Optimum.ipynb

The solution to finding the optimum is to perform an Elbow method that supplies you with the appropriate number of clusters in a given data set.

I have supplied you two data sets: Alpha and Beta. I suggest you plot the Data Set Alpha with K against the sum of squared errors (SSE).

Take note that the calculation for cost is Error Sum of Squares for this example. SSE is the sum of the squared differences between each point and its assigned clusters' means. This is a common measure of variation within a given data cluster. This example uses an acceptable cost comparable for given K values to find an optimum K (Figure 6-14).

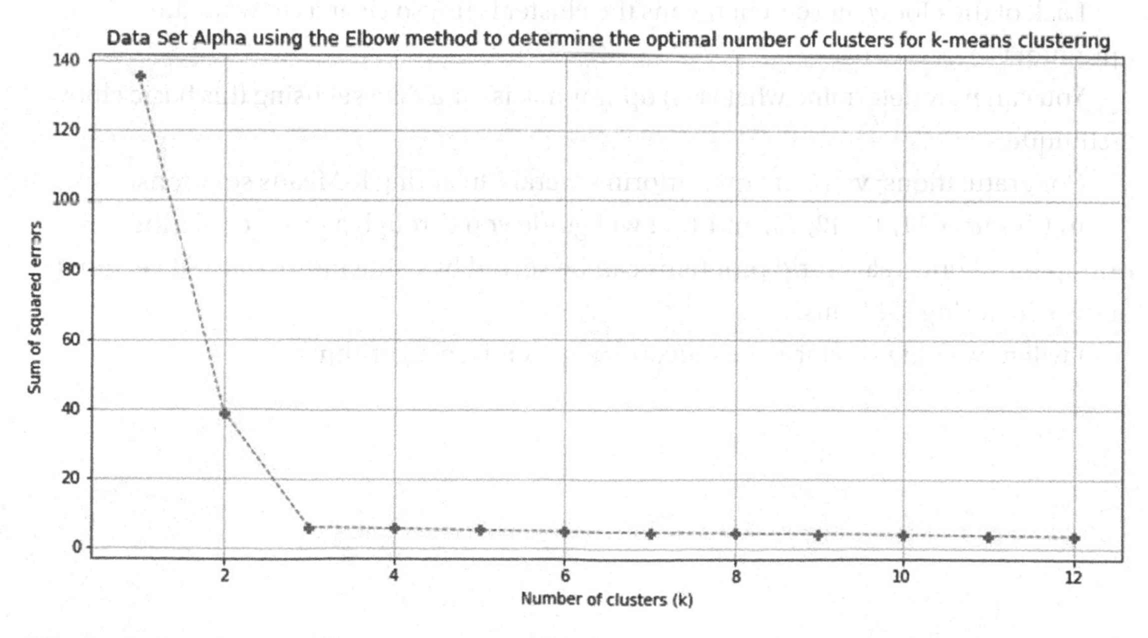

Figure 6-14. *Acceptable cost comparable for given K values to find an optimum K*

The Alpha data set has a k=3 optimum as the elbow movement indicates.

I suggest you look at Data Set Beta with K against the SSE (Figure 6-15).

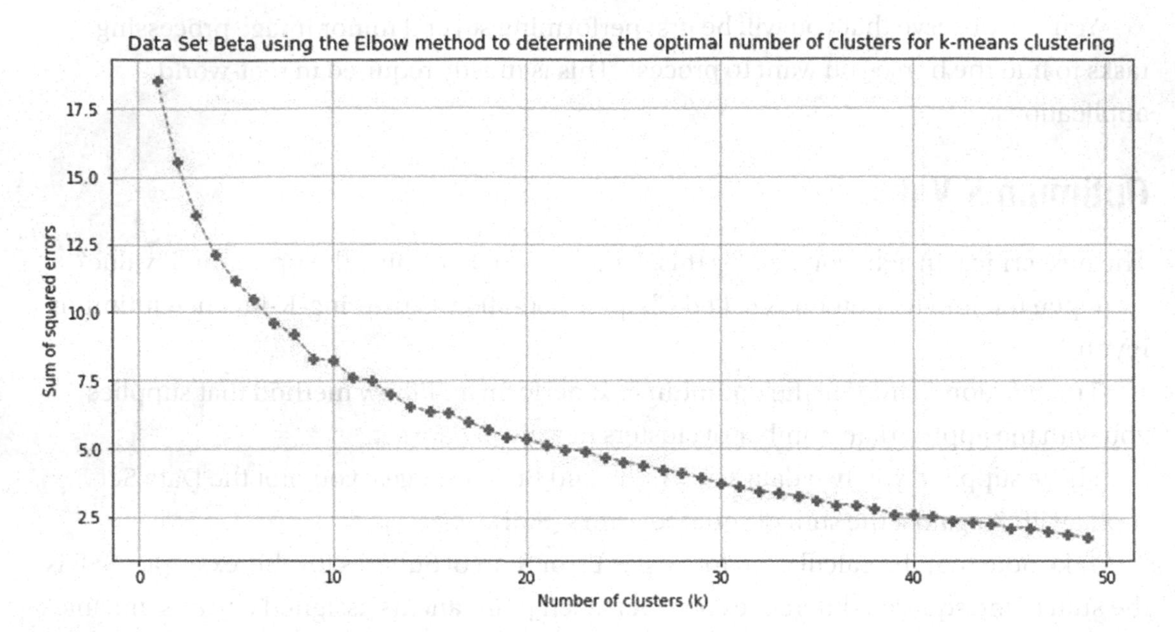

***Figure 6-15.** Data Set Beta with k against SSE*

Lack of the elbow movement means the cluster is not so clear as to what the optimum is.

You can now determine what is an optimum k is for a data set using this basic elbow technique.

Congratulations; you can now perform several Clustering K-Means solutions.

In Chapters 11, 12, 13, 14, and 15, I will guide you through a number of other examples on how real-world problems can be solved by using industrialized versions of the Clustering K-Means.

I will now guide you through a second unsupervised algorithm.

Gaussian Mixture Models

A Gaussian mixture model is a probabilistic model that assumes all the data points are generated from a mixture of a finite number of Gaussian distributions with unknown parameters. One can think of mixture models as generalizing K-Means clustering to incorporate information about the covariance structure of the data as well as the centroids of the latent Gaussians.

Example:

Open the Jupyter notebook called: Chapter-006-006-Gaussian-Mixture-01.ipynb

You will use the Rose series 7 data set to explain the Gaussian model.

The solution cycle through the four different covariance parameters:

- 'full' - each component has its own general covariance matrix.
- 'tied' - all components share the same general covariance matrix.
- 'diag' - each component has its own diagonal covariance matrix.
- 'spherical' - each component has its own single variance.

Your results are shown in Figure 6-16.

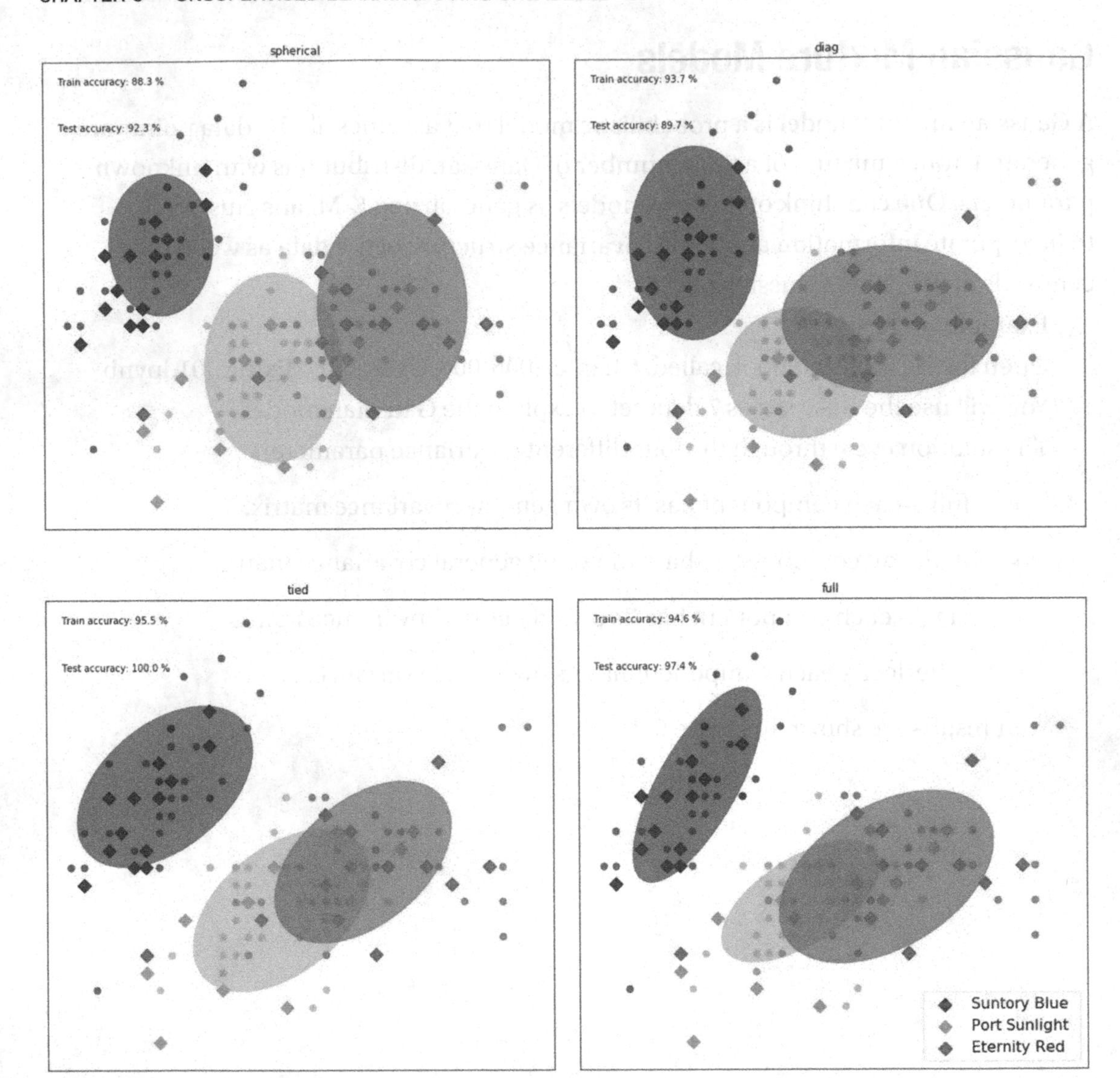

Figure 6-16. *Gaussian mixture model Results*

Can you see how the different covariance type impacts the outcome of the clustering?

Tip I suggest you always test these covariance parameters against your data sets as there are several times I found that only the slightest change supplies better results.

Open the Jupyter notebook called: Chapter-006-007-Gaussian-Mixture-02.ipynb

The process is to find the best covariance parameters hyperparameters for a given data set.

Your results can be seen in Figure 6-17.

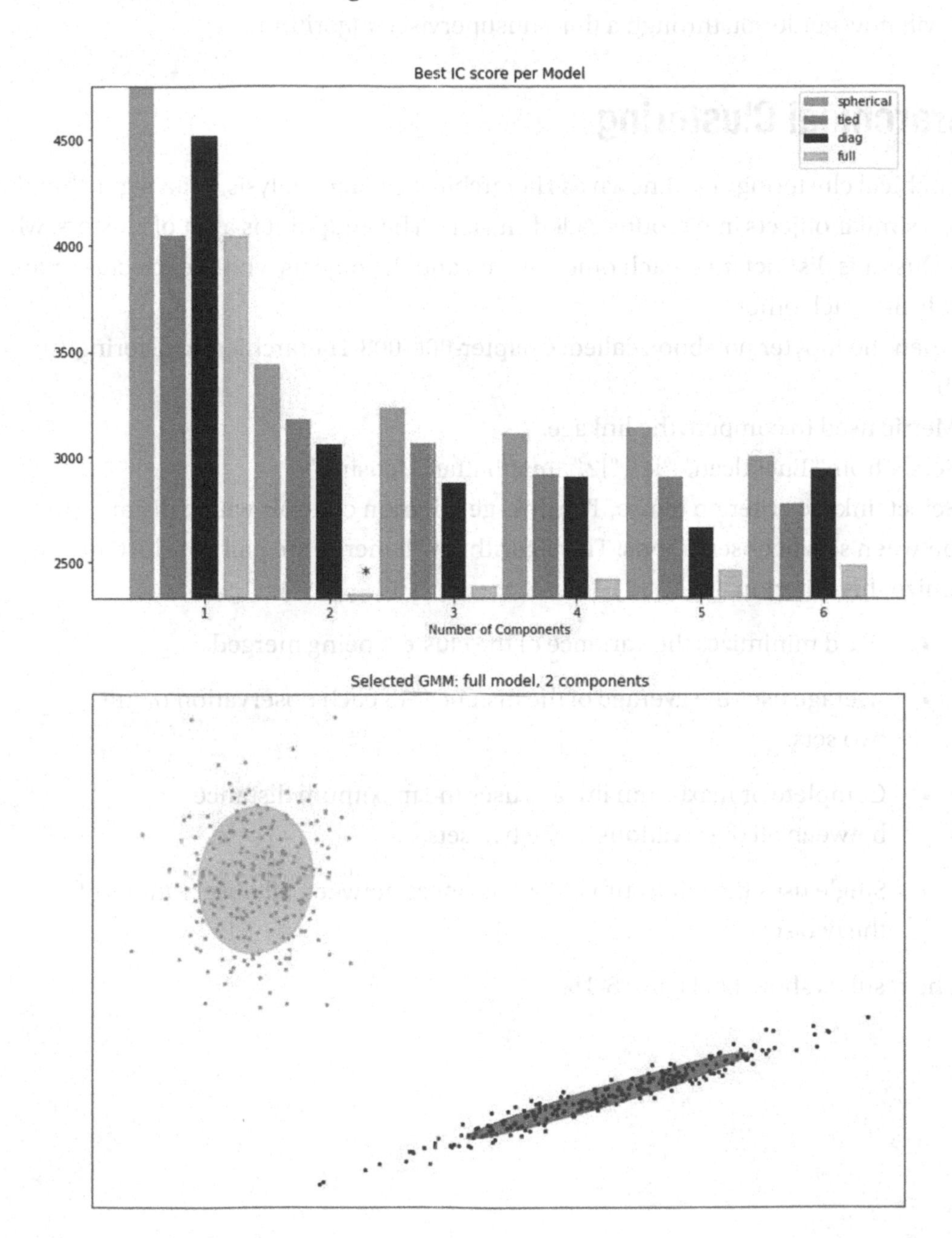

Figure 6-17. *Best Gaussian mixture model Results*

Well done; you can now successfully perform a Gaussian mixture model solution.

In Chapters 11, 12, 13, 14, and 15, I will guide you through several additional examples on how real-world problems can be solved by using industrialized versions of the Gaussian mixture models.

I will now guide you through a third unsupervised algorithm.

Hierarchical Clustering

Hierarchical clustering, also known as Hierarchical cluster analysis, is an algorithm that groups similar objects into groups called clusters. The endpoint is a set of clusters, where each cluster is distinct from each other cluster, and the objects within each cluster are broadly like each other.

Open the Jupyter notebook called: Chapter-006-008-Hierarchical-clustering-01.ipynb

Metric used to compute the linkage.

Select from "Euclidean," "l1," "l2," "manhattan," "cosine"

Select linkage criterion to use. The linkage criterion controls which distance to use between sets of observations. The algorithm will merge the pairs of clusters that minimize this criterion.

- Ward minimizes the variance of the clusters being merged.
- Average uses the average of the distances of each observation of the two sets.
- Complete or maximum linkage uses the maximum distances between all observations of the two sets.
- Single uses the minimum of the distances between all observations of the two sets.

The result is shown in Figure 6-18.

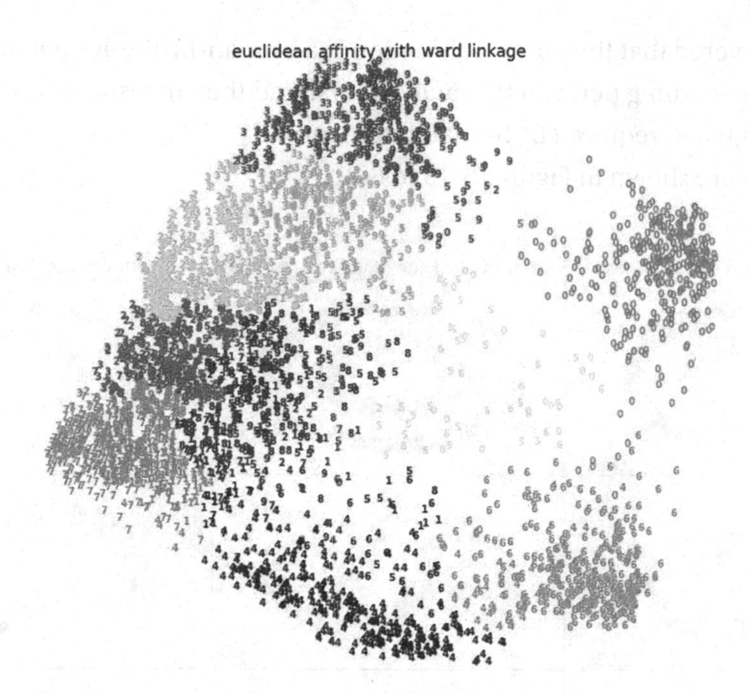

Figure 6-18. *Hierarchical clustering(1 of 8)*

The parameters for the clustering are showing that these contrasts in parameters have either a positive or negative impact on the success of the clustering.

I suggest we look at this impact in a more controlled manner.

Let's investigate the impact of linkage and affinity combinations.

Open the Jupyter notebook called: Chapter-009-009-Hierarchical-clustering-02.ipynb, Chapter-009-010-Hierarchical-clustering-02.ipynb. and Chapter-009-011-Hierarchical-clustering-02.ipynb

You will use noisy circles, noisy moons, blobs. and no structure data configurations to produce a data set in which you can test the behavior of the different hyperparameters.

Tip I personally use this test for my IML model selection as it shows you quickly how specific patterns impact your results.

Execute the notebooks and observe the results.

You should observe that the different parameters have several interesting and counterintuitive outcomes.

This is the main reason humans are still, in general, better at identifying complex patterns.

I have discovered that the best combination is a human-in-the-loop methodology. Let the machine learning perform the analyses first and then investigate the found patterns and adjust as required by human intervention.

Your results are shown in Figures 6-19 to 6-21.

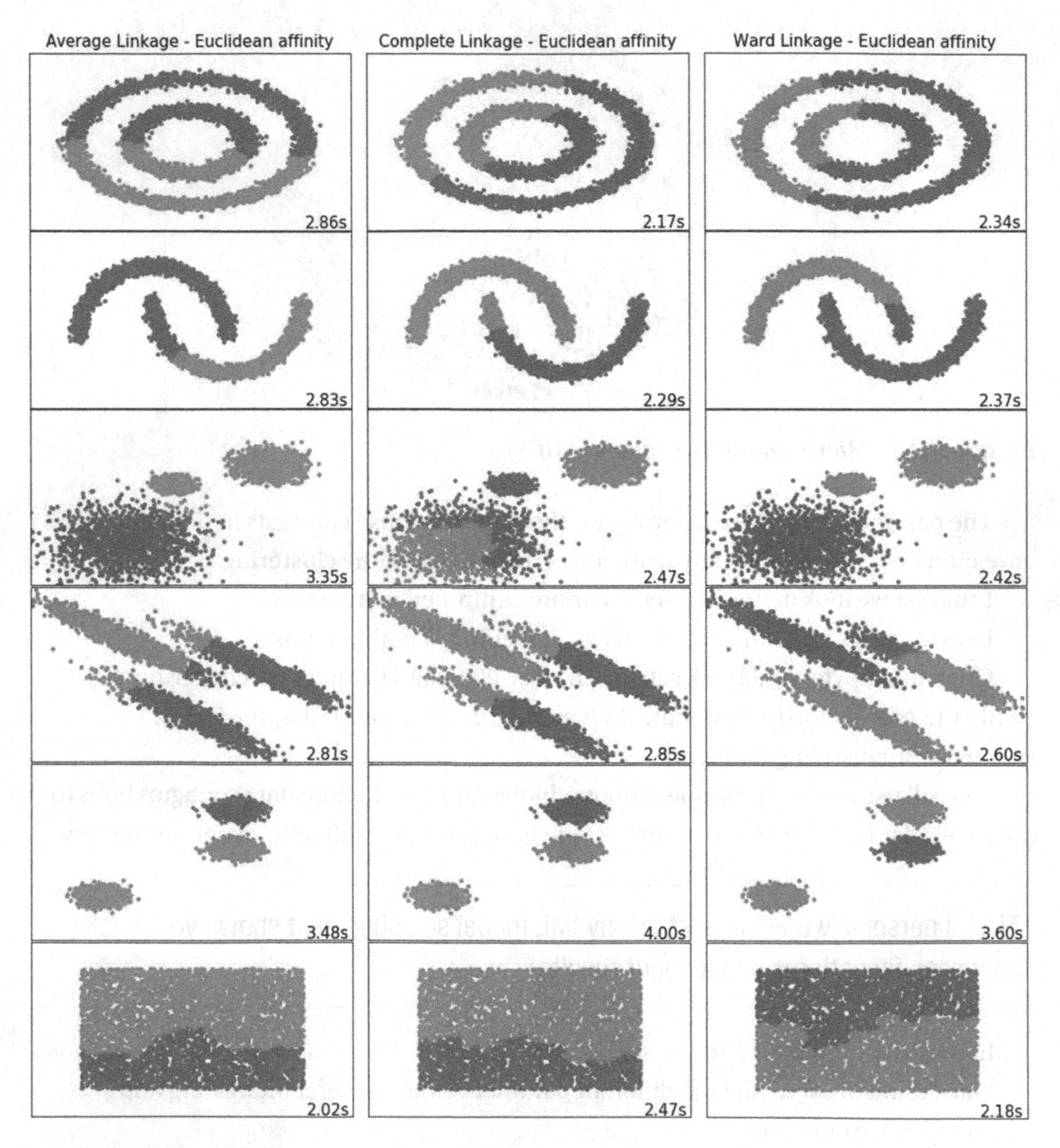

Figure 6-19. *Hierarchical – Linkage Impact*

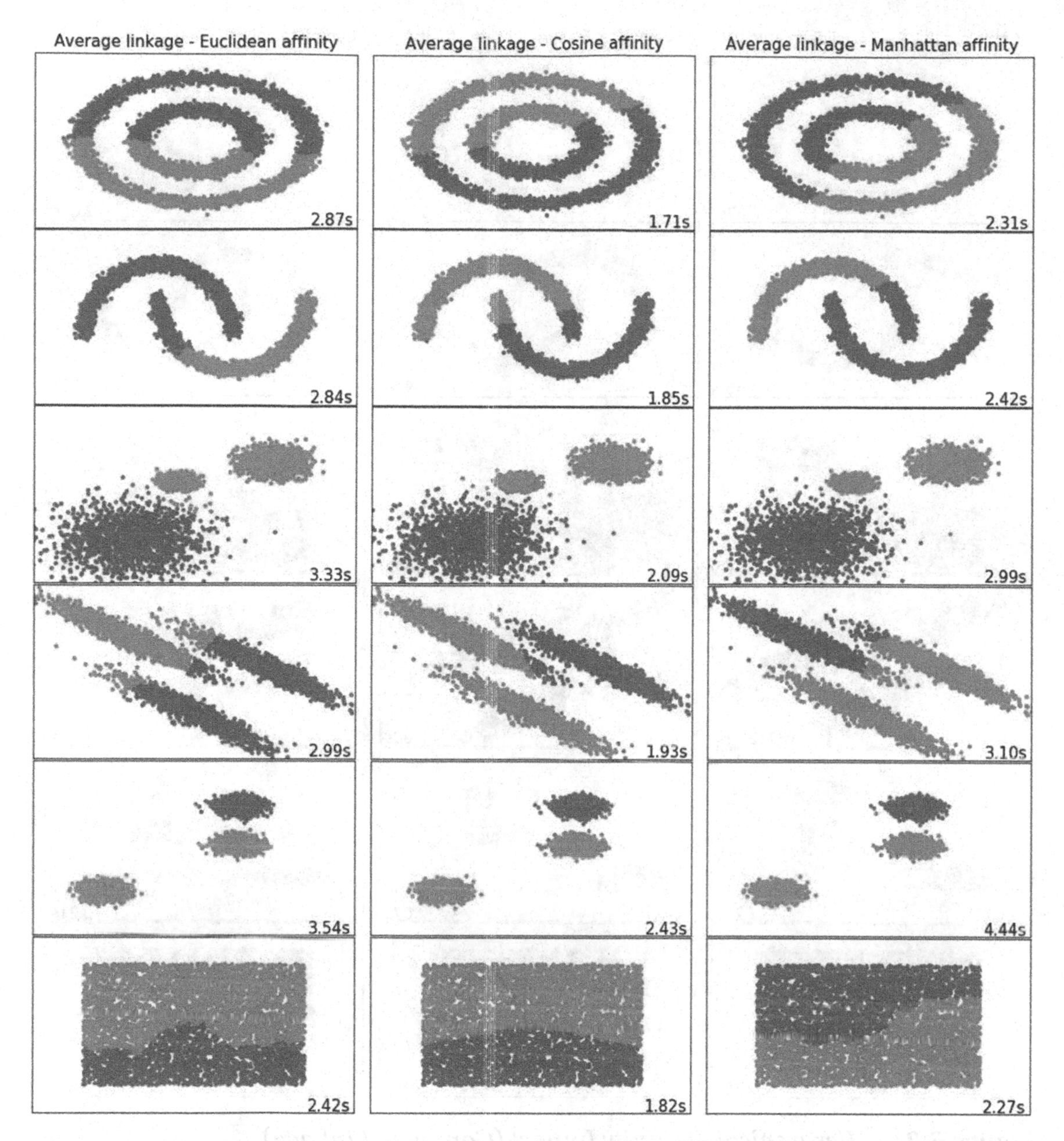

***Figure 6-20.** Hierarchical – Affinity Impact (Average Linkage)*

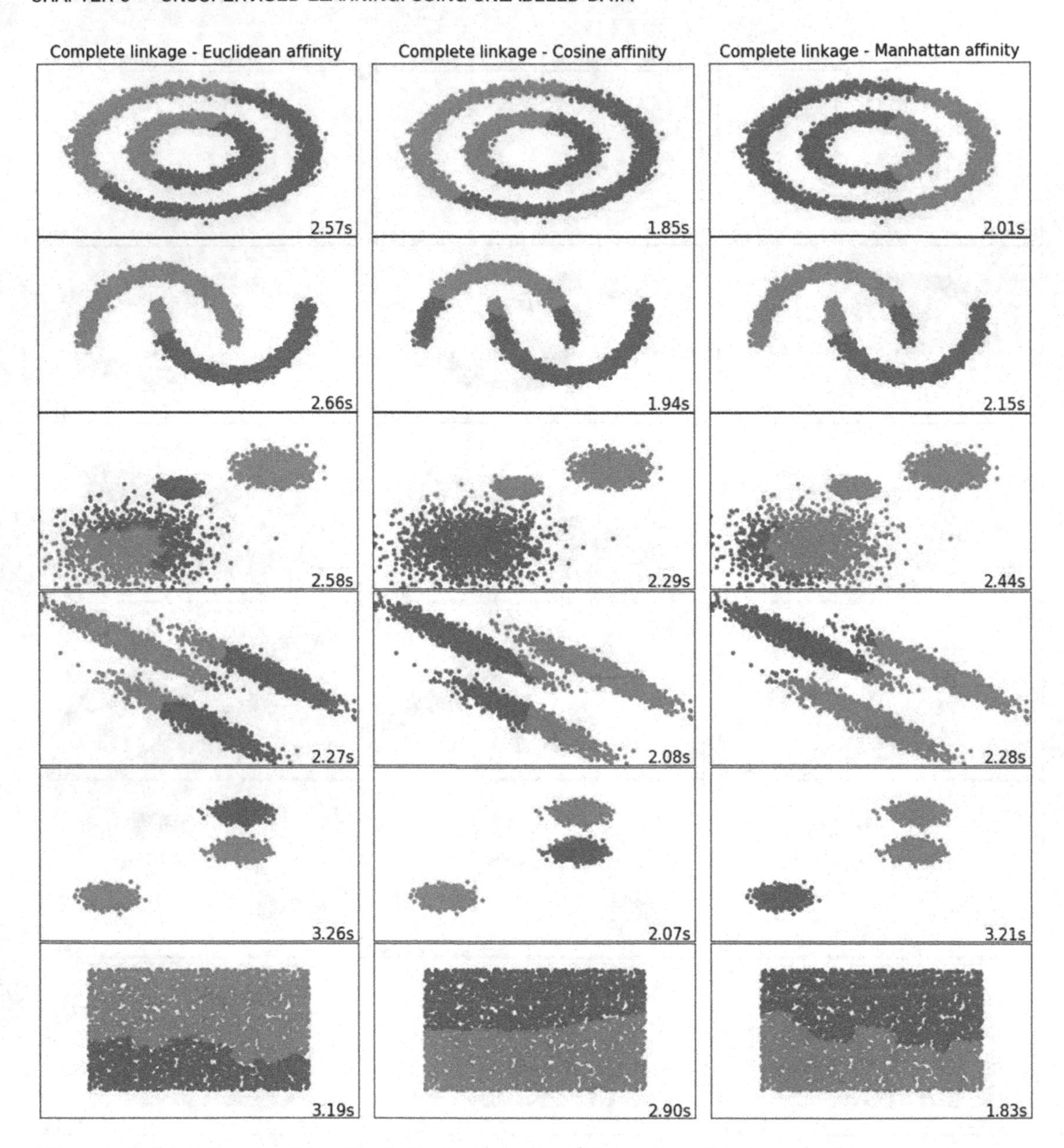

Figure 6-21. *Hierarchical – Affinity Impact (Complete Linkage)*

The results show that the affinity has a clear impact on the clustering algorithms.

Advice The ability to select the precise parameters and turning the algorithms you utilize are the biggest differentiator between an average and an expert data scientist. I recommend you take each algorithm and practice with diverse data sets with numerous patterns to appreciate each algorithm's parameters' impact on the result of the machine learning process.

You are making great progress and can now successfully perform a Hierarchical clustering solution.

In Chapters 11, 12, 13, 14, and 15, I will guide you through several supplementary examples on how real-world problems can be solved by using industrialized versions of the Hierarchical clustering models.

I will now guide you through a major important technique that is useful for real-world solutions.

Anomaly Detection

Anomaly detection is a technique used to identify unusual patterns that do not conform to anticipated actions, called outliers. In simple terms, they do not fit or act as normal.

Anomalies can be generally classified as the following types.

Point Anomalies

A single instance of data is anomalous if it's too far off from the values of the rest of the population.

Business use case:

Detecting credit card fraud based on "amount spent."

Contextual Anomalies

The abnormality is context specific. This type of anomaly is common in time-series data.

Business use case:

Spending $150.00 on food every day during the holiday season is normal but may be odd otherwise.

Suddenly starting to spend money in locations you never did before or in two locations within a time period that is unrealistic will all be contextual anomalies.

Collective Anomalies

A set of data instances collectively helps in detecting anomalies.

Business use case:

Someone is trying to copy data from a remote machine to a local host unexpectedly, an anomaly that would be flagged as a potential cyberattack.

Open the Jupyter notebook called: Chapter-006-012-Anomaly-Detection-01.ipynb

We will investigate where you buy coffee. If somebody is buying coffee using your cards outside these clusters, it is `potential` fraud.

I will show you how to use the following detection procedures:

- Robust covariance
- One-Class SVM
- Isolation Forest
- Local Outlier Factor

Execute the notebook.

The blue dots are not your transactions. I suggest we contact your bank as you are clearly the owner of one of two cards. You have a cloned card issue!!

The results are shown in Figure 6-22.

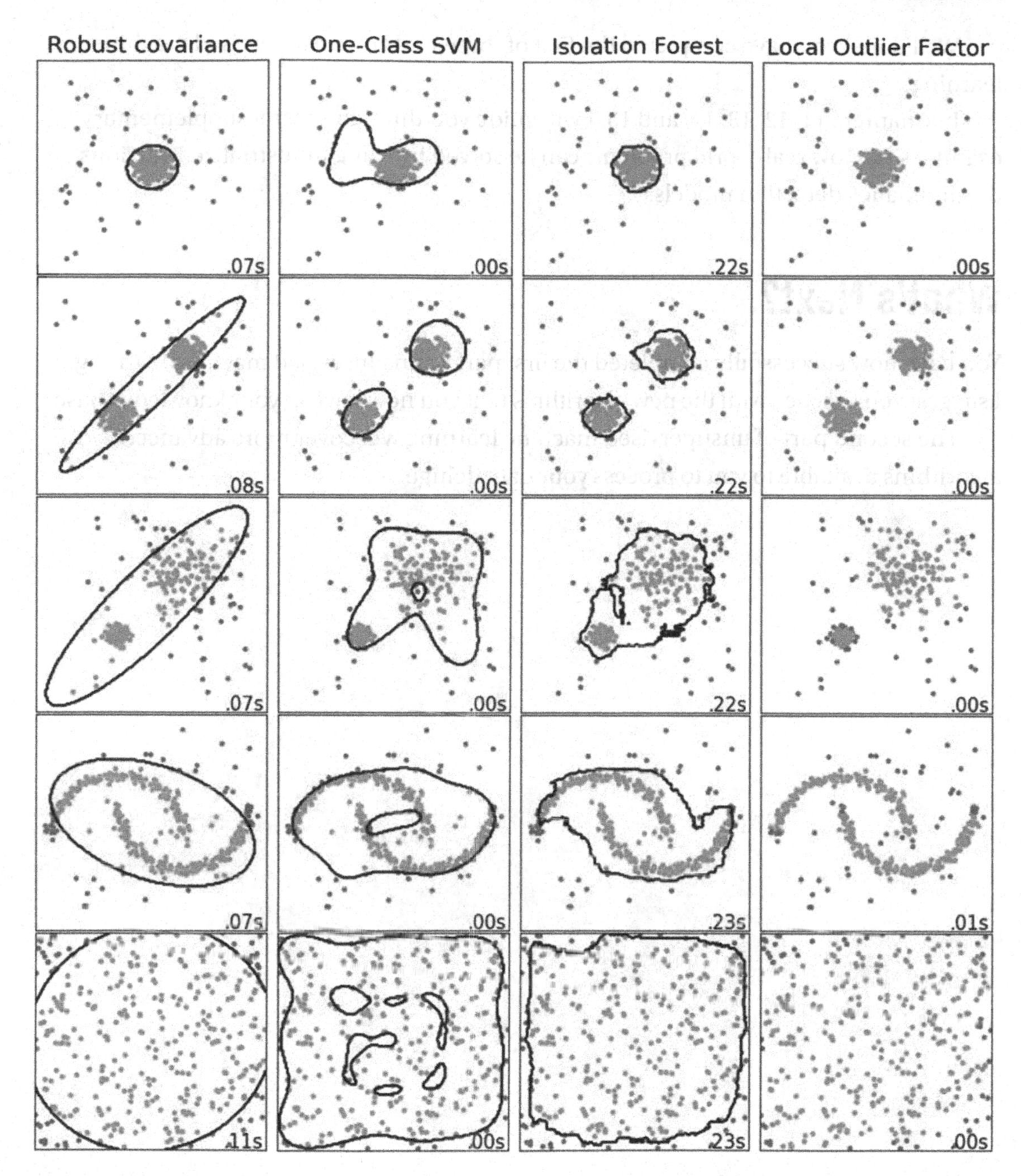

Figure 6-22. *Results*

Well done; you have completed the first of three parts on unsupervised machine learning.

In Chapters 11, 12, 13, 14, and 15, I will guide you through several supplementary examples on how real-world problems can be solved by using industrialized versions of the anomalies detection models.

What's Next?

You have now successfully completed the first part of unsupervised machine learning. I suggest you practice with the new algorithms that you now have in your knowledge base.

The second part of unsupervised machine learning will cover more advanced algorithms available to you to process your data deluge.

CHAPTER 7

Unsupervised Learning: Neural Network Toolkits

The next unsupervised learning technique take the unsupervised process one step further.

I am honored to introduce you to auto encoders. They are the future of discovering and exploring the massive data lakes we are deploying within the industrialized world, formed from the data we generate every day of our lives.

Neural Networks Autoencoders

An autoencoder is a type of artificial neural network used to learn efficient data encoding in an unsupervised manner. The aim of an autoencoder is to learn a representation for a set of data; this is typically used for dimensionality reduction. Recently, the autoencoder concept has become more widely used for learning generative models of data. Some of the most powerful AI achievements in the current business world have involved sparse autoencoders stacked inside of deep neural networks. It is an extremely powerful technique to use against data sets.

Open the Jupyter notebook called: Chapter-007-001 Neural-Networks-Autoencoder-01.ipynb

These examples provide you with a simple autoencoder that explains the principle of the process so that you can use it as a base for other autoencoder solutions.

```
Your input is: [[0 0 1], [0 1 1], [1 0 1], [1 1 1]]
Your required output is: [[0], [0], [1], [1]]
 Output after Training: [[0.00966553], [0.00786406], [0.99359009],
 [0.99211908]]
Output after Stepper function: [[0.], [0.], [1.], [1.]]
```


A. F. Vermeulen, *Industrial Machine Learning*, https://doi.org/10.1007/978-1-4842-5316-8_7

The basic principle is you apply a sigmoid function followed by a stepper function to achieve the desired outcome. These autoencoders can easily be transferred to an integrated circuit (IC) ecosystem to assist with edge processing on solutions with sensors.

You have successfully reduced (autoencoded) three sensors to one output.

Next investigate Chapter-007-002-Neural-Networks-Autoencoder-02.ipynb

This autoencoder uses two layers to process the input into an output. Please run the notebook to observe the process. You are making great progress and can now successfully perform two autoencoder solutions.

In Chapters 11, 12, 13, 14, and 15, I will guide you through several further examples on how the real-world problems can be solved by using industrialized versions of the autoencoders models.

I will now guide you through a new technique that is useful for real-world solutions. Next we will discuss GANs.

Generative Adversarial Networks (GAN)

Generative adversarial networks (GANs) are a class of AI algorithms used in unsupervised machine learning, implemented by a system of two neural networks challenging each other in a zero-sum game framework. The competing ML processes result in a rapid learning cooperative solution that evolves the solution into an optimum solution with ease.

So, let's discuss the zero-sum game:

In game theory and economic theory, a zero-sum game is a mathematical representation of a situation in which each participant's gain or loss of utility is exactly balanced by the losses or gains of the utility of the other participants. If the total gains of the participants are added up and the total losses are subtracted, they will sum to zero.

Therefore, sharing a cup of coffee with your loved one, where drinking a larger share of the coffee may reduce the amount of coffee available for others results still in a zero-sum game if all partakers value each unit of cup of coffee as equally valued. In other words, everyone is happy!

Next I will introduce PyTorch.

PyTorch is a Python-based scientific computing package that uses the power of graphics processing units to assist with processing.

The base difference between TensorFlow based on Theano engine and PyTorch using Torch is the way these frameworks define the computational graphs in the hardware. As a result, the Torch framework is as fast as TensorFlow in most cases, but Torch is faster for Recurrent Neural Networks. It is a known fact that Keras with TensorFlow is slower, but Keras with TensorFlow is winning the race. The main reason is that TensorFlow provides excellent functionalities and services when compared to many other currently popular, deep learning frameworks. The Keras to TensorFlow integration now makes the deep learning deployment easier. Therefore, many new projects are using Keras with TensorFlow as the standard.

You will need the Torch library for CPU use:

```
conda install -c pytorch pytorch-cpu
```

Open the Jupyter notebook called: Chapter-007-003-GAN-01.ipynb and run the notebook.

Your results:

```
Using data [Data and variances]
0: D: 0.7086962461471558/0.6972255110740662 G: 0.6868060827255249 (Real: [3.7825968647003174, 1.2327905192370257], Fake: [0.09708623103797435, 0.009256606334190703])
5000: D: 0.6805199980735779/0.6616045236587524 G: 0.8273075819015503 (Real: [4.117388695478439, 1.274467621457445], Fake: [4.077913434505462, 1.0943474129771693])
10000: D: 0.37435632944107056/0.8348380327224731 G: 0.4132739007472992 (Real: [3.9343803453445436, 1.3165514626757275], Fake: [3.7064030361175537, 1.3515258142353932])
15000: D: 0.024186545982956886/2.411825656890869 G: 0.9955140948295593 (Real: [4.019607482552528, 1.3916774260541982], Fake: [4.100582712888718, 1.3490263347503852])
20000: D: 0.8105971813201904/0.08567392826080322 G: 1.3057440519332886 (Real: [3.8651703649759295, 1.291156659436279], Fake: [4.00323578953743, 1.2979101588489939])
25000: D: 0.00010073692101286724/1.0811676474986598e-05 G: 12.61683464050293 (Real: [4.023330308198929, 1.1151899002588046], Fake: [14.387419891357421, 0.6251815240638156])
```

```
30000: D: 9.059946023626253e-06/-5.972999872483342e-14 G: 26.81680679321289
(Real: [4.0317733579874035, 1.3068207970685468], Fake: [37.9575638961792,
0.9236522286831534])
35000: D: 2.3841761276344187e-07/-9.841016890277388e-13 G:
27.20176887512207 (Real: [4.068414183855057, 1.3843235577079283], Fake:
[62.59007144927978, 1.5635452696977525])
40000: D: 0.0002785713004413992/-1.000088900582341e-12 G: 27.63102149963379
(Real: [4.057045288085938, 1.1723868704134213], Fake: [68.872311668396,
8.993019130772057])
45000: D: 2.5033971269294852e-06/-1.000088900582341e-12 G:
27.63102149963379 (Real: [3.9334994852542877, 1.190026090616771], Fake:
[69.51191040039062, 8.09731438960914])
```

You are making great progress and can now successfully perform a GAN solution.

In Chapters 11, 12, 13, 14, and 15, I will guide you through several more examples on how real-world problems can be solved by using industrialized versions of the GANs.

Note Generative Adversarial Networks is an unsupervised learning algorithm that is currently evolving at an exponential rate as it is generating new insights in the fields of health care, finance, and smart cities.

Interesting Fact A GAN system created the 2018 painting *Edmond de Belamy*, which sold for US$432,500. This proves Machine Learning can learn and master even human-like trades easily.

Convolutional Neural Networks (CNNs)

In machine learning, a convolutional neural network (CNN or ConvNet) is a class of deep, feed-forward artificial neural networks, most commonly applied to analyzing visual imagery. CNNs use a variation of multilayer perceptrons designed to require minimal preprocessing.

ConvNets or CNNs are a category of Neural Networks that have proven very effective in areas such as image recognition and classification. ConvNets have been successful in identifying faces, objects, and traffic signs apart from powering vision in robots and self-driving cars.

Open the Jupyter notebook called: Chapter-007-004A-CNN-01.ipynb

The process uses the traditional MNIST Data to generate a CNN.

Execute the notebook to get the required data sets.

You should have two files called Train-28x28_CNN_text.txt and Test-28x28_CNN_text.txt in data directory.

Now open the Jupyter notebook called: Chapter-03-017B-CNN-01.ipynb

Part A loads the required data.

The conversion to detect the digits is achieved by using a one-hot structure:

```
label:  0  in one-hot representation:  [1 0 0 0 0 0 0 0 0 0]
label:  1  in one-hot representation:  [0 1 0 0 0 0 0 0 0 0]
label:  2  in one-hot representation:  [0 0 1 0 0 0 0 0 0 0]
label:  3  in one-hot representation:  [0 0 0 1 0 0 0 0 0 0]
label:  4  in one-hot representation:  [0 0 0 0 1 0 0 0 0 0]
label:  5  in one-hot representation:  [0 0 0 0 0 1 0 0 0 0]
label:  6  in one-hot representation:  [0 0 0 0 0 0 1 0 0 0]
label:  7  in one-hot representation:  [0 0 0 0 0 0 0 1 0 0]
label:  8  in one-hot representation:  [0 0 0 0 0 0 0 0 1 0]
label:  9  in one-hot representation:  [0 0 0 0 0 0 0 0 0 1]
```

In Part B you create a sigmoid function. In Part C you create the Neural Network using a class.

The following command activates the Neural Network:

```
ANN = NeuralNetwork(no_of_in_nodes = image_pixels,
                    no_of_out_nodes = 10,
                    no_of_hidden_nodes = 100,
                    learning_rate = 0.1)
```

In Part D you train your Neural Network.

The training is done on 60,000 images from the MNIST Data that was uploaded.

In Part E you test your Neural Network against 10,000 images from the MNIST Data that was uploaded.

Part F displays your achievement:

```
Accuracy Train => 0.949
Accuracy Test ==> 0.9477
=====================================================

[[5797    1   43   24   17   46   42   15   15   22]
 [   1 6633   62   38   13   32   21   62  109   14]
 [   5   23 5479   43    4    8    3   45    8    2]
 [   4   23  104 5779    1   87    1   19   82   57]
 [  10   10   45    7 5505   31    7   48   34   64]
 [   5    6    5   44    0 4992   43    1   12    5]
 [  29    2   61   18   67   82 5778    8   36    4]
 [   0    6   40   44    4    9    0 5820    4   28]
 [  51   11   93   52    4   52   23   10 5427   22]
 [  21   27   26   82  227   82    0  237  124 5731]]
=====================================================
Digit: 0 Precision: 0.9787 Recall: 0.9626
Digit: 1 Precision: 0.9838 Recall: 0.9496
Digit: 2 Precision: 0.9196 Recall: 0.9749
Digit: 3 Precision: 0.9426 Recall: 0.9386
Digit: 4 Precision: 0.9423 Recall: 0.9556
Digit: 5 Precision: 0.9209 Recall: 0.9763
Digit: 6 Precision: 0.9763 Recall: 0.9495
Digit: 7 Precision: 0.929 Recall: 0.9773
Digit: 8 Precision: 0.9275 Recall: 0.9446
Digit: 9 Precision: 0.9634 Recall: 0.874
```

Part G tests the process for epochs = 5

```
Epoch: 0
Accuracy Train=> 0.9457
Accuracy Test==> 0.9462
Epoch: 1
Accuracy Train=> 0.9607
Accuracy Test==> 0.956
Epoch: 2
```

```
Accuracy Train=> 0.9684
Accuracy Test==> 0.9603
Epoch: 3
Accuracy Train=> 0.9736
Accuracy Test==> 0.963
Epoch: 4
Accuracy Train=> 0.9726
Accuracy Test==> 0.9613
```

This shows that the CNN is stable and achieves > 94% accuracy.

Recurrent Neural Networks (RNNs)

A recurrent neural network (RNN) is a class of an artificial neural network where connections between nodes form a directed graph along a sequence. This allows it to exhibit temporal dynamic behavior for a time sequence. Unlike feed-forward neural networks, RNNs can use their internal state to process sequences of inputs. This gives them the ability to remember what it did before and how they adapted to this knowledge.

This makes them applicable to tasks like unsegmented, connected handwriting recognition or speech recognition.

Open the Jupyter notebook called: Chapter-007-005-RNN-01.ipynb

Part E is the main engine and processes the following results:

```
==================================================
Cycle: 0
==================================================
Error:[0.3217455]
Pred:[1 0 0 0 1 0 0 1]
True:[1 0 0 0 1 0 0 1]
108 + 29 = 137
==================================================
==================================================
Cycle: 1000
==================================================
Error:[0.19121145]
Pred:[1 0 1 1 1 1 0 1]
```

```
True:[1 0 1 1 1 1 0 1]
76 + 113 = 189
================================================
================================================
Cycle: 2000
================================================
Error:[0.12260766]
Pred:[0 0 1 0 1 0 1 0]
True:[0 0 1 0 1 0 1 0]
10 + 32 = 42
================================================
================================================
Cycle: 3000
================================================
Error:[0.25359378]
Pred:[1 0 0 1 0 0 0 1]
True:[1 0 0 1 0 0 0 1]
108 + 37 = 145
================================================
================================================
Cycle: 4000
================================================
Error:[0.27286803]
Pred:[1 0 0 0 1 0 1 0]
True:[1 0 0 0 1 0 1 0]
123 + 15 = 138
================================================
================================================
Cycle: 5000
================================================
Error:[0.27208995]
Pred:[1 1 0 1 0 0 0 1]
True:[1 1 0 1 0 0 0 1]
83 + 126 = 209
================================================
```

```
================================================
Cycle: 6000
================================================
Error:[0.17898913]
Pred:[1 1 0 1 0 1 0 0]
True:[1 1 0 1 0 1 0 0]
112 + 100 = 212
================================================
================================================
Cycle: 7000
================================================
Error:[0.17206773]
Pred:[0 1 0 1 1 0 0 1]
True:[0 1 0 1 1 0 0 1]
79 + 10 = 89
================================================
================================================
Cycle: 8000
================================================
Error:[0.21744059]
Pred:[1 0 1 0 1 1 1 1]
True:[1 0 1 0 1 1 1 1]
82 + 93 = 175
================================================
================================================
Cycle: 9000
================================================
Error:[0.18515105]
Pred:[0 1 0 0 0 1 1 1]
True:[0 1 0 0 0 1 1 1]
51 + 20 = 71
================================================
================================================
Cycle: 10000
================================================
```

```
Error:[0.12941404]
Pred:[0 1 1 1 1 1 1 0]
True:[0 1 1 1 1 1 1 0]
76 + 50 = 126
================================================
================================================
Cycle: 11000
================================================
Error:[0.22409284]
Pred:[1 1 1 1 0 0 0 0]
True:[1 1 1 1 0 0 0 0]
118 + 122 = 240
================================================
================================================
Cycle: 12000
================================================
Error:[0.16455129]
Pred:[1 0 1 0 1 1 0 0]
True:[1 0 1 0 1 1 0 0]
49 + 123 = 172
================================================
================================================
Cycle: 13000
================================================
Error:[0.20210813]
Pred:[1 0 1 1 1 0 0 0]
True:[1 0 1 1 1 0 0 0]
106 + 78 = 184
================================================
================================================
Cycle: 14000
================================================
Error:[0.10140656]
Pred:[0 1 0 1 1 0 1 1]
True:[0 1 0 1 1 0 1 1]
```

```
18 + 73 = 91
================================================
================================================
Cycle: 15000
================================================
Error:[0.17551984]
Pred:[1 0 0 0 1 0 0 0]
True:[1 0 0 0 1 0 0 0]
111 + 25 = 136
================================================
```

Can you observe how the state is transferred from one cycle to the next?

Note The LSTM (Long Short-Term Memory) Neural Networks used in modern image recognition and characterization work by combining RNN with ConvNet to identify and classify in one process.

Spectral Bi-clustering Algorithm

Bi-clustering, block clustering, co-clustering, or two-mode clustering are data mining techniques that agree to simultaneous clustering of the rows and columns of a matrix.

The data is generated with the make_checkerboard function, then shuffled and passed to the Spectral Bi-clustering algorithm. The rows and columns of the shuffled matrix are rearranged to show the bi-clusters found by the algorithm.

Open the Jupyter notebook called: Chapter-03-019-Spectral-Biclustering-01.ipynb

The core engine is:

```
model = SpectralBiclustering(n_clusters=n_clusters, method='log',
random_state=1968)
```

Your processing results are shown in Figures 7-1 and 7-2.

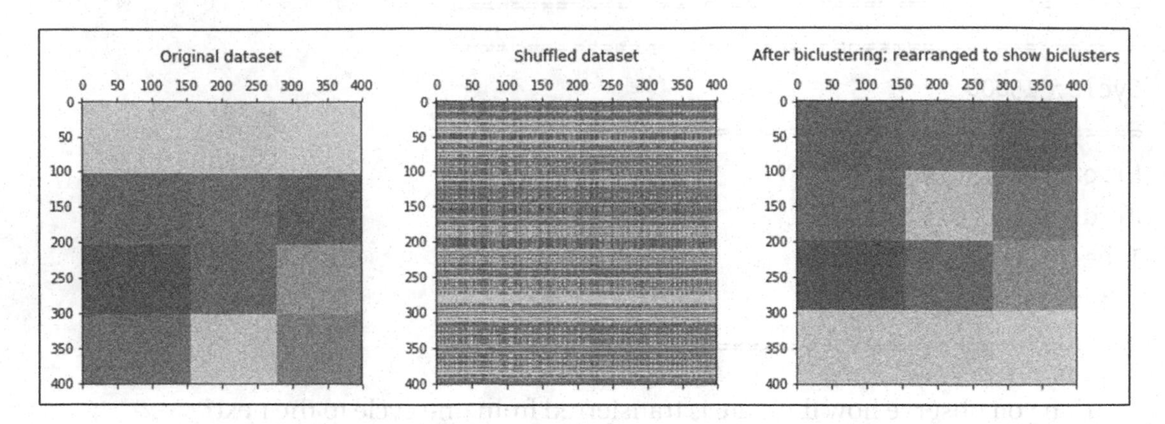

***Figure 7-1.** Bi-clustering Results*

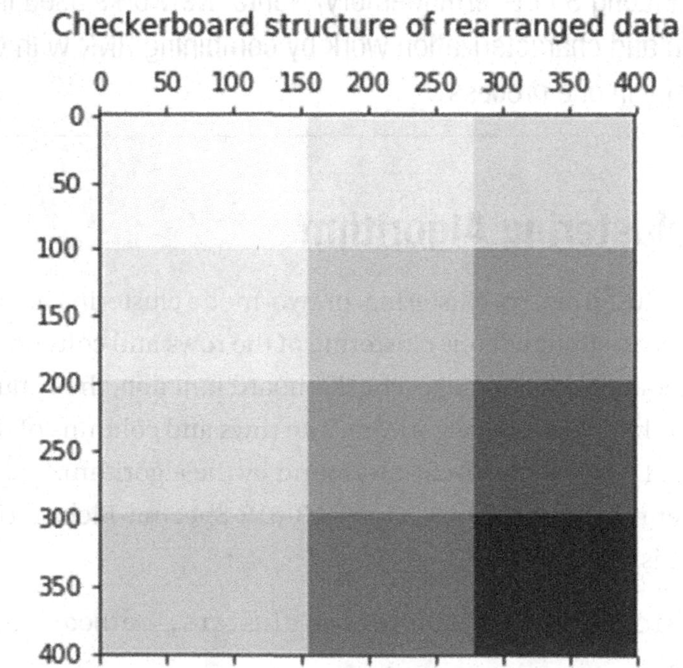

***Figure 7-2.** Bi-clustering Final Result*

You can clearly see the clusters, and this is useful when you need to perform classification of images.

BIRCH Clustering Algorithm

The BIRCH algorithm (balanced iterative reducing and clustering using hierarchies) is an unsupervised data processing algorithm used to accomplish hierarchical clustering over predominantly large data sets.

Our example compares the timing of BIRCH (with and without the global clustering step) and MiniBatchKMeans on a synthetic data set having 100,000 samples and 2 features generated using make_blobs.

If n_clusters is set to None, the data is reduced from 100,000 samples to a set of 158 clusters. This can be viewed as a preprocessing step before the final (global) clustering step that further reduces these 158 clusters to 100 clusters.

Open the Jupyter notebook called: Chapter-007-007-Birch-clustering-01.ipynb

Your results can be seen in Figure 7-3.

```
Birch without global clustering as the final step took 2.08 seconds
n_clusters : 158
Birch with global clustering as the final step took 2.06 seconds
n_clusters : 100
Time taken to run MiniBatchKMeans 2.62 seconds
```

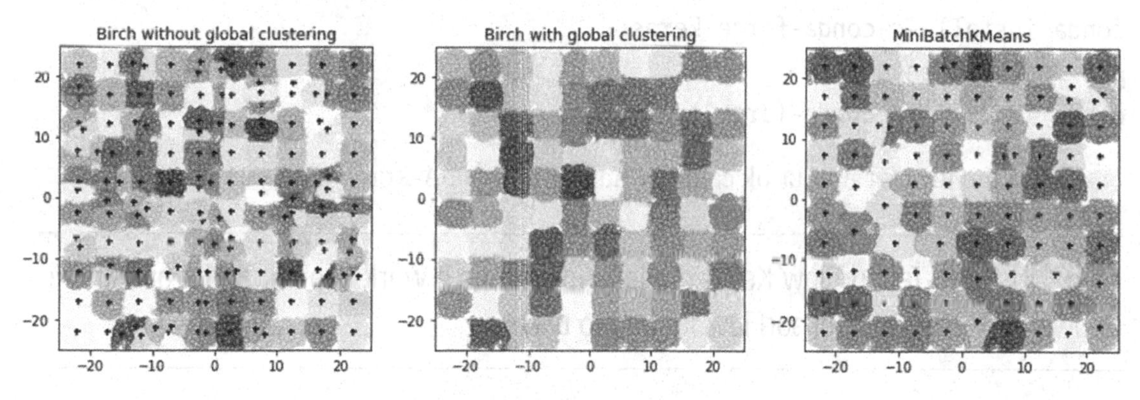

Figure 7-3. *BIRCH clustering Results*

Machine Learning Toolkits

A Machine learning Toolkit is a collection of methods for data analysis that automates analytical model building by supporting a pre-tuned set of techniques that work together seamlessly.

Scikit-Learn

Scikit-learn (formerly scikits.learn) is an open source software machine learning library for Python. It features various classification, regression, and clustering algorithms including support vector machines, random forests, gradient boosting, K-Means, and DBSCAN.

You had been using this library earlier in the book.

Install:

```
conda install -c conda-forge scikit-learn
```

Here is a friendly face I discovered: Chapter-007-008-scikit-learn-01.ipynb

Keras

Keras is an open source neural network library written in Python. It can run on top of TensorFlow, Microsoft Cognitive Toolkit, or Theano. Designed to enable fast experimentation with deep neural networks, it focuses on being user friendly, modular, and extensible.

Install:

```
conda install -c conda-forge keras
pip install tensorflow
conda install -c conda-forge tensorflow
```

Open the Jupyter notebook called: Chapter-007-009A-Keras-01.ipynb

---

Tip Use the TensorFlow Keras implementation as it works better, and it opens the power of Keras and TensorFlow for you to use.

Your results:

```
0 -> 0.9875 =>  1 =>  1 # Yes
1 -> 0.0170 =>  0 =>  0 # Yes
2 -> 0.9954 =>  1 =>  1 # Yes
3 -> 0.0006 =>  0 =>  0 # Yes
4 -> 0.9954 =>  1 =>  1 # Yes
5 -> 0.0113 =>  0 =>  0 # Yes
```

```
6 -> 0.9773 =>  1 =>  1 # Yes
7 -> 0.0070 =>  0 =>  0 # Yes
8 -> 0.0000 =>  0 =>  1 # No
9 -> 0.9984 =>  1 =>  1 # Yes
```

You can observe that this Keras and TensorFlow combination is the future match that will have a major impact on the IML ecosystem.

XGBoost

The Scalable, Portable, and Distributed Gradient Boosting (GBDT, GBRT, or GBM) Library for Python is a common machine learning solution.

XGBoost is an enhanced, distributed gradient boosting library designed to be extremely efficient, flexible, and portable. It implements machine learning algorithms under the gradient boosting framework. XGBoost provides a parallel tree boosting (also known as GBDT, GBM) that solves several data science problems in a fast and accurate way. The similar code runs on major distributed environments (Hadoop, SGE, and MPI) and can solve problems over billions of records.

Install Library:

```
conda install -c conda-forge xgboost
```

Open the Jupyter notebook called: Chapter-007-010- xgboost-01.ipynb

Result:

```
0   -> [9.0, 102.0, 76.0, 37.0, 0.0, 32.9, 0.665, 46.0]  => 0 => 1 # No
1   -> [9.0, 119.0, 80.0, 35.0, 0.0, 29.0, 0.263, 29.0]  => 1 => 1 # Yes
2   -> [7.0, 159.0, 64.0, 0.0, 0.0, 27.4, 0.294, 40.0]   => 1 => 0 # No
3   -> [7.0, 106.0, 92.0, 18.0, 0.0, 22.7, 0.235, 48.0]  => 0 => 0 # Yes
4   -> [2.0, 71.0, 70.0, 27.0, 0.0, 28.0, 0.586, 22.0]   => 0 => 0 # Yes
```

Tip The XGBoost is currently one of the ML algorithms that have a very good reputation for quick results on a data set. I personally use it on a regular basis.

In the wider community, XGBoost is one of the most used algorithms in any data science practice. This is one algorithm you will use!

StatsModels

StatsModels is a Python module that provides classes and functions for the estimation of numerous different statistical models, as well as for conducting statistical tests and statistical data exploration.

Install library:

```
conda install -c conda-forge statsmodels
```

Open the Jupyter notebook called: Chapter-007-011-statsmodels-01.ipynb

```
Execute the notebook to see the variety of the capability of this library.
```

Tip Using the StatsModel library supports a great transition for most statisticians into the ecosystem of machine learning.

LightGBM

LightGBM is a gradient boosting framework that uses tree-based learning algorithms.

Install library:

```
conda install -c conda-forge lightgbm
```

Open the Jupyter notebook called: Chapter-007-012-LightGBM-01.ipynb

Execute the notebook.

Your results:

```
Confusion Matrix
[[64  4]
 [ 5 27]]
====================
```

Accuracy: 0.91

The library is performing well in general, and I foresee it further evolving over the next five years into a contender for one of the ten algorithms you should know.

CatBoost

Gradient boosting on decision trees library.

"CatBoost" name comes from two words "Category" and "Boosting".

Install library:

```
conda install -c conda-forge catboost
```

Open the Jupyter notebook called: Chapter-007-013-CatBoost-01.ipynb

Execute the notebook.

Result:

```
bestTest = 0.8358208955
bestIteration = 428
```

That proves that the algorithm found the best result at 428 out of 500 cycles. That is nearly a 20% saving in processing.

That is the end of the machine learning toolkits. The knowledge to master these will serve you well in the future to handle new data solutions.

What's Next?

You have now successfully completed Part 2 of unsupervised machine learning.

The two biggest achievements in this chapter are you learning how to use TensorFlow and XGBoost.

The use of some form of Neural Network algorithm is now a must-have skill. So please ensure you understand the last two chapters as this is where the biggest growth of work for machine learning is happening.

You are doing well as you have now covered the most common types of machine learning. The next chapter is Part 3 where I will introduce you to Deep Learning algorithms.

CHAPTER 8

Unsupervised Learning: Deep Learning

The next set of unsupervised learning techniques take the unsupervised process into neural networks.

Artificial neural networks (ANN) are ecosystems that compute solutions by mimicking the activities similar to the process in brains, that is, a chain of activities with a common outcome.

Deep Learning

Deep Learning is a subfield of machine learning concerned with algorithms to generate an artificial neural network.

TensorFlow

TensorFlow (`https://www.tensorflow.org/`) is an open source software library for numerical computation using data flow graphs. Nodes in the graph represent mathematical operations, while the graph edges represent the multidimensional data arrays (tensors) communicated between them. The flexible architecture allows you to deploy computation to one or more 'CPUs' or 'GPUs' in a desktop, server, or mobile device with a single 'API'.

To install it, you can use any of the following:

conda install -c anaconda tensorflow (I used this one on my laptop)

conda install -c conda-forge tensorflow

conda install -c anaconda tensorflow-gpu (I use this in my GPU-enabled Desktop Workstation)

A. F. Vermeulen, *Industrial Machine Learning*, https://doi.org/10.1007/978-1-4842-5316-8_8

You have used this processing before in earlier chapters.

Open the Jupyter notebook called: Chapter-008-001-Tensorflow-01.ipynb

Execute the notebook to solve the mathematic formula of $y = m * x + k$ using Tensorflow.

Solution: $y = (0.196810 * x) + 0.201904$

The processing learns the values by trying a match to the train data set.

Open the Jupyter notebook called: Chapter-008-002-Tensorflow-02.ipynb

---

Warning On a Mac, there is a known issue with certificate verify failing when loading the mnist data set. You need to perform an "/Applications/Python/3.6/Install/Certificates.command" in your Python installation to fix the issue.

Your results: Loss: 0.149082 and Accuracy: 0.957300. A loss of 14.9% and an accuracy of 95.7% is a good result, and the process is working well.

PyTorch

PyTorch is an open source machine learning library for Python, based on Torch, used for applications such as natural language processing.

PyTorch is an optimized tensor library for deep learning using GPUs and CPUs.

PyTorch enables fast, flexible experimentation and efficient production through a hybrid front end, distributed training, and ecosystem of tools and libraries.

Look at: `https://pytorch.org/` for more details on this engine.

Install:

```
conda install pytorch torchvision cudatoolkit=10.0 -c pytorch
```

If you only want or can use the CPU install: `conda install pytorch-cpu torchvision-cpu -c pytorch`

Now open the Jupyter notebook: Chapter-008-003-Torch-01.ipynb

We will be comparing NumPy against PyTorch to introduce you to the machine learning techniques.

NumPy provides an n-dimensional array object and numerous functions for manipulating these arrays.

One one hand, NumPy is a generic framework for scientific computing; it does not know anything about computation graphs, or deep learning, or gradients. On the other hand, we can easily use NumPy to fit a two-layer network to random data by manually implementing the forward and backward passes through the network using NumPy operations.

PyTorch is an open source machine learning library for Python, based on the Torch ecosystem that supports two types of processing:

- Tensor computation (comparable NumPy) with strong GPU acceleration.
- Deep neural networks built on a tape-based autodiff system.

So, I suggest we investigate how much better it is than a pure NumPy solution.

I have set the max to 1968 and then 19680 for two runs.

Torch (10.753 seconds) is 1.38 times faster than NumPy (14.803 seconds) on 1968 records.

Torch (104.135 seconds) is 1.43 times faster than NumPy (149.317 seconds) on 19680 records.

Tip It proves that Torch will outperform NumPy by 40% on average. When you enable the GPUs, this number gets even higher.

Theano

Theano is a Python library that allows you to define, optimize, and efficiently evaluate mathematical expressions involving multidimensional arrays. This is an optimizing compiler for evaluating mathematical expressions on CPUs and GPUs. The engine has been in use since 2007 and therefore is used widely in companies. The core developer Montreal Institute for Learning Algorithms (MILA) has in September 2017 informed the machine learning community that the major development will cease.

I have since assisted several customers to convert their Theano solutions to Keras TensorFlow solutions.

Install Library:

```
conda install -c conda-forge theano
```

Open the Jupyter notebook called: Chapter-008-004-Theano.ipynb

Run the notebook. This will show you how the Theano engine performs the machine learning.

Your results should be as follows:

```
print('Difference: ')
print(result[0])

Difference:
[[ 1.  0.]
 [-1. -2.]]

print('Absolute Difference: ')
print(result[1])

Absolute Difference:
[[1. 0.]
 [1. 2.]]

print('Squared Difference: ')
print(result[2])

Squared Difference:
[[1. 0.]
 [1. 4.]]

print('Powered Difference: ')
print(result[3])

[[ 1.    1.  ]
 [-1.    0.25]]
```

Tip I suggest you perform a collection of data sources through the engine to ensure you understand the techniques required to use this engine successfully.

I also suggest you spend some time studying the engine to enable you to maintain existing code or to port the code to another engine.

Compare Clusters

The following compares within the cluster example will test your understanding of the whole clustering process.

Open the Jupyter notebook called: Chapter-008-005-Compare-clustering-01.ipynb

The display (Figure 8-1) shows how the most minor shift of the profile of the data has diverse outcomes on the same machine learning algorithms and hyperparameters.

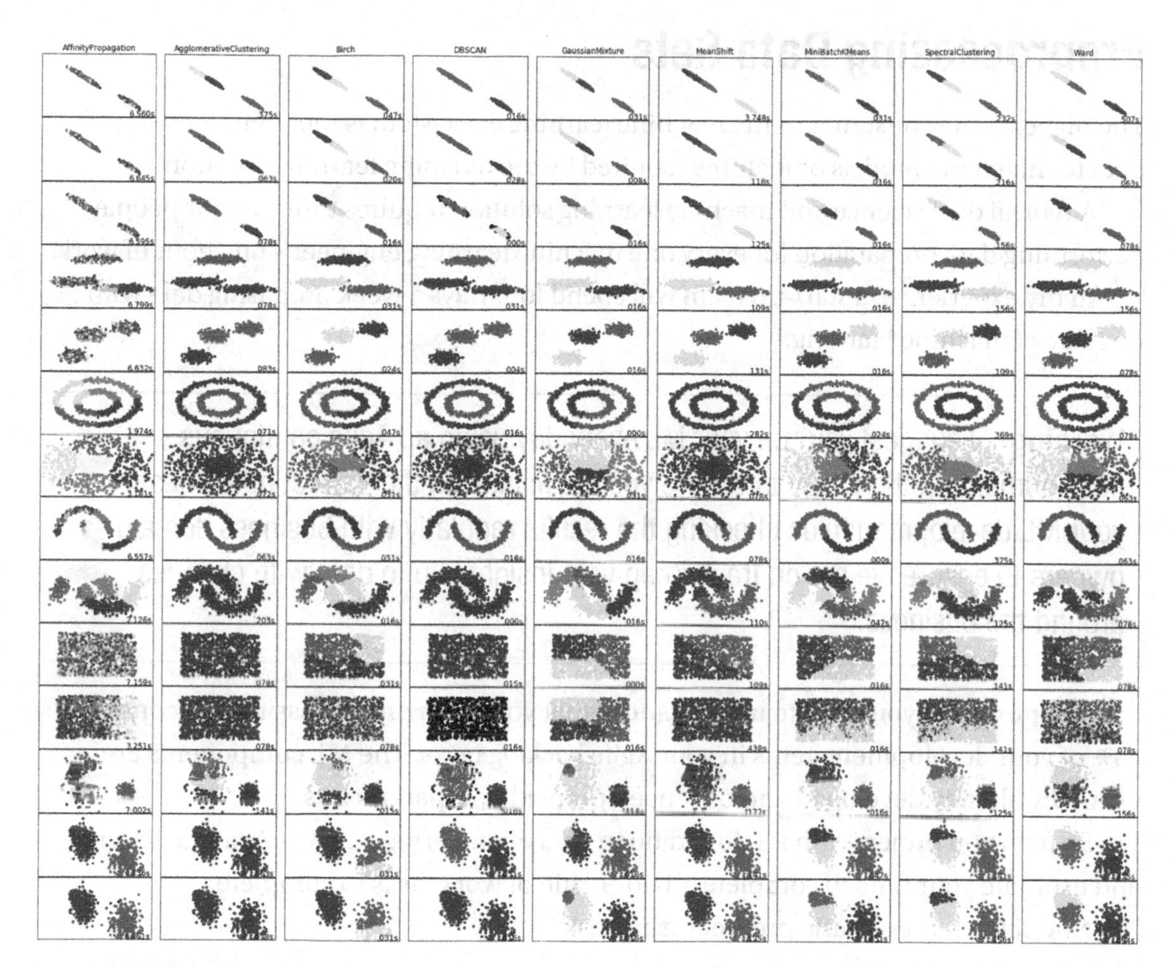

Figure 8-1. *Cluster Compare Results*

Tip Investigate the machine learning algorithms and their hyperparameters by creating a standard test bed with diverse data models to ensure you understand the behavior of the machine learning.

The result is shown in Figure 8-1.

Preprocessing Data Sets

The biggest time consumer in the machine learning ecosystem is the preprocessing of data to match the models or features required by the machine learning solution.

A typical data science and machine learning solution requires three to four people performing data preparation for every one machine learning engineer's one hour of work.

In my experience, a start-up team will spend four days a week preparing data with one day of machine learning.

Warning I suggest you run your ML every day using an Agile continuous development methodology to ensure you are achieving the correct outcomes from your ML development plus checking the results regularly with business domain owners to ensure we are on track Keep your insights up to date with changes around the business.

The next ratio you need to understand is that your data preprocessing will consume 80% of your development items in your Agile backlog tasks. The ML components are normally already developed and only need proper hyperparameters.

So plan you project with a 1-to-4 ratio in data engineering staff vs. data science staff, and estimate your time to complete a 1-to-9 ratio of work tasks to complete.

Now let's start our basic preprocessing tasks.

Preprocessing Data

Open the Jupyter notebook called: Chapter-008-006-Preprocessing-Data-01.ipynb Part C is a scale processor:

This processor standardizes a data set along any axis by centering to the mean and a component-wise scale to the unit variance of each item in the data set.

```
X_scaled = preprocessing.scale(X_train,
                               axis=0,
                               with_mean=True,
                               with_std=False,
                               copy=True
                              )
```

Part D is a RobustScaler processor:

This processor standardizes a data set along any axis with centering to the median and component-wise scale according to the interquartile range.

```
scaler = preprocessing.RobustScaler(with_centering=True,
                                    with_scaling=True,
                                    quantile_range=(25.0, 75.0),
                                    copy=True
                                   ).fit(X_train)
```

The sklearn.preprocessing library has several useful preprocessing engines. I suggest you investigate them all to determine what they support you with for your preprocessing requirements.

Processor	Description
preprocessing.Binarizer([threshold, copy])	Binarize data (set feature values to 0 or 1) according to a threshold.
preprocessing.FunctionTransformer([func, ...])	Constructs a transformer from an arbitrary callable.
preprocessing.KBinsDiscretizer([n_bins, ...])	Bin continuous data into intervals.
preprocessing.KernelCenterer()	Center a kernel matrix.
preprocessing.LabelBinarizer([neg_label, ...])	Binarize labels in a one-vs.-all fashion.
preprocessing.LabelEncoder	Encode labels with value between 0 and n_classes-1.
preprocessing.MultiLabelBinarizer([classes, ...])	Transform between iterable of iterables and a multi-label format.
preprocessing.MaxAbsScaler([copy])	Scale each feature by its maximum absolute value.
preprocessing.MinMaxScaler([feature_range, copy])	Transforms features by scaling each feature to a given range.
preprocessing.Normalizer([norm, copy])	Normalize samples individually to unit norm.
preprocessing.OneHotEncoder([n_values, ...])	Encode categorical integer features as a one-hot numeric array.
preprocessing.OrdinalEncoder([categories, dtype])	Encode categorical features as an integer array.
preprocessing.PolynomialFeatures([degree, ...])	Generate polynomial and interaction features.
preprocessing.PowerTransformer([method, ...])	Apply a power transform featurewise to make data more Gaussian-like.

preprocessing.QuantileTransformer([...])	Transform features using quantiles information.
preprocessing.RobustScaler([with_centering, ...])	Scale features using statistics that are robust to outliers.
preprocessing.StandardScaler([copy, ...])	Standardize features by removing the mean and scaling to unit variance.
preprocessing.add_dummy_feature(X[, value])	Augment data set with an additional dummy feature.
preprocessing.binarize(X[, threshold, copy])	Boolean thresholding of array-like or scipy.sparse matrix.
preprocessing.label_binarize(y, classes[, ...])	Binarize labels in a one-vs.-all fashion.
preprocessing.maxabs_scale(X[, axis, copy])	Scale each feature to the [-1, 1] range without breaking the sparsity.
preprocessing.minmax_scale(X[, ...])	Transforms features by scaling each feature to a given range.
preprocessing.normalize(X[, norm, axis, ...])	Scale input vectors individually to unit norm (vector length).
preprocessing.quantile_transform(X[, axis, ...])	Transform features using quantiles information.
preprocessing.robust_scale(X[, axis, ...])	Standardize a data set along any axis.
preprocessing.scale(X[, axis, with_mean, ...])	Standardize a data set along any axis.
preprocessing.power_transform(X[, method, ...])	Power transforms are a family of parametric, monotonic transformations that are applied to make data more Gaussian-like.

This is only the standard set. I have also found that a daisy chain of preprocessors can generally improve the data additionally with improved outcomes in the final ML model's performance and tuning capabilities.

Tip I suggest you devote a few weeks into basically matching known data sets against these processors to understand how they convert and modify your ML outcomes. In the real world I have found the use of appropriate preprocessing saves you hours of complex hypertuning of your models.

So let's look at a real data set to demonstrate a few of the interesting insights you can achieve.

Open the Jupyter notebook called: Chapter-008-007-Preprocessing-Data-02.ipynb

Run the notebook and track your progress. The basic concept is to visually get insights of the relationship and distribution of two selected features out of the given data set. I personally perform this process for any data set I first receive to get insights on the data distribution and characteristics.

So here we go ... Let's investigate the data. See Figure 8-2.

```
make_plot(1)
```

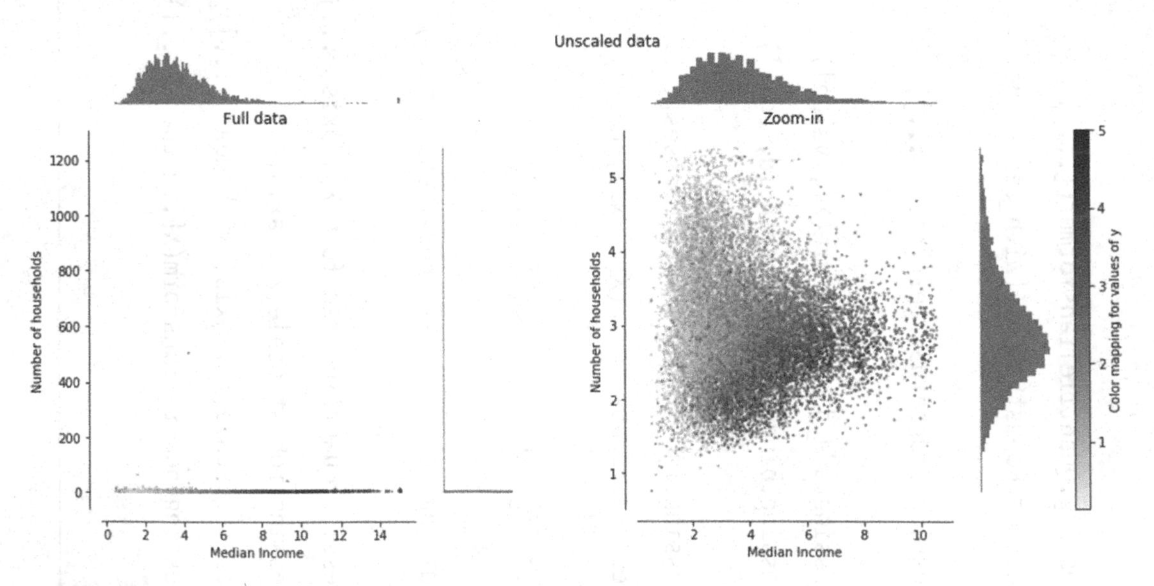

Figure 8-2. *Preprocessing Run 0 Results*

Note that the number of households are all distributed on the lower part of the value scale.

If you apply a standard scaler you observe a better distribution. See Figure 8-3.

```
make_plot(1)
```

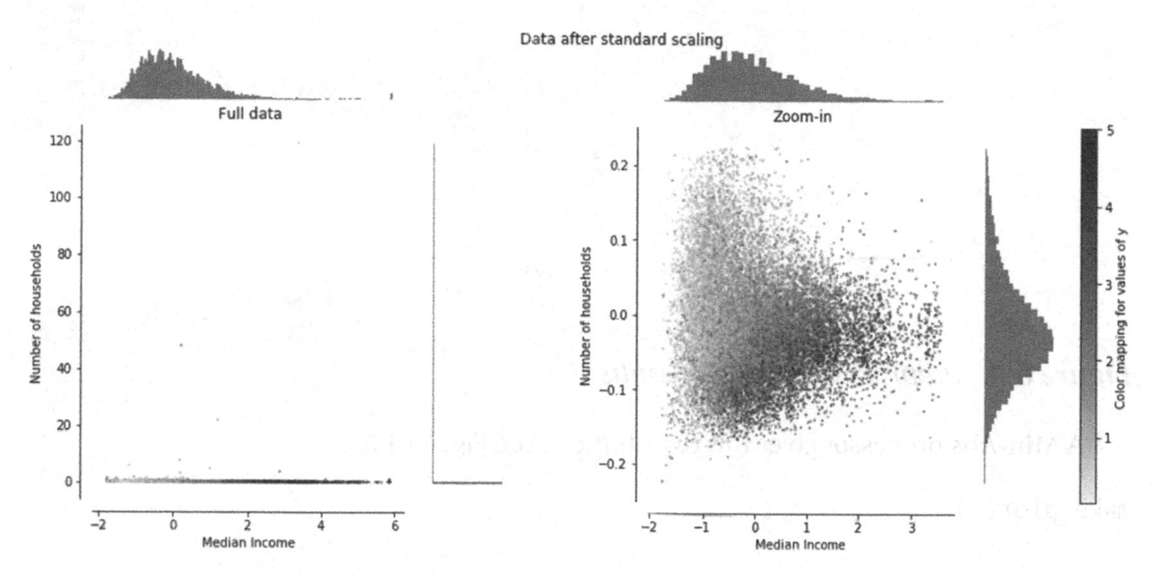

***Figure 8-3.** Preprocessing Run 1 Results*

If you apply a Min-Max Scaler, you discover the data is spread mostly around a mean. See Figure 8-4.

```
make_plot(2)
```

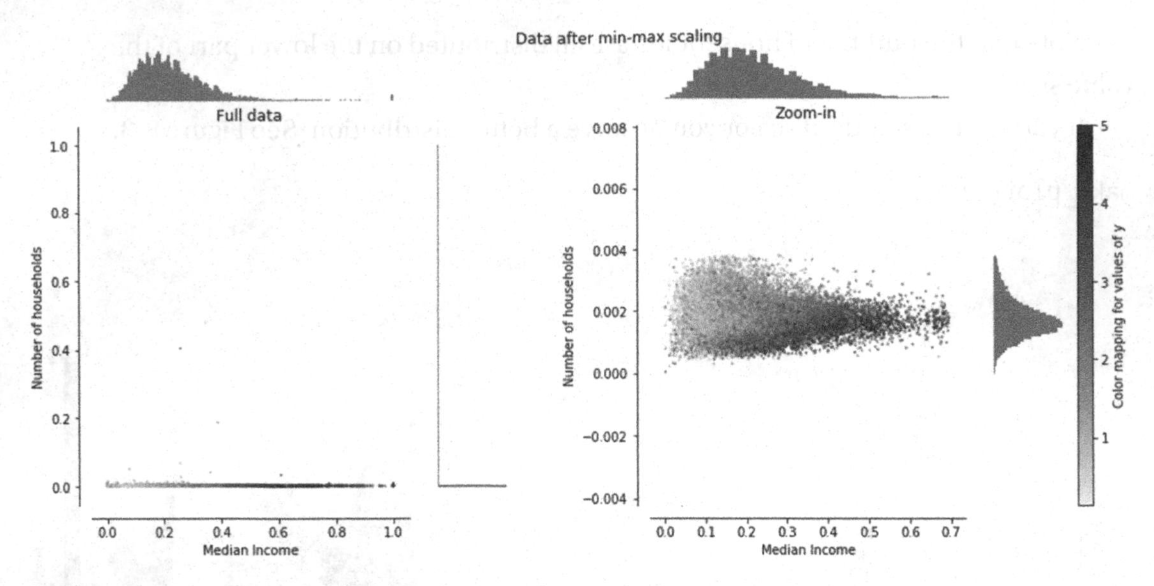

Figure 8-4. *Preprocessing Run 2 Results*

A Min-Abs processor gives this distribution. See Figure 8-5.

```
make_plot(3)
```

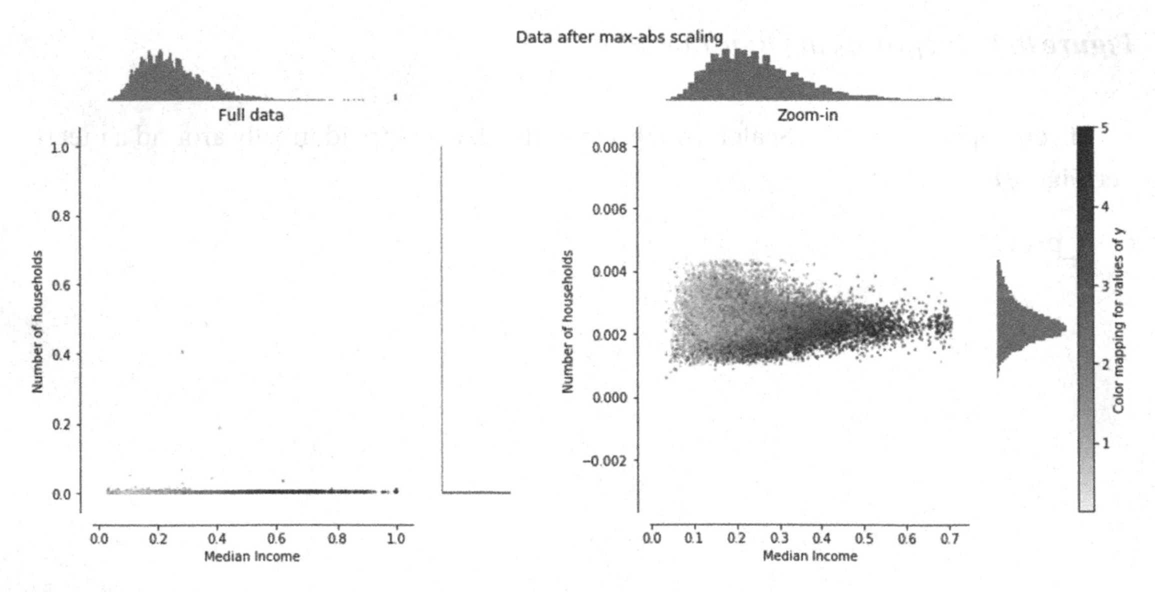

Figure 8-5. *Preprocessing Run 3 Results*

A robust Scaler gives this distribution. See Figure 8-6.

```
make_plot(4)
```

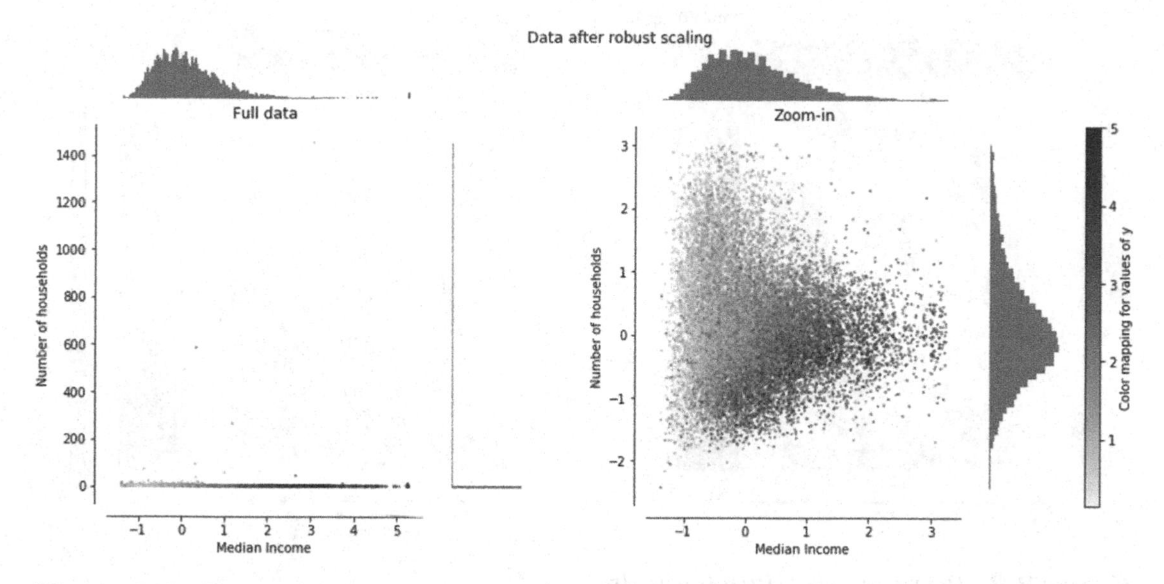

Figure 8-6. *Preprocessing Run 4 Results*

See Figure 8-7 for more results.

```
make_plot(5)
```

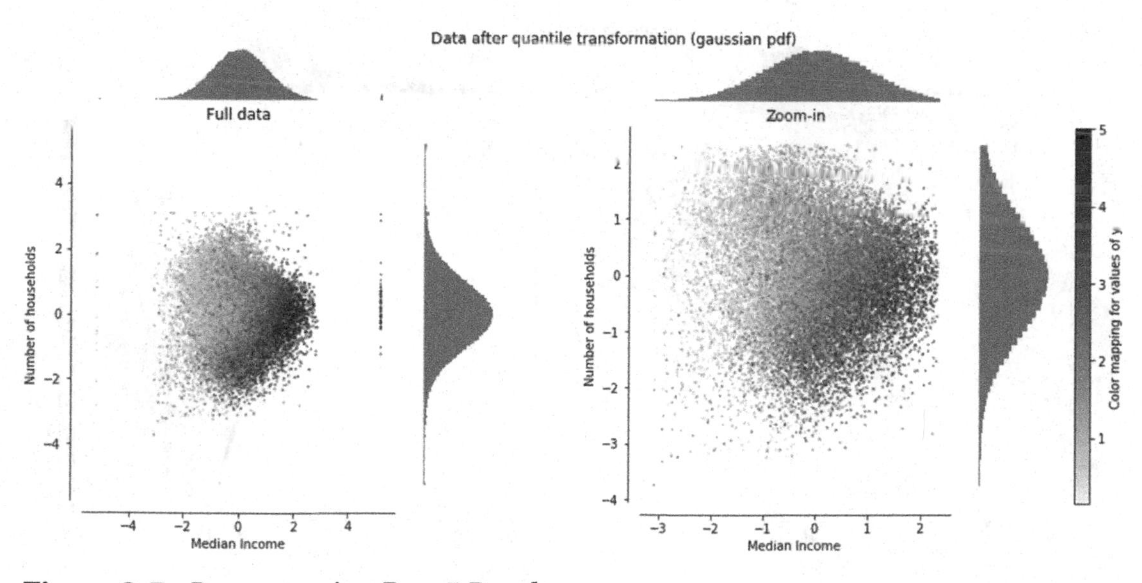

Figure 8-7. *Preprocessing Run 5 Results*

See Figure 8-8 for more results.

```
make_plot(6)
```

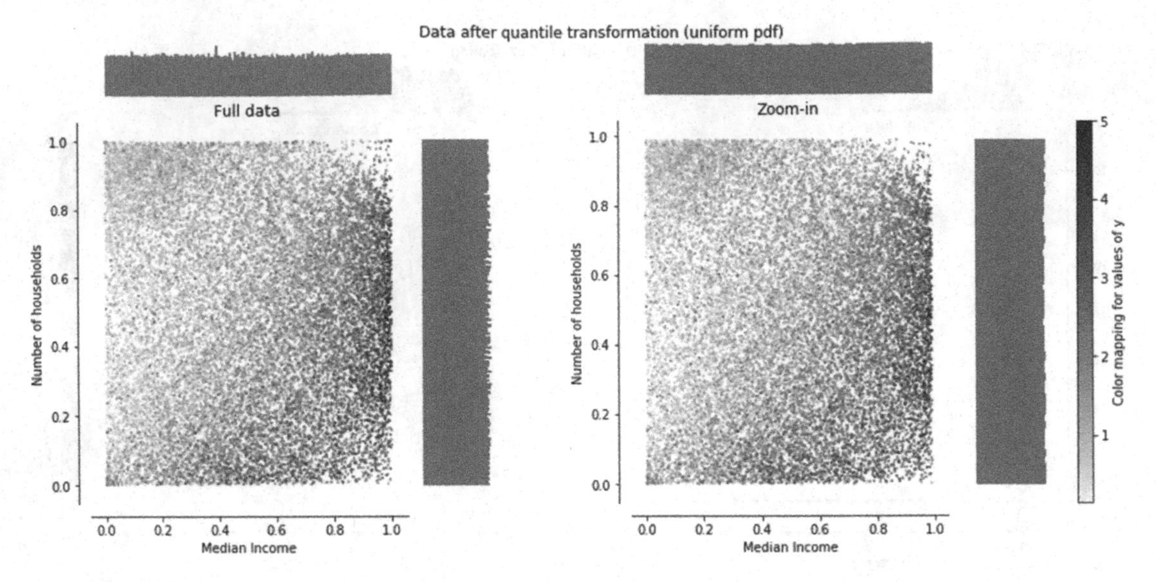

Figure 8-8. *Preprocessing Run 6 Results*

See Figure 8-9 for more results.

```
make_plot(7)
```

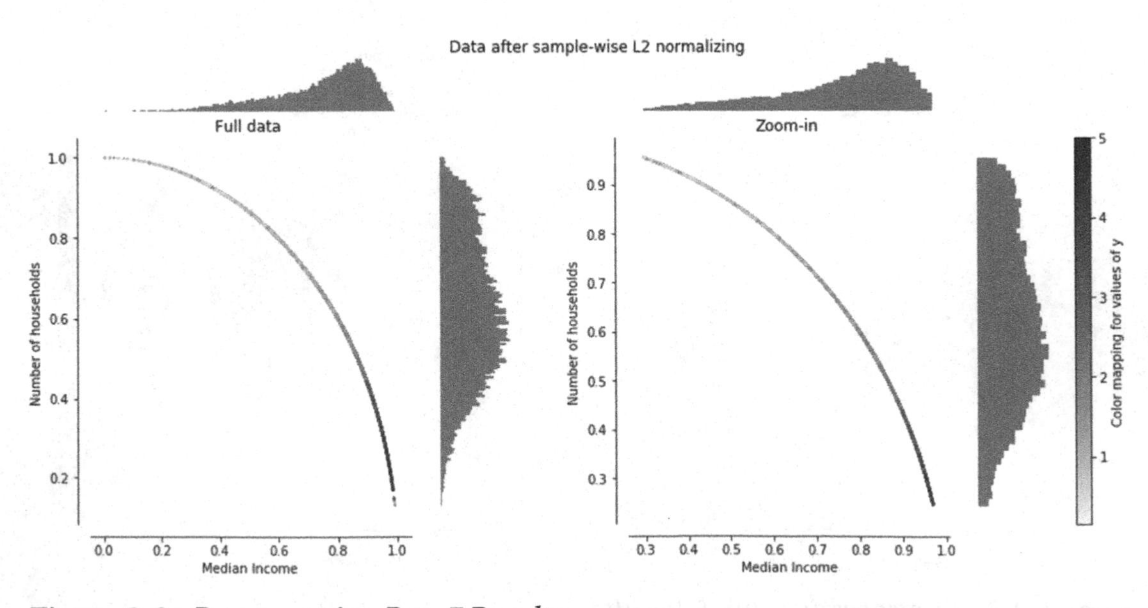

Figure 8-9. *Preprocessing Run 7 Results*

Features

Features are the backbone of the machine learning process; without you having been able to identify the patterns in the data, that is, features, your ML will not perform well. I suggest you investigate the following notebooks to achieve insights into the capabilities of features and techniques you can perform to enhance their capabilities.

Open the Jupyter notebooks called:

- Chapter-008-008-Features-01.ipynb
- Chapter-008-009-Model-Bagging.ipynb
- Chapter-008-010-Model-Boosting.ipynb
- Chapter-008-011-Model-Stacking.ipynb

Applications

I will guide you through an unsupervised learning application next to demonstrate how to use the knowledge you have acquired in a real-world solution.

Stock Market

The application is a stock market trading company that trades specific blue-chip customer portfolios. You will basically mine for specific blue-chip stocks and track their values.

Open the Jupyter notebook called: Chapter-008-012-Stock-Market.ipynb. See Figure 8-10.

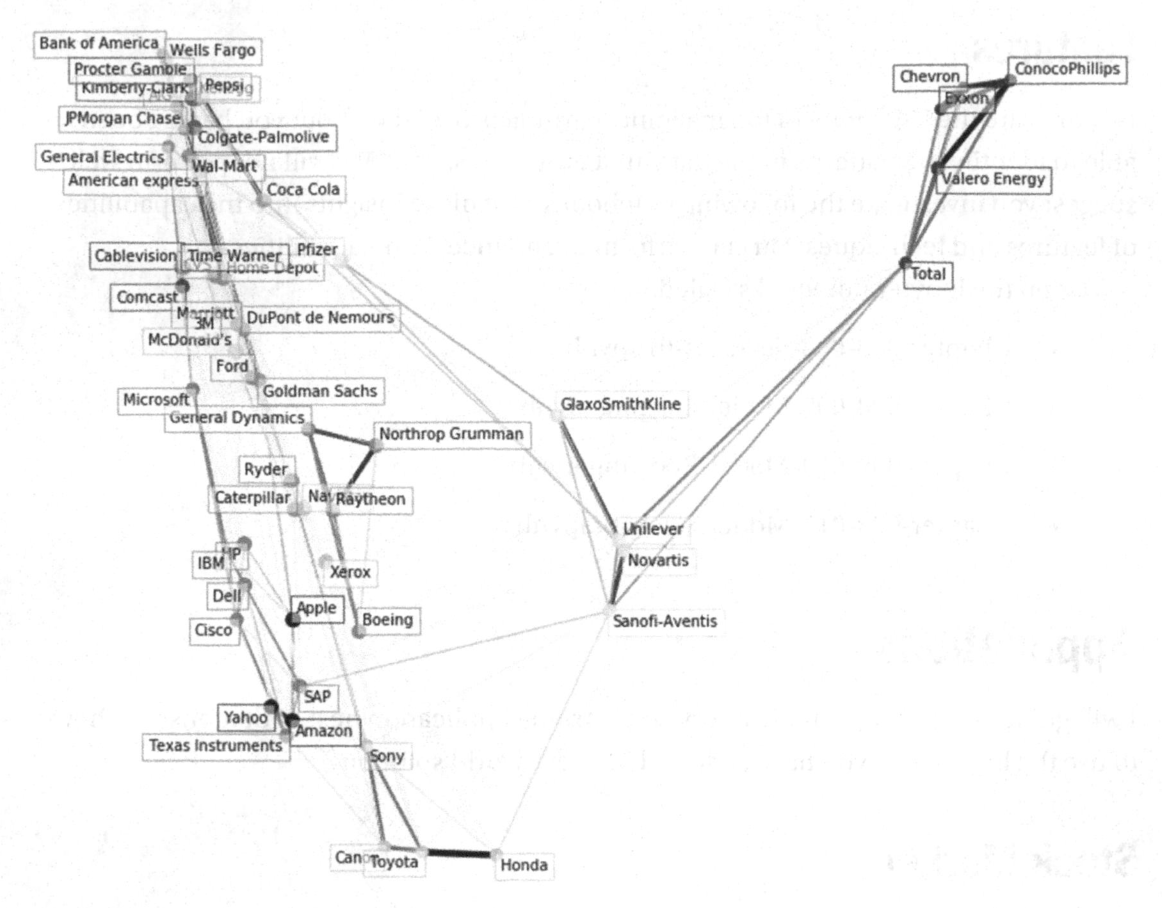

Figure 8-10. *Stock Market Results*

Good progress! You have completed the stock market solution.

Warning ML suggests major amounts of coffee and tea need to be supplied to the reader.

I will now share a number of non-machine learning tools that you can use to industrialize the delivery of your machine learning. Remember that machine learning is part of the decision sciences ecosystem, and any solutions in the extended field of study are applicable to your data science.

Tip I suggest you study solutions that apply to decision science, operational research, and business management to find the correct and appropriate solutions for your ML solutions.

What's Next?

You have now successfully completed the unsupervised machine learning. You are now prepared to handle practically any data with ease. Because you can now process data without any prior flagging, this unsupervised method will support just about any data requirements.

You are doing well as you have now covered the most common types of machine learning. The next chapter is about Reinforcement Learning.

CHAPTER 9

Reinforcement Learning: Using Newly Gained Knowledge for Insights

In reinforcement learning an action is executed for each data point. The action's execution generates a reward signal that indicates how good the decision was.

This algorithm then modifies its action strategy to enhance the possibility of the highest reward against the specific environment.

Due to the missing training data, the agent learns from repeating experience. It collects the knowledge by keeping record ("this action was good, that action was bad"). It learns via trial and error as it attempts modifying and executing for better reward tasks, with the sole goal of maximizing long-term reward from the applied environment.

The chapter also covers Deep Learning, which enhances the concepts of deep reinforcement learning to solve current business solutions.

Markov Decision Process (MDP)

A Markov decision process (MDP) is a discrete-time stochastic control process. It provides a mathematical framework for modeling decision-making in situations where outcomes are partly random and partly under the control of a decision-maker. MDPs are useful for studying optimization problems solved via dynamic programming and reinforcement learning.

The basic data flow and interaction to make the transitions work is in Figure 9-1.

A. F. Vermeulen, *Industrial Machine Learning*, https://doi.org/10.1007/978-1-4842-5316-8_9

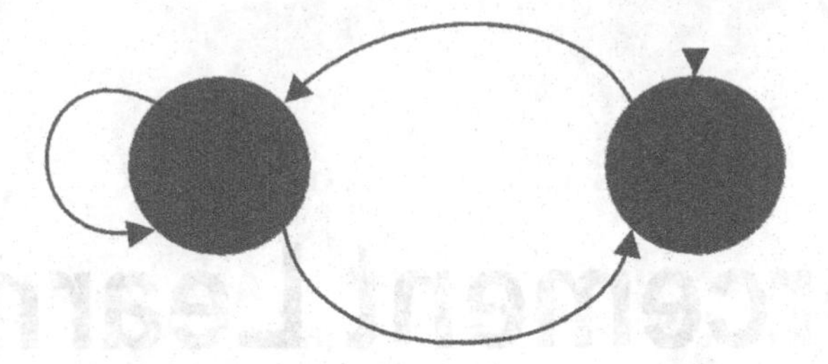

***Figure 9-1.** Transition Diagram*

We will convert these transitions into code by opening the Jupyter Notebook: Chapter-009-001-Markov-Decision-Process-01.ipynb

Robot Walk

The following transitional diagram explains how a robot (1, 1) gets to (3, 4) by avoiding (2, 4).

The robot has an 80% likelihood of traveling in the direction it is facing but can also with a 10% likelihood turn too far and end up going left or right. See Figure 9-2.

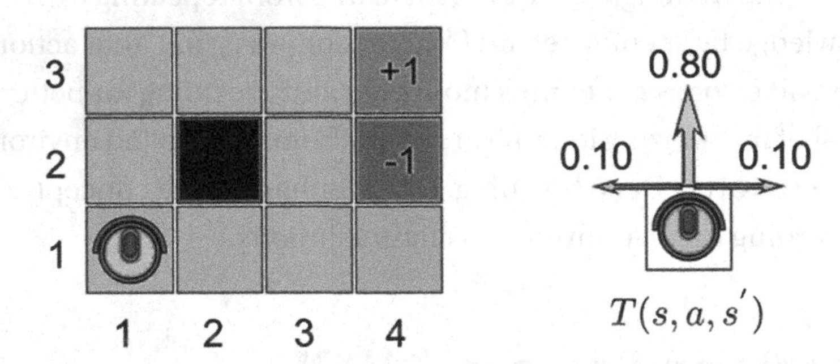

***Figure 9-2.** Robot Walk Transitions*

To create these robot transitions, you need Jupyter Notebook: Chapter-009-002-Robot-Walk-01.ipynb

Please follow the code as it transforms the transitions diagram into code to solve the solution of getting "Robbie" home.

Dynamic Programming

Dynamic programming (DP) is both a mathematical optimization method and a computer programming method.

DP is deployed against "complex problem" problems by breaking them into subsets and then solving discovered subset(s) of the problem while using the discovered information to solve the more difficult original complex problem.

You simply break the problem into smaller solvable problems and consolidate the results to resolve the bigger problem. With DP, you store your results in a table generally to keep track of your progress.

DP is a powerful technique that allows you to solve different types of problems in time O(n 2) or O(n 3) for which a naive approach would take exponential time.

Dijkstra's Algorithm

Dijkstra's algorithm for the shortest path problem can be solved by following a series of smaller cycles that then together achieve a solution.

To investigate how the algorithm works, open the notebook: Chapter-009-003-Dijkstra-algorithm-01.ipynb

The code demonstrates how the algorithm is used to solve the problem of finding the shortest route between various nodes.

```
print (Graph.dijkstra(graph,'a','e'))
```

```
Result:
deque(['a', 'b', 'd', 'e'])
```

The next solution demonstrates how to solve a more complex problem.

Open the Jupyter Notebook called: Chapter-009-004-Dijkstra-algorithm-02.ipynb

The results are:

```
Name:
Type: Graph
Number of nodes: 6
Number of edges: 9
Average degree:   3.0000

Dijkstra path:

['a', 'b', 'd', 'e']
```

I also suggest you investigate the other graphs this code produces.

Tip The networkx library is extremely useful for handling relationships between nodes, that is, agents.

Activity Selection Problem

The problem requires you to select a list of activities you can perform within a given time period.

Note The following implementation assumes that the activities are already sorted according to their finish time.

The results should show a maximum set of activities that can be done by a single person, one at a time.

Report back:

- n --> Total number of activities
- s[]--> An array that contains start time of all activities
- f[] --> An array that contains finish time of all activities

Open the Jupyter Notebook called: Chapter-009-005-Activity-Selection-Problem-01.ipynb

```
The following activities are selected:
Activity: A Start: 1h00 Finish: 2h00
Activity: B Start: 3h00 Finish: 4h00
Activity: D Start: 5h00 Finish: 7h00
Activity: E Start: 8h00 Finish: 9h00
```

Tower of Hanoi

The game "Towers of Hanoi" uses three rods. A number of disks are stacked in decreasing order from the bottom to the top of one rod, that is, the largest disk at the bottom and the smallest one on top. The disks build a conical tower. The aim of the game is to move the tower of disks from one rod to another rod. See Figure 9-3.

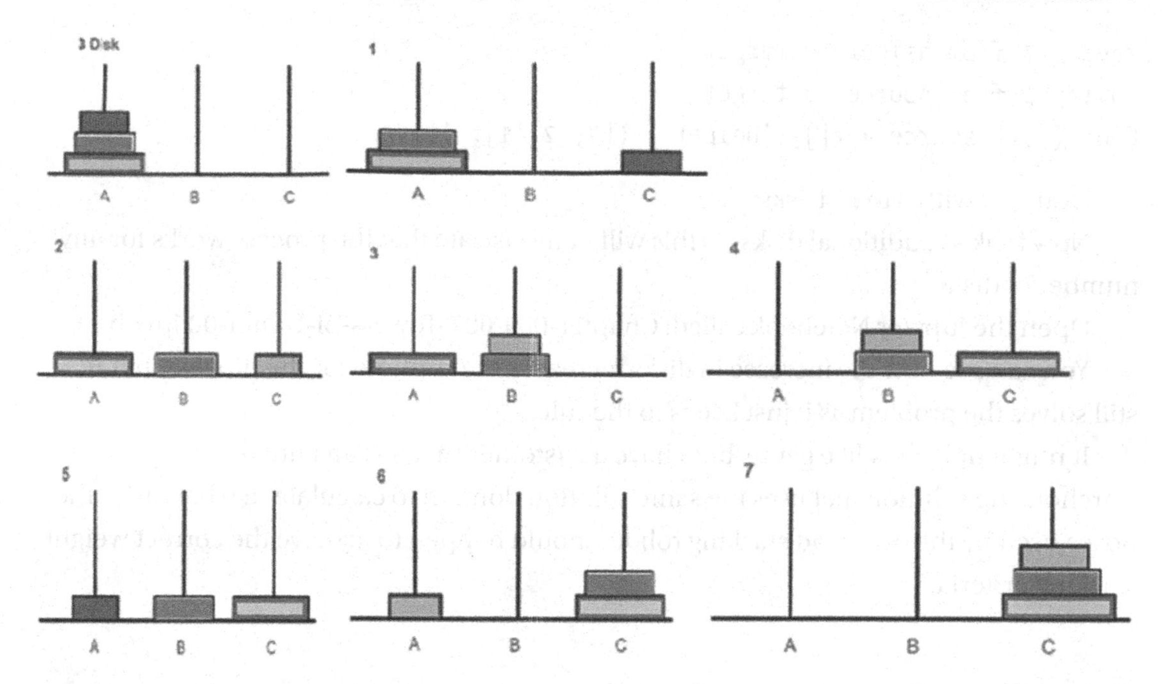

Figure 9-3. *Towers of Hanoi*

The following rules must be obeyed:

- Only one disk may be moved at a time.
- Only the most upper disk from one of the rods can be moved in a move.
- It can be put on another rod, if this rod is empty or if the uppermost disk of this rod is larger than the one which is moved.

Example – 3 Disks only:
Open the Jupyter Notebook called: Chapter-009-006-Towers-of-Hanoi-01.ipynb

```
Results:

Begin: ([3, 2, 1], 'source') ([], 'helper') ([], 'target')
moving 1 from source to target
moving 2 from source to helper
moving 1 from target to helper
moving 3 from source to target
moving 1 from helper to source
```

```
moving 2 from helper to target
moving 1 from source to target
End: ([], 'source') ([], 'helper') ([3, 2, 1], 'target')
```

Example with 3 to 11 Disks:

Now look at additional disks, as this will demonstrate that the process works for any number of disks.

Open the Jupyter Notebook called: Chapter-009-007-Towers-of-Hanoi-02.ipynb

You can see how the increase in disks increases the runtime for the process, but it still solves the problem as it just keeps to the rules.

It might only look like game, but I have a customer that has an automatic warehousing solution that uses the same solution domain to calculate in what order the boxes used by the overhead stacking robots should happen to achieve the correct weight stacking criteria.

Traveling Salesman Problem

The traveling salesman problem (TSP) asks the following question: "Given a list of cities and the distances between each pair of cities, what is the shortest possible route that visits each city and returns to the origin city?" It is an NP-hard problem in combinatorial optimization.

Open the Jupyter Notebook called: Chapter-009-008-Travelling-Salesman-01.ipynb

Your result is shown in Figure 9-4.

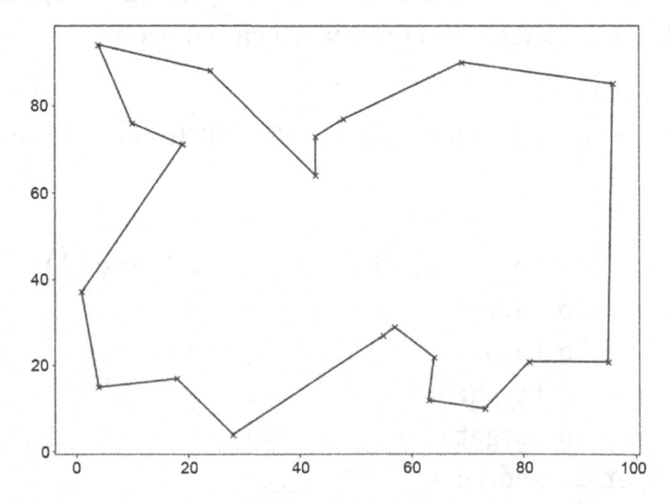

Figure 9-4. *Traveling salesman problem*

The traveling purchaser problem and the vehicle routing problem are equally generalizations of TSP. It is commonly used for solving problems of routing of public transport or bus routes.

I will now show you the use of a better optimized version of the same problem.

Open the Jupyter Notebook called: Chapter-009-009-Travelling-Salesman-02.ipynb

Your results are shown in Figure 9-5.

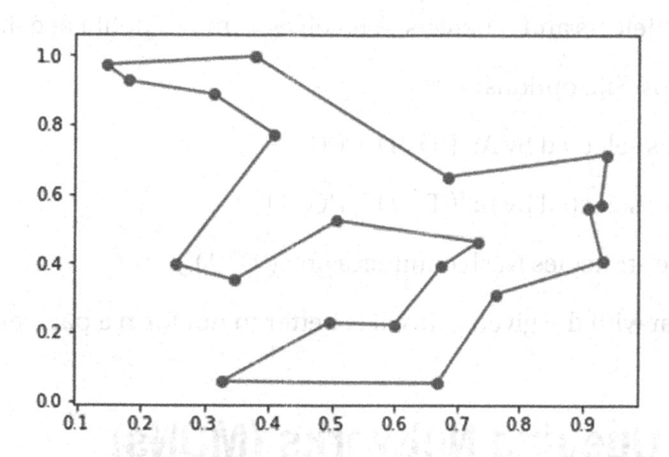

Figure 9-5. *Traveling Route Problem*

Route: [0 13 12 4 17 7 11 15 18 2 6 14 8 3 1 5 10 16 9 19]
Distance: 4.033827398325876

Prisoner's Dilemma

The prisoner's dilemma is a standard illustration of a game analyzed in game theory that shows why two completely sensible persons may not cooperate, even if it appears that it is in their best interests to react.

Open the Jupyter Notebook called: Chapter-009-010-Prisoner-Dilemma-01.ipynb

Results:

The results are investigating if Members of Parliament will cooperate forming a partnership (C) or defect (D). The test is common when parties at an election did not achieve clear majority rule.

Say Party A is rewarded as follows:

When B cooperates and A cooperates, A receives 3 units of political gains

When B cooperates and A defects, A receives 5 units of political gains

When B defects and A cooperates, A receives 0 units of political gains

When B defects and A defects, A receives 1 unit of political gains

Here are the possible options:

Strategies selected by A: [('D', 'D'), ('C', 'D')]

Strategies selected by B: [('D', 'D'), ('C', 'D')]

The pure strategies Nash equilibria are: ('D', 'D')

This shows that with the given gains, it is better to not form a partnership.

Multiclass Queuing Networks (MQNs)

Multiclass Queuing Networks are networks that models the typical network structure created by real-world processing situations where there is a dependency between different processing points with queues acting as a buffer between the points.

Open the Jupyter Notebook called: Chapter-009-011-MQN-01.ipynb

---

Tip Rerun the process until you get a combination that converges.

The solution converged in 5,572 steps.

Queue A:

```
[[1.47025168 0.83294556]
 [0.68592644 0.36197673]
 [0.2433538  1.72692295]
 [2.0728354  1.49116639]
 [0.64876782 1.38615141]
 [0.30623    0.64057238]
 [1.35141951 0.78225212]
 [0.64335072 1.63394717]]
```

Queue B:

```
[[2.17254805 3.64517471]
 [3.16230687 2.97824253]
 [2.79010516 4.11028026]
 [2.38418286 4.05329159]
 [2.69742024 3.57219974]
 [3.03045585 2.64732508]
 [1.74762308 3.58107411]
 [3.09587674 2.65190328]]
```

Queue C:

```
[[ 2.83518099e+00  1.67734127e-03]
 [ 3.08180062e+00  1.68267411e-01]
 [ 2.71948131e+00 -3.15558079e-01]
 [ 2.79336116e+00 -4.95018881e-01]
 [ 2.52133725e+00  8.35166780e-01]
 [ 2.95958847e+00  3.69114538e-01]
 [ 2.79235514e+00  5.09964946e-01]
 [ 4.46361008e+00  4.24490087e-01]]
```

Recommender Systems

A recommender system or a recommendation system is a type of information filtering system that searches for a valid prediction of the "rating" or "preference" any user would give to an item. It tries to foresee the future by looking at previous behaviors of a specific user or finding common matching patterns with other users using the same services. This is a common process on e-commerce systems. It will, for example, suggest new items for you to buy or suggest similar items somebody else bought. In an industry, this is called "upselling" or "cross-selling."

Movie Recommender System

Open the Jupyter Notebook called: Chapter-009-012-Recommender-01.ipynb

Your results are shown in Figure 9-6.

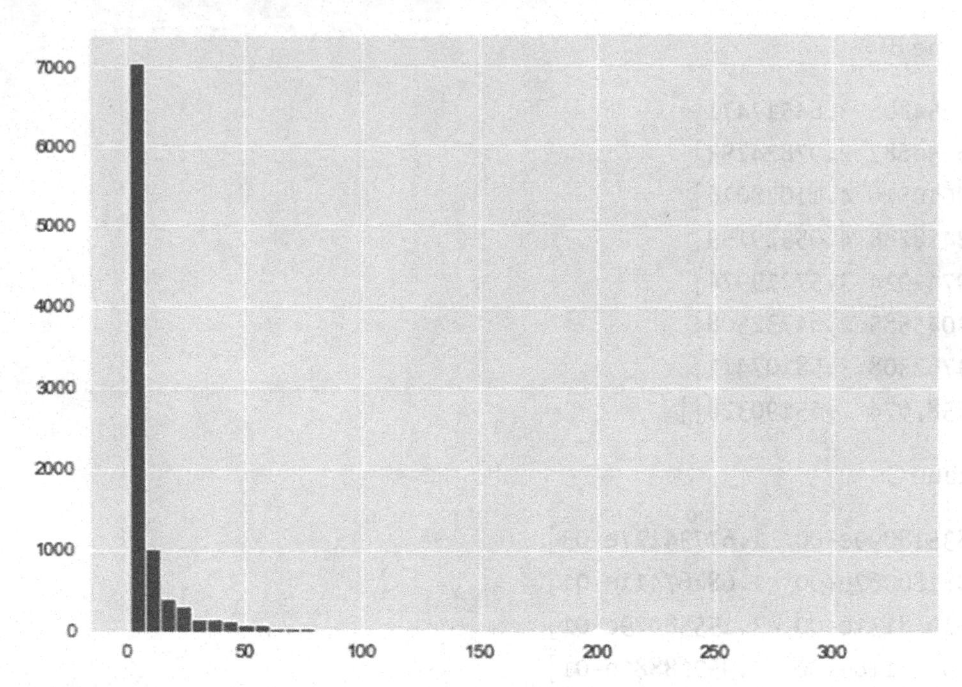

***Figure 9-6.** Recommender System – Rating Count Histogram*

Your results are shown in Figure 9-7.

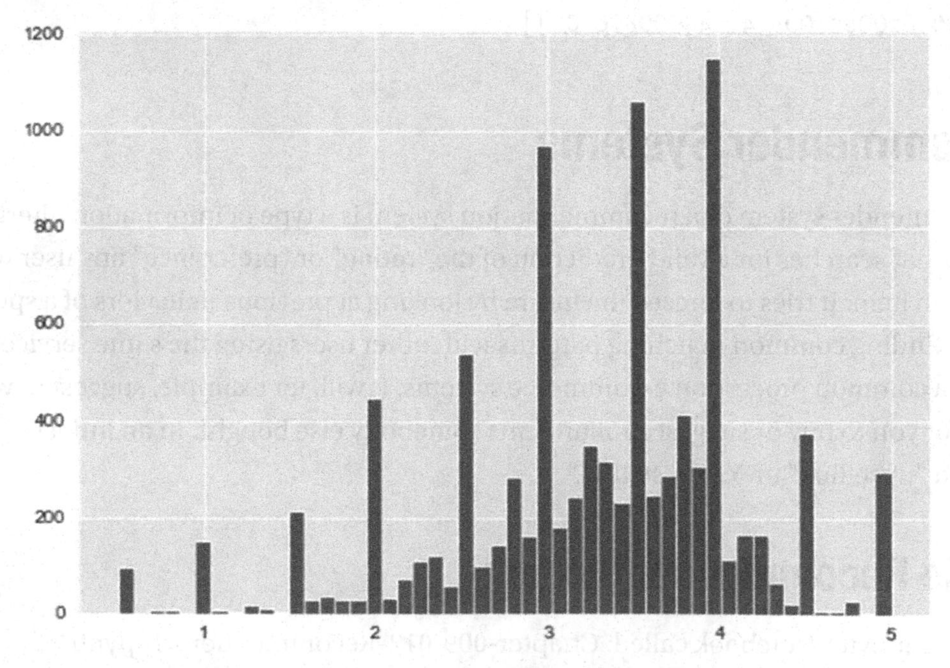

***Figure 9-7.** Recommender System – Rating Histogram*

Your results are shown in Figure 9-8.

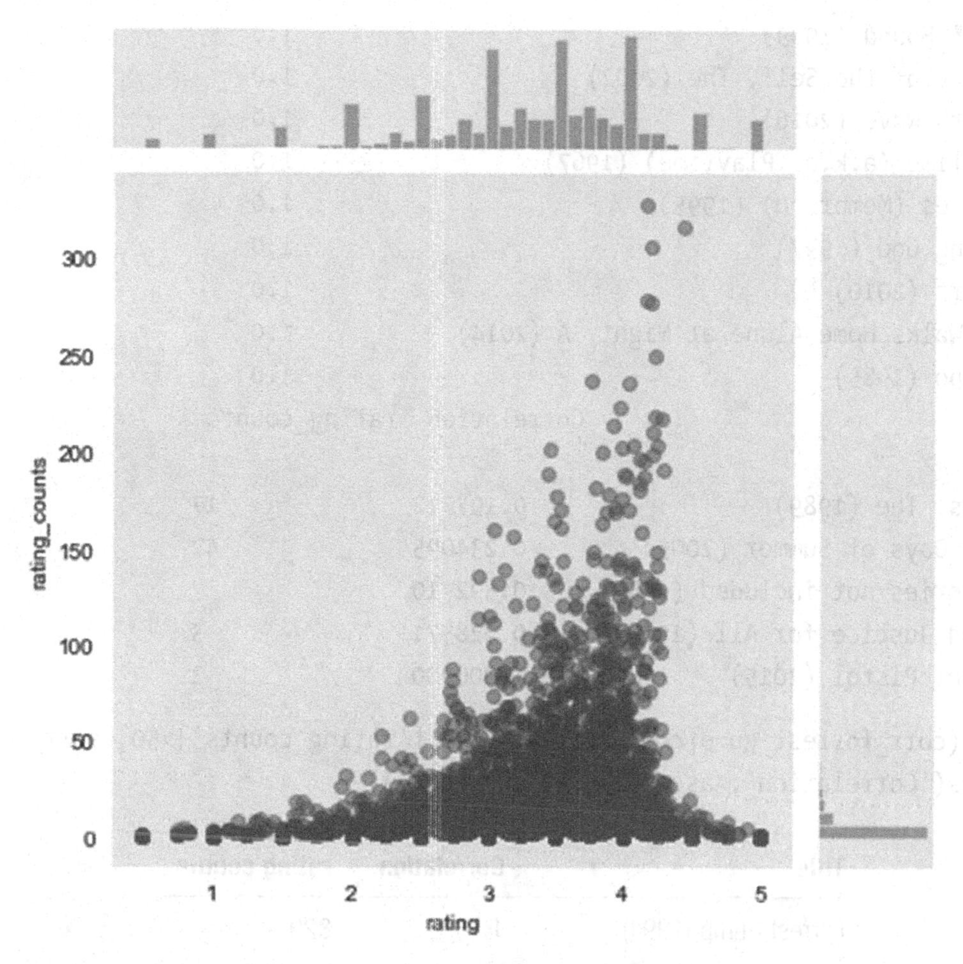

***Figure 9-8.** Recommender System – Joint Plot*

```
                                  Correlation
title
'burbs, The (1989)                   0.197712
(500) Days of Summer (2009)          0.234095
*batteries not included (1987)       0.892710
...And Justice for All (1979)        0.928571
10 Cent Pistol (2015)               -1.000000
```

```
                                          Correlation
title
Lost & Found (1999)                               1.0
Century of the Self, The (2002)                   1.0
The 5th Wave (2016)                               1.0
Play Time (a.k.a. Playtime) (1967)                1.0
Memories (Memorîzu) (1995)                        1.0
Playing God (1997)                                1.0
Killers (2010)                                    1.0
Girl Walks Home Alone at Night, A (2014)          1.0
Tampopo (1985)                                    1.0
                             Correlation  rating_counts
title
'burbs, The (1989)              0.197712             17
(500) Days of Summer (2009)     0.234095             42
*batteries not included (1987)  0.892710              7
...And Justice for All (1979)   0.928571              3
10 Cent Pistol (2015)          -1.000000              2
```

```
print(corr_forrest_gump[corr_forrest_gump ['rating_counts']>50].sort_
values('Correlation', ascending=False).head())
```

Title	Correlation	rating counts
Forrest Gump (1994)	1	329
Mr. Holland's Opus (1995)	0.652144	80
Pocahontas (1995)	0.550118	68
Grumpier Old Men (1995)	0.534682	52
Caddyshack (1980)	0.520328	52

The next recommender system uses some shared articles.

Open the Jupyter Notebook called: Chapter-009-013-Recommender-02.ipynb

```
Loaded 3122 articles with 13 columns

Loaded 72312 articles with 8 columns
```

Your results are shown in Figure 9-9.

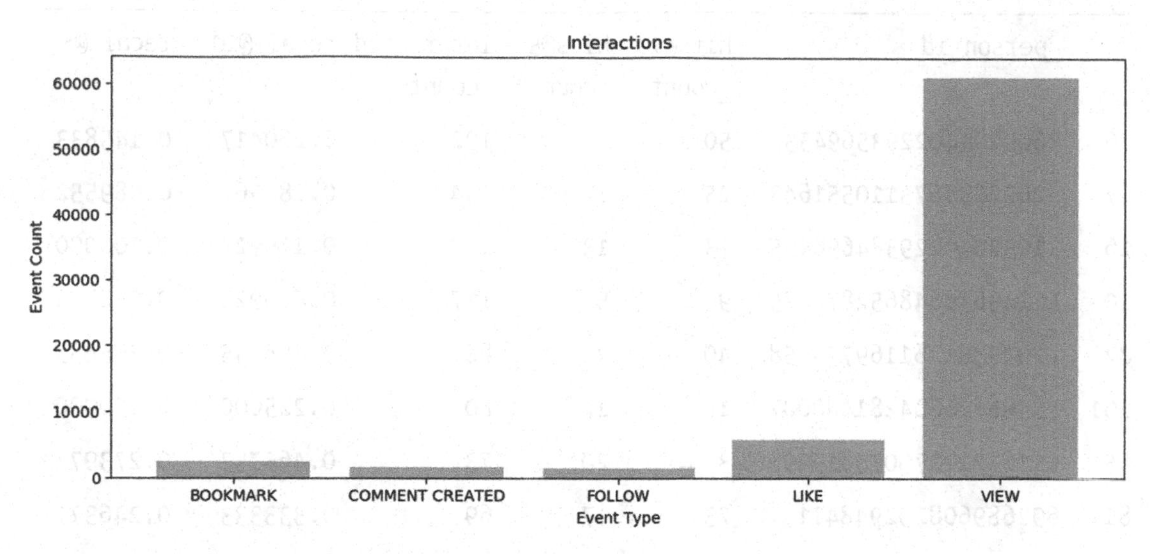

***Figure 9-9.** Event Spread - Interactions*

Number of interactions: 72312

Number of users: 1895

Number users with at least 5 interactions: 1140

Number of interactions from users with at least 5 interactions: 69868

Number of unique user/item interactions: 39106

Number of interactions on Train set: 31284

Number of interactions on Test set: 7822

Your results:

```
Evaluating Popularity recommendation model...
1139 users processed

Global metrics:
{'modelName': 'Popularity', 'recall@5': 0.2418818716440808, 'recall@10':
0.3725389925850166}

print(pop_detailed_results_df.head(10))
```

Your results:

	_person_id	hits@10_count	hits@5_count	interacted_count	recall@10	recall@5
76	3609194402293569455	50	28	192	0.260417	0.145833
17	-2626634673110551643	25	12	134	0.186567	0.089552
16	-1032019229384696495	23	13	130	0.176923	0.100000
10	-1443636648652872475	9	5	117	0.076923	0.042735
82	-2979881261169775358	40	26	88	0.454545	0.295455
161	-3596626804281480007	18	12	80	0.225000	0.150000
65	1116121227607581999	34	20	73	0.465753	0.273973
81	692689608292948411	23	17	69	0.333333	0.246377
106	-9016528795238256703	18	14	69	0.260870	0.202899
52	3636910968448833585	28	21	68	0.411765	0.308824

Content-Based Filtering Model

Content-based filtering approaches leverage descriptions or attributes from items the user has interacted with to recommend similar items.

It depends only on the user's previous choices, making this method robust to avoid the cold-start problem.

The "cold start" happens when the collected data set are not optimal for the engine to provide the best possible recommendation results.

Ignoring stopwords (words with no semantics) from English and Portuguese (as you have a corpus with mixed languages):

```
get_ipython().system('pip install -U nltk')
```

Let's set up the NLTK content on your system.

```
import nltk
nltk.download
```

Now you can start the program to demonstrate the ways that NLTK can be used.

```
from nltk.tokenize import sent_tokenize, word_tokenize
from nltk.corpus import stopwords
from sklearn.feature_extraction.text import TfidfVectorizer
stopwords_list = stopwords.words('english') + stopwords.words('portuguese')
```

Now you train a model with a vector size of 5000, composed by the core unigrams and bigrams found in the corpus, ignoring stopwords.

```
vectorizer = TfidfVectorizer(analyzer='word',
                             ngram_range=(1, 2),
                             min_df=0.003,
                             max_df=0.5,
                             max_features=5000,
                             stop_words=stopwords_list)
item_ids = articles_df['contentId'].tolist()
tfidf_matrix = vectorizer.fit_transform(articles_df['title'] +
"" + articles_df['text'])
tfidf_feature_names = vectorizer.get_feature_names()
print(tfidf_matrix.shape)
```

Your result:

```
(3047, 5000)
```

You can now loop through the results:

I only loop the first 20 records.

```
for i in range(20):
    print(' %2.0d => %0.4f - %0.4f' % (i, tfidf_matrix[i,0], tfidf_
    matrix[i,1]))
```

Your results:

```
 0 => 0.0000 - 0.0000
 1 => 0.0957 - 0.0000
 2 => 0.0000 - 0.0000
 3 => 0.0000 - 0.0000
```

```
 4 => 0.0000 - 0.0398
 5 => 0.0000 - 0.0000
 6 => 0.0000 - 0.0000
 7 => 0.0000 - 0.0134
 8 => 0.0000 - 0.0151
 9 => 0.0000 - 0.0000
10 => 0.0000 - 0.0000
11 => 0.0104 - 0.0000
12 => 0.0000 - 0.0000
13 => 0.0000 - 0.0145
14 => 0.0000 - 0.0000
15 => 0.0000 - 0.0000
16 => 0.0000 - 0.0276
17 => 0.0000 - 0.0000
18 => 0.0000 - 0.0000
19 => 0.0000 - 0.0300
```

Well done ... you have the basic reinforcement learning model.

Now that you have an introduction into the technique, I suggest you investigate these more complex models.

Tip I suggest you take various types and combinations of data and distributions to practice your ML models. This is the only way to achieve IML that is effective and efficient.

Framework for Solving Reinforcement Learning Problems

Shortest Path Problem

Let me take you through an example (Figure 9-10) to make it clear.

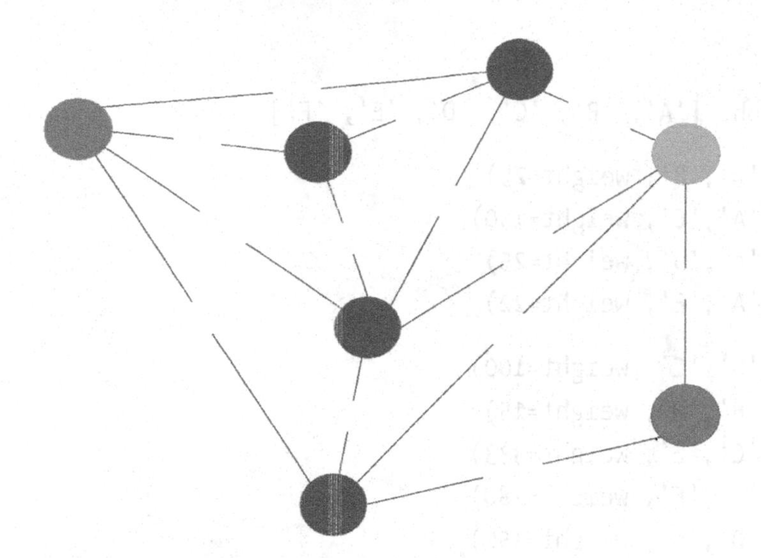

Figure 9-10. *Shortest Path Problem*

This is a representation of a shortest path problem. The task is to go from place A to place F, with as low a cost as possible. The numbers at each edge between two places represent the cost taken to traverse the distance. The negative cost is some incomes on the way. You must define Value as the total cumulative reward when you use a given policy.

Here,

The set of states are the nodes {A, B, C, D, E, F, G}

The action to take is to go from one place to another {A -> B, C -> D, etc.}

The reward function is the value represented by edge, that is, cost

The policy is the "way" to complete the task {A -> ? -> F}

Can you solve the route?

Open the Jupyter Notebook called: Chapter-009-014-Recommender-Network-03.ipynb

```
n = ['A','B','C','D','E','F']
import networkx as nx
G=nx.Graph()
G.add_nodes_from(n)
print ("Nodes of graph: ")
print(G.nodes())
```

Results:

```
Nodes of graph: ['A', 'B', 'C', 'D', 'E', 'F']

G.add_edge ('A','B', weight=25)
G.add_edge ('A','C', weight=150)
G.add_edge ('A','D', weight=25)
G.add_edge ('A','E', weight=22)

G.add_edge ('B','C', weight=100)
G.add_edge ('B','D', weight=15)
G.add_edge ('C','E', weight=333)
G.add_edge ('C','F', weight=280)
G.add_edge ('D','F', weight=150)
G.add_edge ('E','F', weight=60)
G.add_edge ('E','G', weight=180)
G.add_edge ('F','G', weight=390)

print ("Nodes of graph: ")
print(G.nodes())
print ("Edges of graph: ")
print(G.edges())
Nodes of graph:
['A', 'B', 'C', 'D', 'E', 'F', 'G']
Edges of graph:
[('A', 'B'), ('A', 'C'), ('A', 'D'), ('A', 'E'), ('B', 'C'), ('B', 'D'),
('C', 'E'), ('C', 'F'), ('D', 'F'), ('E', 'F'), ('E', 'G'), ('F', 'G')]

%matplotlib inline
import matplotlib.pyplot as plt

pos=nx.circular_layout(G, dim=2, scale=2)

nx.draw(G, pos=pos, node_size=800, node_shape='D')
labels=nx.draw_networkx_labels(G,pos=pos)
edge_labels=nx.draw_networkx_edge_labels(G,pos=pos)
plt.draw()
```

Your results are shown in Figure 9-11.

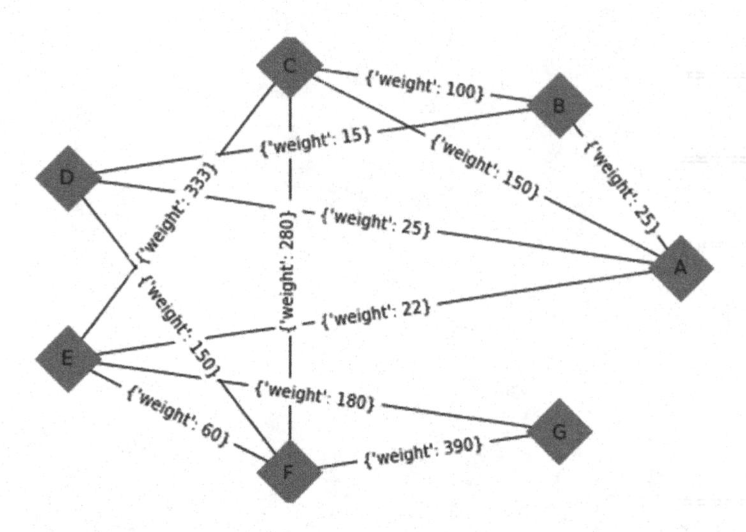

Figure 9-11. *Shortest Path Solution*

```
print (nx.shortest_path(G, source='A',target='F', weight='weight'))
```

Result: ['A', 'E', 'F']

```
print (nx.shortest_path_length(G,source='A',target='F', weight='weight'))
```

Result: 82

```
print (nx.shortest_path(G,source='A',target='G', weight='weight'))
```

Result: ['A', 'E', 'G']

```
print (nx.shortest_path_length(G,source='A',target='G', weight='weight'))
```

Result: 202

```
for c in range(3):
    print ('Cut-off:', c)
    print ('=================')
    p = nx.single_source_shortest_path_length(G, source='A', cutoff=c)
    for key, value in p.items():
        print(key, ' => ', value)
    print('=================')
```

Your Results:

```
Cutoff: 0
=================
A => 0
=================
Cutoff: 1
=================
A => 0
B => 1
C => 1
D => 1
E => 1
=================
Cutoff: 2
=================
A => 0
B => 1
C => 1
D => 1
E => 1
F => 2
G => 2
=================
```

So here, your policy was to take ['A', 'E', 'F'] and your value is 82.

The other route ['A', 'E', 'G'] with your value is 202.

Congratulations! You have just implemented a reinforcement learning algorithm.

An Implementation of Reinforcement Learning

You will be using a Deep Q-learning algorithm. Q-learning is a policy-based learning algorithm with the function estimated as a neural network.

You will first see what a Cart Pole problem is then go on to coding up a solution.

When I was a kid, I remember that I would pick a stick and try to balance it on one hand. My friends and I used to have this competition where whoever balanced it for more time would get a "reward," a chocolate!

Let's code it up!

To set up our code, you need to first install a few things. Open the Jupyter Notebook called: Chapter-009-015-Q-Learn-01.ipynb

You need to have the following installed:

```
pip install keras-rl
pip install h5py
pip install gym
```

Let's start to build your Q-learning process:

```
import numpy as np
import gym

from keras.models import Sequential
from keras.layers import Dense, Activation, Flatten
from keras.optimizers import Adam

from rl.agents.dqn import DQNAgent
from rl.policy import EpsGreedyQPolicy
from rl.memory import SequentialMemory
ENV_NAME = 'CartPole-v0'
```

You must get the environment and extract the number of actions available in the CartPole problem

```
env = gym.make(ENV_NAME)
np.random.seed(20)
env.seed(20)
nb_actions = env.action_space.n
```

You need to create a single hidden layer neural network model to support the environment's actions.

```
model = Sequential()
model.add(Flatten(input_shape=(1,) + env.observation_space.shape))
model.add(Dense(16))
model.add(Activation('relu'))
model.add(Dense(nb_actions))
model.add(Activation('linear'))
```

Now you can look at your model.

```
print(model.summary())
```

Next you configure and compile our agent.

I suggest you use the policy as Epsilon Greedy and you set the memory as Sequential Memory since you must store the result of actions that your CartPole performed and the rewards it gets for each action.

```
policy = EpsGreedyQPolicy()
memory = SequentialMemory(limit=50000, window_length=1)
dqn = DQNAgent(model=model, nb_actions=nb_actions, memory=memory, nb_steps_
warmup=10, target_model_update=1e-2, policy=policy)
dqn.compile(Adam(lr=1e-3), metrics=['mae'])
```

It is time to perform the training process.

Warning This can take some time demanding on the compute capabilities of your ecosystem.

When running this on cloud, switch off the visualize.

```
try:
  dqn.fit(env, nb_steps=5000, visualize=True, verbose=2)
except:
  dqn.fit(env, nb_steps=5000, visualize=False, verbose=2)
```

Your result:

Training for 5000 steps...

```
10/5000: episode: 1, duration: 0.005s, episode steps: 10, steps per
second: 1825, episode reward: 10.000, mean reward: 1.000 [1.000, 1.000],
mean action: 0.000 [0.000, 0.000], mean observation: 0.135 [-1.986, 3.027],
loss: --, mean_absolute_error: --, mean_q: --
/usr/local/lib/python3.6/dist-packages/rl/memory.py:39: UserWarning: Not
enough entries to sample without replacement. Consider increasing your
warm-up phase to avoid oversampling!
  warnings.warn('Not enough entries to sample without replacement. Consider
increasing your warm-up phase to avoid oversampling!')
```

(Removed steps 2 to 4999 to save space in book. See notebook for details.)

```
4976/5000: episode: 234, duration: 0.143s, episode steps: 50, steps per
second: 349, episode reward: 50.000, mean reward: 1.000 [1.000, 1.000],
mean action: 0.460 [0.000, 1.000], mean observation: -0.167 [-0.756,
0.225], loss: 2.869326, mean_absolute_error: 10.015974, mean_q: 19.497936
done, took 14.771 seconds
```

Or with visualize, see Figure 9-12.

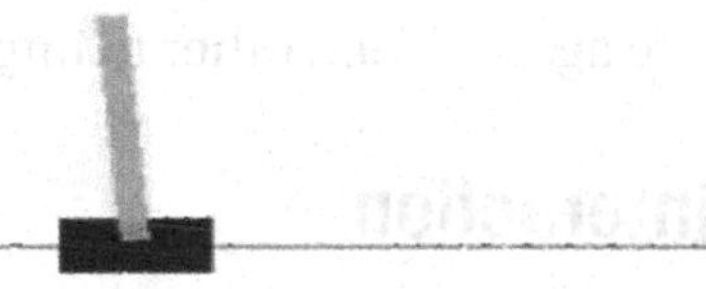

Figure 9-12. *CartPole*

You can now test the reinforcement learning model against the learnings.

```
try:
  dqn.test(env, nb_episodes=5, visualize=True, verbose=2)
except:
  dqn.test(env, nb_episodes=5, visualize=False, verbose=2)
```

Results:

```
Episode 1: reward: 200.000, steps: 200
Episode 2: reward: 82.000, steps: 82
Episode 3: reward: 119.000, steps: 119
Episode 4: reward: 187.000, steps: 187
Episode 5: reward: 83.000, steps: 83
```

Note The five tests yield diverse results. The results are driven by the start point that is random, but they all balance at the end.

It you have a CartBot like the above one trying to balance itself, you have successfully gotten a Q-learning solution.

Increasing the Complexity

The following domains are where Reinforced Learning has been applied with great success:

Game Theory

Top-end artificial intelligence (AI) systems at the moment are constructed on a concept of a single agent engaging in a given task or, in the case of adversarial models, a couple of agents that compete against each other to improve the overall behavior of a system.

Multi-agent Interaction

A multi-agent system is common where we simulate any system that consists of a team interacting within the same ecosystem. Some actions enhance other agent's effort positively or obstruct agents negatively.

- Swarms of Robotics
- Swarms
- Vehicular Navigation
- Medicine Research
- Industrial Logistic for process control

- Computer Networking
- Traffic Control
- Smart Cities

If you look around your own environment, can you see applications for this technology?

The following industrialized versions of these environments are highly interesting:

AlphaGo - `https://deepmind.com/research/alphago/`

Do you remember seeing this on the news ... AlphaGo is the first computer program to defeat a professional human Go player?

This processing engine is now also trying to solve some critical real-world problems.

Here is a small subset of the reinforcement learning currently active in the world (Table 9-1).

Table 9-1. *Subset of Reinforcement Learning*

Algorithm	Description	Model	Policy	Action Space	State Space	Operator
Monte Carlo	Monte Carlo Simulations	Model-Free	Off-policy	Discrete	Discrete	Sample-means
Q-learning	State-action-reward-state	Model-Free	Off-policy	Discrete	Discrete	Q-value
SARSA	State-action-reward-state-action	Model-Free	On-policy	Discrete	Discrete	Q-value
Q-learning - Lambda	State-action-reward-state with eligibility traces	Model-Free	Off-policy	Discrete	Discrete	Q-value
SARSA - Lambda	State-action-reward-state-action with eligibility traces	Model-Free	On-policy	Discrete	Discrete	Q-value
DQN	Deep Q-Network	Model-Free	Off-policy	Discrete	Continuous	Q-value
DDPG	Deep Deterministic Policy Gradient	Model-Free	Off-policy	Continuous	Continuous	Q-value
A3C	Asynchronous Actor-Critic Algorithm	Model-Free	Off-policy	Continuous	Continuous	Q-value
NAF	Q-Learning with Normalized Advantage Functions	Model-Free	Off-policy	Continuous	Continuous	Advantage
TRPO	Trust Region Policy Optimization	Model-Free	On-policy	Continuous	Continuous	Advantage
PPO	Proximal Policy Optimization	Model-Free	On-policy	Continuous	Continuous	Advantage

I can debate numerous books of information on these reinforcement learning processes as there are countless different options currently being developed by the data scientist and machine learning community worldwide.

This is the most active research and development ecosystem in the field of science.

It is the one ecosystem that I predict will have the biggest impact on humanity in the next 200 years. Reinforcement learning is the first process that truly has the capability to replace or displace people on a massive scale.

Tip If there is one technology in this book that you learn well, it should be reinforcement learning.

I am now proceeding to a more general trend for the rest of the chapter.

Modeling Environment

I want to explain a general term called "simulator."

"Simulation is the imitation of the operation of a real-world set of processes within a system over time."

Essentially, there is a system that has several inputs that applies approximated mathematical functions to selected inputs, and returns an output in the form of data that can be visualized to observe what the system is doing at that moment.

Think back to the CartPole. It was a simple visualization, but you could see the little cart trying to balance the pole.

The problem you will find is the scope of your model or simulation. As most events in the real world are related in one way or another, you would need to work out how wide your projects scope is before you start.

> The question is: How do you determine the scope?
>
> So, you have a Cart with a pole ... Is that the scope?

Let's think bigger ... What about a complete car? See Figure 9-13.

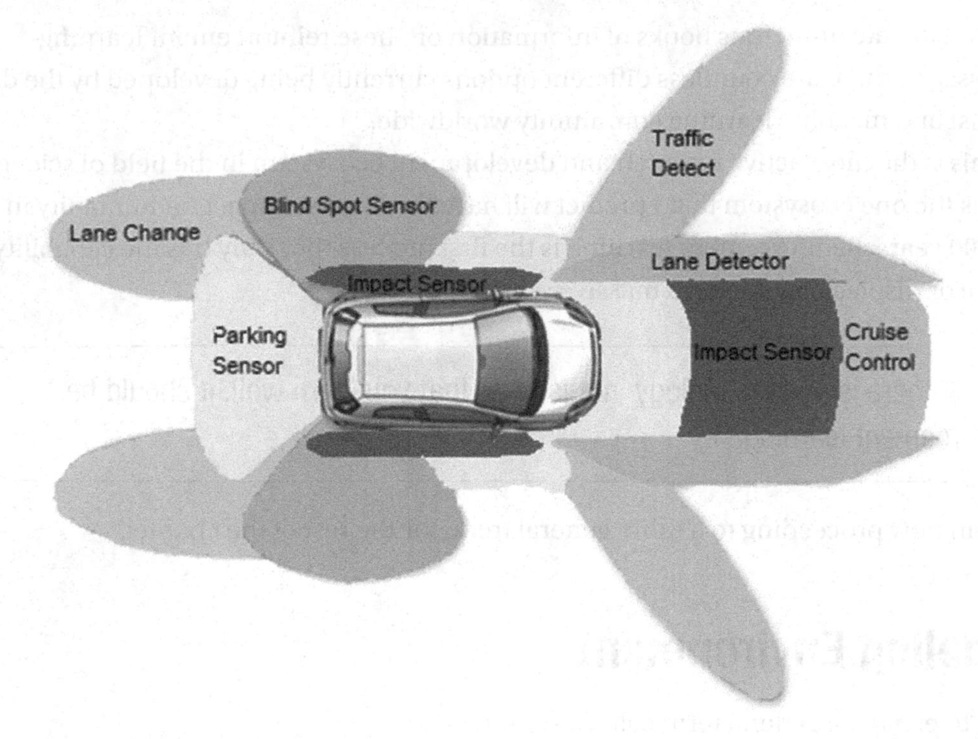

***Figure 9-13.** Scope of car*

Let's go bigger again ... Numerous vehicles?

Every vehicle that travels through a town can map the city and then share the information (Figure 9-14).

Figure 9-14. *All vehicles in tow*

Even bigger ... Entire earth's population of vehicles in the world? See Figure 9-15.

Figure 9-15. *All vehicles in the world*

What if each of the vehicles in the world was connected? Possible but improbable scope?

So, what I am saying is the larger your scope, the larger the volume of data to process and the greater the probability you are could not resolve the model on a single given system.

Status Feature Creation

The next step is to identify the features in the model.

I suggest you take an image of your model and then tag it with a tag for every unique feature.

Here is the vehicle model tagged (Figure 9-16).

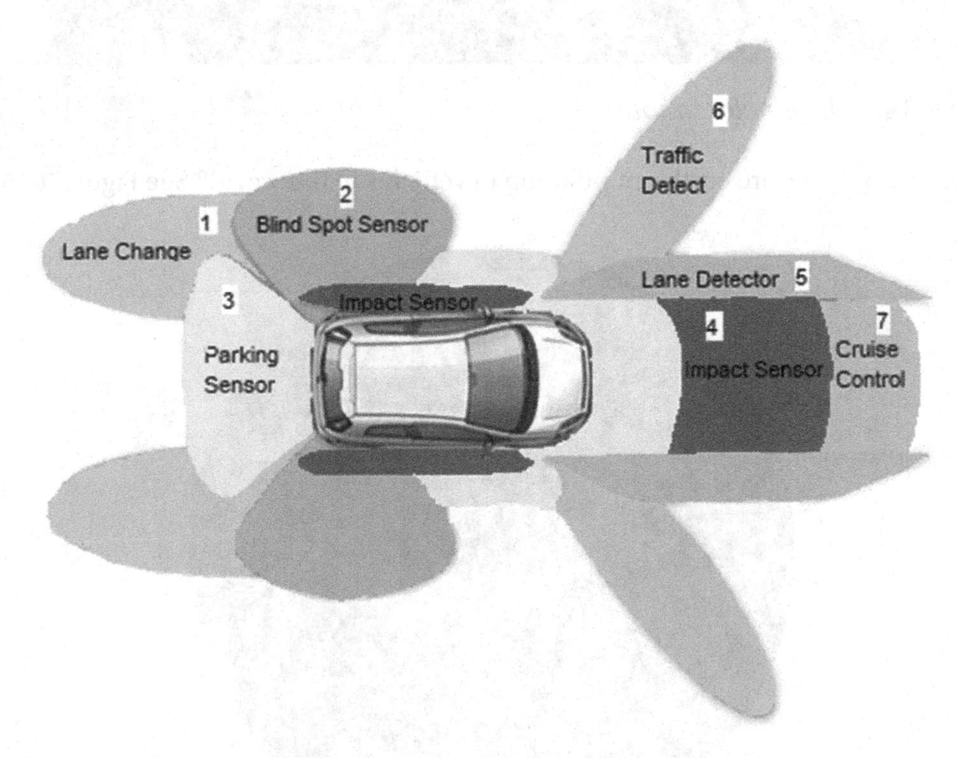

Figure 9-16. *Vehicle model*

I will now guide you through the features I identified.

Feature 1:

Lane Changing Feature is a binary sensor as the sensor only reports a value of 0 or 1.

Feature 2:

Blind Spot Sensor Feature is a binary sensor with also a value of 0 or 1.

Feature 3:

Parking Sensor Feature is a five-level sensor and also has a value of 0.0 between 1.0.

It reports inside the vehicle as a set of LEDs that show relatively how close you are to the object behind.

The second output is that when weighted correctly, the 1.0 is five meters +/- tenth meters in distance.

Feature 4:

Impact Sensor Feature is a ten-level sensor and is also a value of 0.0 between 1.0.

The second output is that when weighted corrected the 1.0 is fifteen meters +/- quarter meter in distance.

Feature 5:

Lane Detector Feature is a four-level sensor and is also a value of 0.0 between 1.0.

The second output is that when weighted correctly, the 1.0 is fifteen meters +/- quarter meter in distance.

Feature 6:

Traffic Detector Feature is a five-level sensor and is also a value of 0.0 between 1.0.

The second output is that when weighted correctly, the 1.0 is fifteen meters +/- quarter meter in distance.

Feature 7:

> Cruise Sensor Feature is a five-level sensor and is also a value of -1.0 between 1.0
>
> The sensor reports a 0.0 if the cruise control detects a < 0.0, and it starts to reduce the speed of the vehicle to return to 0.0; and if > 0.0 the speed will increase until it reaches the correct speed.

You now have seven features and therefore seven dimensions feed information about your model's overall status.

Reward Functions

The reward is a seven-stage reward formula. The seven features reward back separately. You can use a seven-by-seven correlation algorithm here to check if there are significant correlations between sensors.

Warning In the real world, this process has been going on for years with millions of hours of driving training data, and they still get outliers.

The prediction is that a sudden algorithm that solves this issue of unexpected results on unknown inputs will immediately cause a major disruption in the capability of most RL systems' capabilities.

Action Generation

The actions that are applied range from warning lights to actively changing the speed of the vehicle or even applying the brakes. The lane detection will even return the car to the road.

Warning More than one possible action could cause the same result or cause a catastrophic negative reward of total failure of the environment.

I would suggest keeping your actions to small incremental steps, not big radical changes.

Final Models

I hope you now understand that the model for some real-world projects can rapidly evolve into hundreds or even thousands of features, states. and actions.

Advice I create a minimum of three models with diverse base features and then check for correlations in the actions suggested to validate the overall system's actions. That way you do not easily reach a catastrophic failure state.

Inverse Reinforcement Learning

Inverse reinforcement learning (IRL) is the process of deriving a reward function from observed behavior. While ordinary "reinforcement learning" involves using rewards and punishments to learn behavior, in IRL the direction is reversed.

This modeling process allows the system to try any new actions and then calculates a binary result to repeat that suggested action.

This is a simple model used when working with massive lists of actions that are possible.

For example, the vehicle to prevent the vehicle leaving the road the car could apply the brakes or turn the steering wheel. In IRL the vehicle will try each individually option and in combination with each other to learn what the result is of that single action or combination of actions. The sum of all actions, that is, the behavior is rewarded.

This reduces the amount of reward measurements. Work well in big scale real-life projects.

IRL is also used in Evolutionary Computing to develop actions and rewards that can automatically discover the modeling for the environment. This will be discussed in detail in Chapter 10.

Deep Reinforcement Learning

Deep learning is a new research pathway within the field of machine learning. The core knowledge behind deep learning is to create architectures consisting of multiple layers of representations in order to learn high-level abstractions. The base of the deep learning is that the model can handle complex feature sets as it processes the complex feature interactions as a series of smaller interactions and returns a net result.

A typical deep learning artificial neural network uses IRL models that cause a Backpropagation to determine which actions achieved the most effect. Backpropagation is a method used in artificial neural networks to calculate a gradient that is needed in the calculation of the weights to be used in the network. Backpropagation is shorthand for "the backward propagation of errors," since an error is computed at the output and distributed backward throughout the network's layers. It is commonly used to train deep neural networks.

This training achieved by deep learning of neural networks is pioneering the combination of numerous approaches with deep reinforcement learning – to produce the first artificial agents to achieve human-level performance across several challenging domains.

You can now create agents that can continually make value judgements by selecting good actions over bad. This knowledge is represented by a Q-network that estimates the total reward that an agent can expect to receive after taking a particular action.

The area is developing at an accelerating pace as more and more real-world problems are now solved by these massive deep learning models that consist of tens of thousands of input features and thousands of hidden layers with a single or only a few systems' behaviors on the output side.

This is an area that will deliver many great results over the coming years.

Multi-agent Reinforcement Learning

Multi-agent reinforcement learning is the concept where one system's observation gets passed to another system model to enhance their state and reward collection. This could deploy where, for example, two vehicles communicate to each other that they are going to take an action. The other option is that the environment passes statuses by default to the individual from their point of view. An example would be a pressure activated part of the road to prevent vehicles driving of the road by activating a signal that the other agent can detect.

For example, in one smart city, the "vehicle's cruise control is interacting with the city's traffic system to ensure a smoother traffic flow through the city. Or heavy good vehicles get routed away from bridges that are too low for their loads. The multi-agent world is the core world of the Internet of things (IoT). The Internet-of-Things revolution will connect billions of agents into a single cluster of sensors and state recorders.

The multi-agent environment, within the next five years, be the norm for systems.

Your machine learning will need to be able to handle these industrialized scale IoT systems combine with the new robotic capability that I discussed in Chapters 6 and 7 will change the world physically over the next ten years.

What Have You Achieved?

At this point in the book, you have now completed the core machine learning methodologies.

You should now be able to perform supervised learning, unsupervised learning, and reinforcement learning.

You should also now understand that the optimal solution is an ensemble of techniques from all three of these core methodologies.

Tip I would suggest you try as many options as you can for a given modeling requirement as the complexity of the solution you will require will demand more creative solutions.

What's Next?

The later Chapters 14 and 15 will explain to you how to convert the knowledge you acquired up to now in the book into industrialized-level actions to generate in IML disruptor style solutions.

We will next cover the world of Evolutionary Computing in Chapter 10.

Tip I suggest you take a few hours to just revisit the previous chapters to ensure you understand the information you now have in your knowledge base.

Now get ready for the next phase of the book.

The next chapter will enable you to automatically generate parts of models via Evolutionary Computing.

CHAPTER 10

Evolutionary Computing

Evolutionary computation is a family of algorithms for global optimization inspired by the process of biological evolution, and the subfield of artificial intelligence (AI) and soft computing studying these algorithms. In technical terms, they are a family of population-based trial-and-error problem solvers with a metaheuristic or stochastic optimization character.

The area of evolutionary computing is supporting many of the new generation AI algorithms to generate new processing rules in a dynamic manner to enable the algorithm to automatically adapt to the new data set. This enables the machine learning to adapt to the data lake it is required to process.

Evolutionary Process

The basic evolutionary process is three steps:

Step One

Generate the initial population of individuals randomly (First generation).

Step Two

Evaluate the fitness of each individual in that population (time limit, sufficient fitness achieved, etc.).

A. F. Vermeulen, *Industrial Machine Learning*, https://doi.org/10.1007/978-1-4842-5316-8_10

Step Three

Repeat the following regeneration steps until termination:

1. Select the best-fit individuals for reproduction (Parents).
2. Breed new individuals through cross-over and mutation operations to give birth to offspring.
3. Evaluate the individual fitness of new individuals.
4. Replace least-fit population with new individuals.

Evolutionary computing techniques mostly involve metaheuristic optimization algorithms. Broadly speaking, the field includes the algorithms and methodologies I will discuss in the rest of this chapter. However, this is a field that is expanding at a massive rate.

My prediction is that within five years 90%+ of the products around us will either be designed by or tested by a type of evolutionary computing ecosystem.

Ant Colony Optimization

The ant colony optimization algorithm (ACO) is a probabilistic technique for solving computational problems that can be reduced to finding good paths through graphs. Artificial Ants stand for multi-agent methods inspired by the behavior of real ants. The pheromone-based communication of biological ants is often the predominant paradigm used.

Combinations of Artificial Ants and local search algorithms have become a method of choice for numerous optimization tasks involving some sort of graph, for example, vehicle routing and Internet routing.

Load the Jupyter Notebook from the example directory in Chapter 10 called: Chapter-10-01-Ant-Colony.ipynb

Run the solution:

```
ant_colony = AntColony(distances, 1, 1, 100, 0.95, alpha=1, beta=1)

shortest_path = ant_colony.run()
```

Output is:

The results are:

```
([(0, 1), (1, 2), (2, 3), (3, 4), (4, 0)], 16.0)
([(0, 1), (1, 3), (3, 4), (4, 2), (2, 0)], 17.0)
([(0, 1), (1, 2), (2, 3), (3, 4), (4, 0)], 16.0)
```

(Removed some results to save space in book. Please see notebook results for complete result set.)

```
 ([(0, 1), (1, 2), (2, 3), (3, 4), (4, 0)], 16.0)
([(0, 2), (2, 3), (3, 4), (4, 1), (1, 0)], 9.0)
```

The next step prints the result for the Shortest Path:

```
print ("\n Shortest Path: {}".format(shortest_path))
```

Result is:

```
Shortest Path: ([(0, 1), (1, 4), (4, 3), (3, 2), (2, 0)], 9.0)
```

Cultural Algorithms

Cultural algorithms (CA) are a branch of evolutionary computation where there is a knowledge component that is called the belief space in addition to the population component. In this sense, CAs can be seen as an extension of a predictable genetic algorithm.

The list of belief space categories are as follows.

Normative Knowledge

A collection of desirable value ranges for the individuals in the population component, for example, acceptable behavior for the agents in population. Normativity is the process in human societies of designating some actions or outcomes as good or desirable or permissible and others as bad or undesirable.

The normative process evolves the environment it controls by using precise guides for the machine learning-generated action to cause an evolutionary process within the solution the machine learning model is managing. The ML is forcing the environment to match the norm as prescribed by the rules.

Domain-Specific or Domain-General Knowledge

Domain-specific learning theories of development hold that machine learning has numerous independent, specialized knowledge structures and not one cohesive knowledge structure. This machine learning training covers only one domain and will not impact another independent knowledge domain. The core knowledge theorists believe machine learning should have highly specialized machine learning models and features that are independent from the next machine learning models and features.

The domain-general theories see evolutionary knowledge as a co-opted cohesive knowledge structure within the machine learning model and features, favoring the domain-general learning model created by the resulting sum of all the evolutionary knowledge since the seed model of the learning was activated.

Situational Knowledge

Specific examples of important events, for example, successful/unsuccessful solutions will now be discussed.

Situational awareness is an important part of cyber, war, economics, business, relationships etc., and usually it is not feasible to have all the possible information in use, therefore scenario-based training helps to make decisions with incomplete information and being able to sense more than what is seen.

Machine learning is typically achieved using categorical learning. This means that in general there are two different ways of categorical learning.

Rule based – the machine learning formulates sets of rules to categorize situations and stores the rules as the category of the situation observed.

Information based – the machine learning formulates sets of features against the information to successfully categorize the situation and stores the binary existence of these features as the category of the situation observed.

Temporal Knowledge

The temporal knowledge as relating to time or order the machine learning has observed in the state of the features.

Tip Read about Allen's Interval algebra. The calculus defines possible relations between time intervals and provides a composition table that can be used as a basis for reasoning about temporal descriptions of events.

An example follows.

Spatial Knowledge

Spatial knowledge is the insight of all the possible states within the topography of the search space of the machine learning's model space. It can also be used to describe the spatial distribution of population of the information the machine learning can learn.

An example follows.

Distributed Evolutionary Algorithms

Distributed Evolutionary Algorithms in Python (DEAP) are an evolutionary computation framework for rapid prototyping and testing of ideas. It incorporates the data structures and tools required to implement most common evolutionary computation techniques such as genetic algorithms, genetic programming, evolution strategies, particle swarm optimization, differential evolution, traffic flow, and estimation of distribution algorithm. It has been developed at Université Laval since 2009.

Install library:

```
conda install -c conda-forge deap
```

Load the Jupyter Notebook from the example directory in Chapter 10 called: Chapter-10-02-Distributed-Evolutionary-Algorithms.ipynb

I suggest you execute the notebook step by step to gain insight into how it works technically.

Take note due to the nature of the random function in this evolutionary process, you may get slightly different answers than those printed below.

Results:

```
### Population ###########################################
[[9.521215056326895, 7.345527981285888], [4.8329426999153355,
6.20495203951839], [10.466145449758365, 7.299748993495174],
[9.756259353772032, 4.805180288174137], [8.9728086099149,
6.435624543518433]]
### Log Book #############################################
gen    evals  avg          std          min          max
0      5      0.0629536    0.0591294    0.00108562   0.16031
1      5      0.081051     0.0361767    0.0180054    0.126552
2      5      0.0916668    0.0385386    0.0312551    0.141644
3      5      0.0663885    0.0164441    0.0541062    0.0985648
4      5      0.0759395    0.0385668    0.0148436    0.136325
5      5      0.0851483    0.0419408    0.00816126   0.133337
```

<< Lines 6 to 996 - Removed due to volume of answer >>>

```
997    5      0.291057     0.150004     0.106485     0.44331
998    5      0.545407     0.481211     0.196919     1.4909
999    5      0.356074     0.108138     0.153049     0.476002
```

The best score is:

```
[8.647687147637413, 6.772257466226377]
```

Open the Jupyter Notebook in Chapter 10 called: Chapter-10-03-Distributed-Evolutionary-Algorithms-Loop.ipynb

Note the change in the main () function:

```
for i in range(10):
    pop, logbook, best = main_proc()
    print('### Population ########################################')
    print(pop)
    print('### Best ##############################################')
    print(best)
```

Execute the notebook to proof the randomness of the example and the speed of the resolution of the evolution process.

Evolutionary Algorithms

I will walk you through the following evolutionary algorithm example to demonstrate to you how this process works in detail.

Open the Jupyter Notebook in the Chapter 10 directory called: Chapter-10-04-Evolutionary-Algorithm-01.ipynb

This evolutionary algorithm uses an extinct rule that only keeps the single best member of the population to survive to the next generation.

Load the libraries

```
import numpy as np
import matplotlib.pyplot as plt
```

```
Setup the parameters
```

```
DNA_SIZE = 1                # DNA (real number)
DNA_BOUND = [0, 5]          # solution upper and lower bounds
N_GENERATIONS = 200
POP_SIZE = 100              # population size
N_KID = 50                  # n kids per generation
```

To find the maximum of this function:

```
def F(x): return np.sin(10*x)*x + np.cos(2*x)*x
```

Find nonzero fitness for selection:

```
def get_fitness(pred): return pred.flatten()
```

Spawn new population member:

```
def make_kid(pop, n_kid):
    # generate empty kid holder
    kids = {'DNA': np.empty((n_kid, DNA_SIZE))}
    kids['mut_strength'] = np.empty_like(kids['DNA'])
    for kv, ks in zip(kids['DNA'], kids['mut_strength']):
        # crossover (roughly half p1 and half p2)
        p1, p2 = np.random.choice(np.arange(POP_SIZE), size=2, replace=False)
        cp = np.random.randint(0, 2, DNA_SIZE, dtype=np.bool)  # crossover points
        kv[cp] = pop['DNA'][p1, cp]
        kv[~cp] = pop['DNA'][p2, ~cp]
```

```
        ks[cp] = pop['mut_strength'][p1, cp]
        ks[~cp] = pop['mut_strength'][p2, ~cp]

        # mutate (change DNA based on normal distribution)
        ks[:] = np.maximum(ks + (np.random.rand(*ks.shape)-0.5), 0.)
        # must > 0
        kv += ks * np.random.randn(*kv.shape)
        kv[:] = np.clip(kv, *DNA_BOUND)    # clip the mutated value
    return kids
```

Extinct unfit population member:

```
def kill_bad(pop, kids):
    for key in ['DNA', 'mut_strength']:
        pop[key] = np.vstack((pop[key], kids[key]))

    fitness = get_fitness(F(pop['DNA']))
    idx = np.arange(pop['DNA'].shape[0])
    good_idx = idx[fitness.argsort()][-POP_SIZE:]
    for key in ['DNA', 'mut_strength']:
        pop[key] = pop[key][good_idx]
    return pop
```

Generate a population:

```
pop = dict(DNA=5 * np.random.rand(1, DNA_SIZE).repeat(POP_SIZE, axis=0),
           mut_strength=np.random.rand(POP_SIZE, DNA_SIZE))
```

Evolve through the generations:

```
%matplotlib inline
plt.ion()
x = np.linspace(*DNA_BOUND, 200)
for g in range(N_GENERATIONS):
    plt.plot(x, F(x))
    plt.scatter(pop['DNA'], F(pop['DNA']), s=200, lw=0, c='red', alpha=0.5)
    plt.text(0, -8, 'Generation=%.0f' % (g+1))
    plt.pause(0.05)

    kids = make_kid(pop, N_KID)
    pop = kill_bad(pop, kids)   # good parent for elitism
```

```
plt.ioff()
plt.show()
```

Results:

The first generation is not doing too well (Figure 10-1).

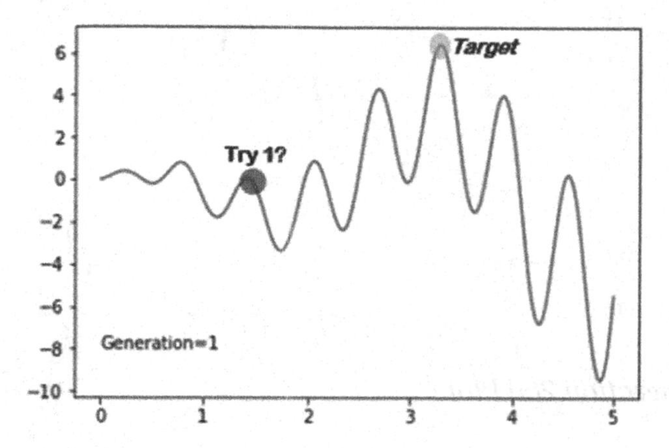

Figure 10-1. *Generation 1 Plot*

The red guess is not even close to the green target. Luckily that is where evolution starts working.

Generation 1:

Let's monitor a few generations.

Generation 10:

During generation 10 we now have more than a few potential next-generation candidates (Figure 10-2).

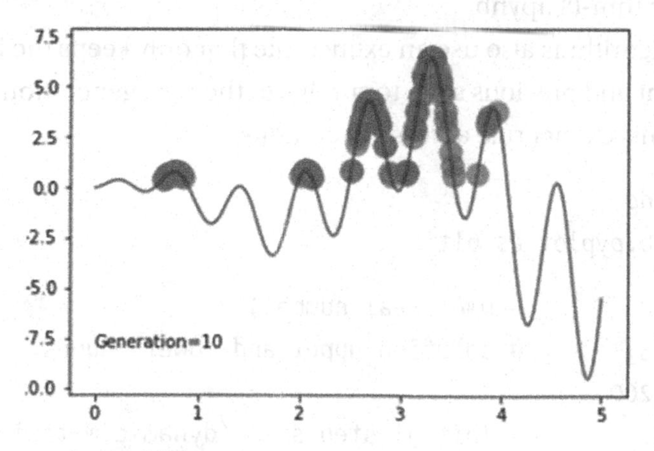

Figure 10-2. *Generation 10 Plot*

The process is getting improved via further evolution.

Generation 200:

Sole survivor! See Figure 10-3.

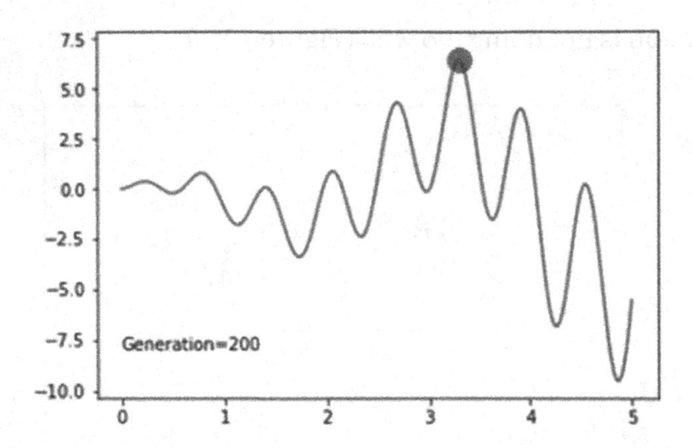

Figure 10-3. *Generation 200 Plot*

Jackpot – Solved the problem within 200 cycles or generations.

Tip Evolutionary Computing in general gives the impression of chaos, but from my experience the algorithm either converges quickly or keeps on running. The latter is normally due to an unknown feature you are not measuring or haven't discovered yet.

Open the Jupyter Notebook in the Chapter 10 directory called: Chapter-10-05-Evolutionary-Algorithm-02.ipynb

Evolutionary algorithms also use an extinct rule that only keeps the best member between the current and previous state to survive to the next generation. I will now show you the impact of this extinct rule across generations.

```
import numpy as np
import matplotlib.pyplot as plt

DNA_SIZE = 1                # DNA (real number)
DNA_BOUND = [0, 5]          # solution upper and lower bounds
N_GENERATIONS = 200
MUT_STRENGTH = 5.           # initial step size (dynamic mutation strength)
```

```
def F(x): return np.sin(10*x)*x + np.cos(2*x)*x     # to find the maximum
of this function

# find non-zero fitness for selection
def get_fitness(pred): return pred.flatten()

def make_kid(parent):
    # no crossover, only mutation
    k = parent + MUT_STRENGTH * np.random.randn(DNA_SIZE)
    k = np.clip(k, *DNA_BOUND)
    return k

def kill_bad(parent, kid):
    global MUT_STRENGTH
    fp = get_fitness(F(parent))[0]
    fk = get_fitness(F(kid))[0]
    p_target = 1/5
    if fp < fk:      # kid better than parent
        parent = kid
        ps = 1.      # kid win -> ps = 1 (successful offspring)
    else:
        ps = 0.
    # adjust global mutation strength
    MUT_STRENGTH *= np.exp(1/np.sqrt(DNA_SIZE+1) * (ps - p_target)/
    (1 - p_target))
    return parent

parent = 5 * np.random.rand(DNA_SIZE)   # parent DNA

plt.ion()       # something about plotting
x = np.linspace(*DNA_BOUND, 200)

for g in range(N_GENERATIONS):
    kid = make_kid(parent)
    py, ky = F(parent), F(kid)       # for later plot
    parent = kill_bad(parent, kid)

    plt.cla()
    plt.plot(x, F(x))
```

```
    plt.scatter(parent, py, s=200, lw=0, c='red', alpha=0.5,)
    plt.scatter(kid, ky, s=200, lw=0, c='blue', alpha=0.5)
    plt.text(0, -7, 'Mutation strength=%.2f' % MUT_STRENGTH)
    plt.text(0, -8, 'Generation=%.0f' % (g+1))
    plt.pause(0.05)

plt.ioff()
plt.show()
```

Results:

Generation 1 (Figure 10-4).

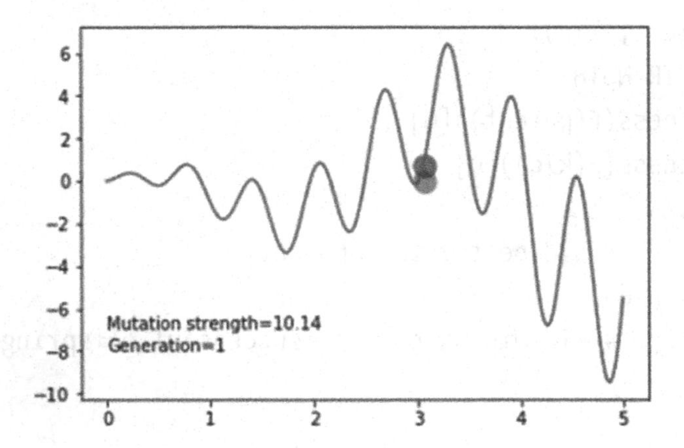

Figure 10-4. *Generation 1 Plot – Multi-generations*

Let's monitor a few selective generations.

Generation 10 (Figure 10-5):

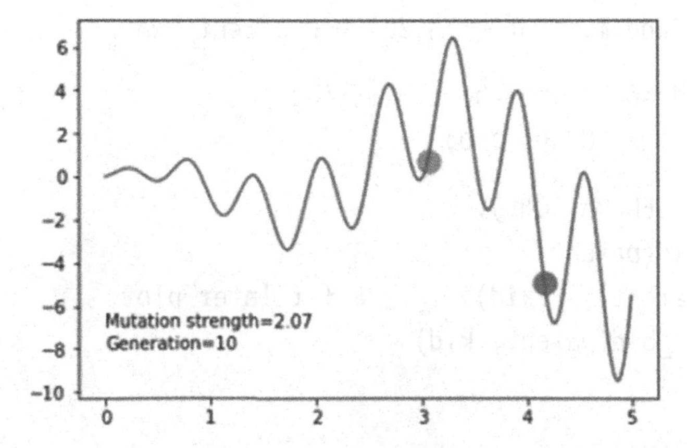

Figure 10-5. *Generation 10 Plot - Multi-generations*

It looks worse!!!

Did something go wrong ...?

No, the random mutation is helping the generation to overcome any local maximums that may cause the generation to converge on a less than optimum solution. This is a major advantage of this version of the algorithm.

This algorithm will settle on an optimum solution because the generation that caused the negative impact will get reset to a previous better parent within one generation.

Generation 200 (Figure 10-6).

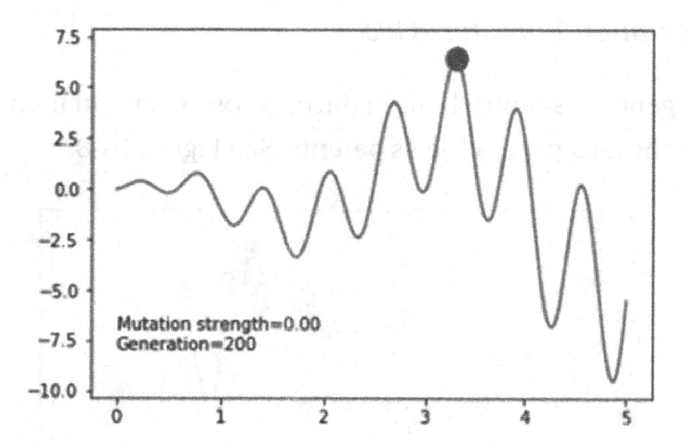

Figure 10-6. *Generation 200 Plot – Multi-generations*

The solution found is an optimum!

Tip Try the early stop option to see how you save a clock cycle with a 'Good Enough' solution. Set mtype='B' with Mutation strength > 0.001

It took 76 generations, and used only 38.000 % total effort for 200 generations.

Next, we will show a different algorithm to the same problem. The solution creates a concept called cross-over.

This means the new generation is a mix of the surviving parents, that is, a cross-over.

Open the Jupyter Notebook in the Chapter 10 directory called: Chapter-10-06-Evolutionary-Algorithm-03.ipynb

Your results:

Generation 1 (Figure 10-7).

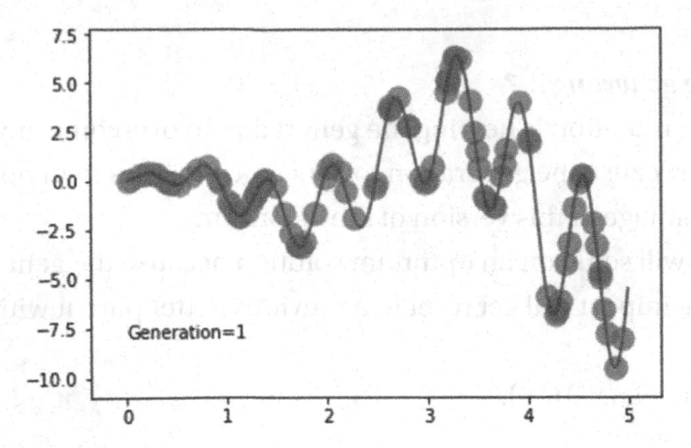

***Figure 10-7.** Generation 1 on Cross-Over*

The cross-over generates more children during cross-over and then only the fittest survive to generate the next generation as parents. See Figure 10-8.

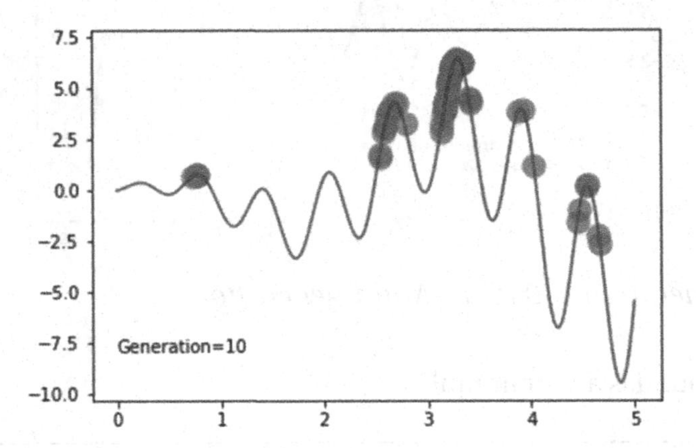

***Figure 10-8.** Generation 10 on Cross-Over*

Note the reduction in children as the cross-over now propagates common strong qualities from one generation to the next.

Let's watch it converge. See Figure 10-9.

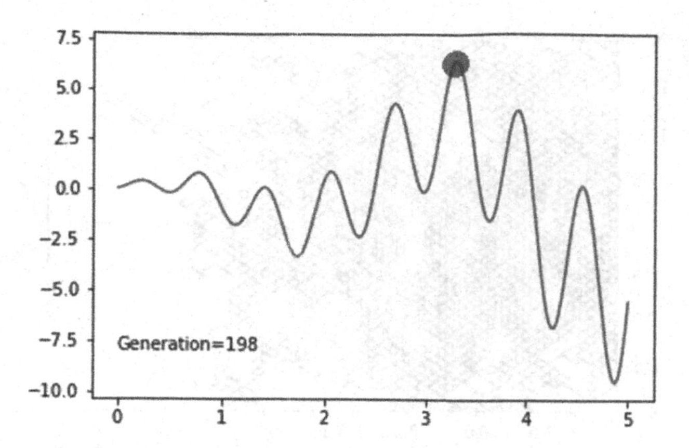

Figure 10-9. *Generation 198 on Cross-Over*

Success!

I suggest you look at the next option yourself:

Open the Jupyter Notebook in the Chapter 10 directory called: Chapter-10-07-Evolutionary-Algorithm-04.ipynb

The basic requirement is your generation needs to move between two points.

This is a common problem that these algorithms can solve easily.

I suggest you now evaluate your results.
Check for the following:

- Check when the best path was already found.
- Check when 95% of the generation is within the significant close-to target.

Here are my findings (Figure 10-10).

```
Gen: 0 | best fit: 0.023423059056103604
```

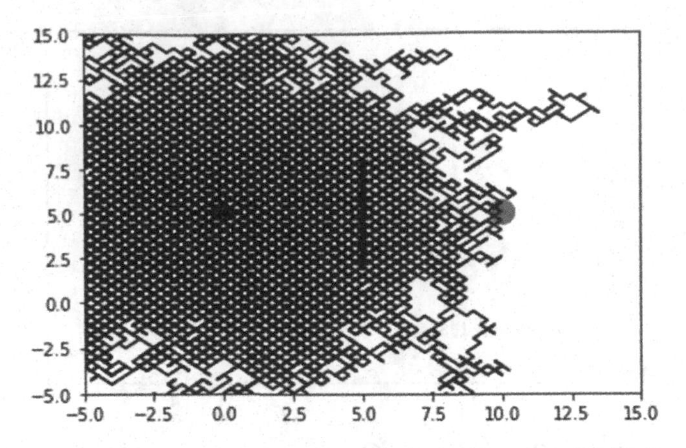

Figure 10-10. *Generation 0*

Only 2 of the 300 generations crossed the endpoint during the random walk. That is a 0.667 % success. See Figure 10-11.

```
Gen: 20 | best fit: 0.545819714368591
```

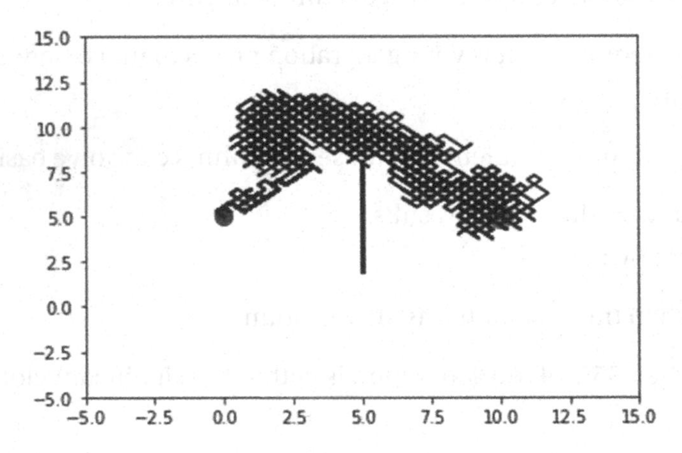

Figure 10-11. *Generation 20*

By generation 20, 95+% of the generation ends up on the endpoint.

This is good but due to the design that completes a fixed number of loops, I wasted another 279 cycles for no gain!

If you look at final generation 299 (Figure 10-12):

```
Gen: 299 | best fit: 0.545819714368591
```

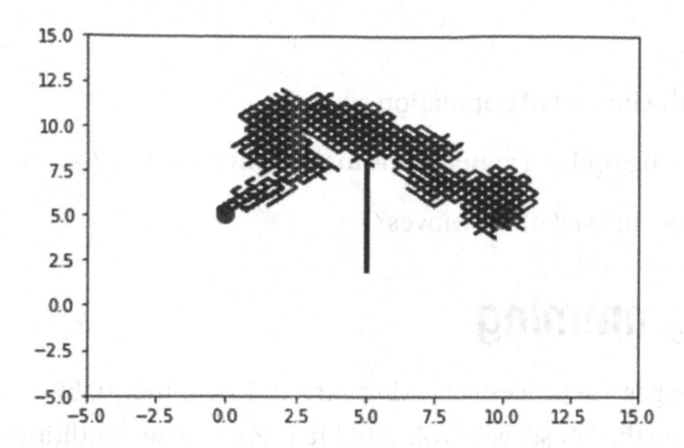

Figure 10-12. *Generation 299*

There is a 0% gain between generation 20 and generation 299.

Warning Investigate always for an early exit strategy in evolutionary programming because you are aiming for a good enough solution, not an optimum; and if you are trying for an optimum, typically an evolutionary solution will not converge on the target 100% in a high-dimensional space due to the curse of dimensionality. You simply do not have enough data or in this case a big enough population to find every possible mutation or cross-over for a 100% solution.

Open the Jupyter Notebook in the Chapter 10 directory called: Chapter-10-08-Evolutionary-Algorithm-05.ipynb

Experiment with the following parameters by changing them:

```
POP_SIZE_MIN = 30
POP_SIZE_MAX = 31
POP_SIZE_STEP = 1

N_GENERATIONS_MIN = 30
N_GENERATIONS_MAX = 31
N_GENERATIONS_STEP = 1

N_MOVES_MIN = 300
N_MOVES_MAX = 301
N_MOVES_STEP = 1
```

Questions:

What is the impact of population size?

What are the optimal generations for the environment?

What is the impact of the moves?

Linear Programming

Linear programming is the process of taking various linear inequalities relating to some situation and finding the "best" value obtainable under those conditions. A characteristic example takes the limitations of materials and labor, and then determines the "best" production levels for maximal profits under those conditions. It is an important area of mathematics called "optimization techniques." This applied field of study is used every day in an organization for allocation of resources.

Consider the following problem:

Minimize:

```
f = -1*x[0] + 4*x[1]
Subject to:
-3*x[0] + 1*x[1] <= 6
1*x[0] + 2*x[1] <= 4
x[1] >= -3
where: -inf <= x[0] <= inf
```

Open the Jupyter Notebook in the Chapter 10 directory called: Chapter-10-09-Linear-programming-01.ipynb

```
import scipy.optimize as opt

c = [-1, 4]
A = [[-3, 1], [1, 2]]
b = [6, 4]
x0_bounds = (None, None)
x1_bounds = (-3, None)
res = opt.linprog(c, A_ub=A, b_ub=b, bounds=(x0_bounds, x1_bounds),
              options={"disp": True})
print(res)
```

Result:

```
Optimization terminated successfully.
         Current function value: -22.000000
         Iterations: 1
     fun: -22.0
 message: 'Optimization terminated successfully.'
     nit: 1
   slack: array([39.,  0.])
  status: 0
 success: True
       x: array([10., -3.])
```

The Linear programming shows that for X0=10 and X1=-3, and the formula is at an optimal solution.

Did you notice that the solution was quick to solve as it has only a series of mathematical formulas to apply? This is characteristic of linear programming.

Warning The formulas for linear programming can get extremely complex if the limits are precise or multifaceted.

I will suggest that you look at SymPy (`www.sympy.org`) if you need any formula optimization requirements.

Open the Jupyter Notebook in the Chapter 10 directory called: Chapter-10-10-SymPy.ipynb

```
from sympy import *
solve(Eq(f, -1*x + 4*y),f) => [-x + 4*y]
solve(3*x + 1*y <= 6,x) => (-oo < x) & (x <= -y/3 + 2)
solve(3*x + 1*y <= 6,y) => (-oo < y) & (y <= -3*x + 6)
solve(1*x + 2*y <= 4,x) => (-oo < x) & (x <= -2*y + 4)
solve(1*x + 2*y <= 4,y) => (-oo < y) & (y <= -x/2 + 2)
solve(y >= -3) => (-3 <= y) & (y < oo)
```

That gives you a quick way to determine the ranges for the ***scipy opt.linprog*** programming.

The area of Linear programming is used extensively in many of my customers' systems as there are many requirements for simple calculations that assist the creation of machine learning.

Congratulations; you can now do Linear programming.

Particle Swarm Optimization

Particle Swarm Optimization (PSO) is a computational method that optimizes a problem by iteratively trying to improve a candidate solution about a given measure of quality. It solves a problem by having a population of candidate solutions, here labeled particles, and moving these particles around in the search space according to simple mathematical formulae over the particle's position and velocity.

Each particle's movement is influenced by its local best-known position but is also guided toward the best-known positions in the search space, which are updated as better positions are found by other particles. This is expected to move the swarm toward the best solutions.

Example:

Open the Jupyter Notebook in the Chapter 10 directory called: Chapter-10-11-Particle-Swarm-01.ipynb

Results:

```
Running PSO ...
Iteration 0: Best Cost = 132.48185745142655
Iteration 1: Best Cost = 132.48185745142655
Iteration 2: Best Cost = 132.48185745142655
Iteration 3: Best Cost = 132.48185745142655
Iteration 4: Best Cost = 132.48185745142655
Iteration 5: Best Cost = 132.48185745142655

<Removed part of results>

Iteration 46: Best Cost = 1.8639730250806885
Iteration 47: Best Cost = 1.8639730250806885
Iteration 48: Best Cost = 1.8471628852322097
Iteration 49: Best Cost = 1.845248848436588
```

```
Global Best:
{'position':
array([-0.11522734, -0.04638722,  0.41577969,  0.21634509, -0.4200067 ,
       -0.84273085, -0.08707212,  0.83846401, -0.10769234, -0.03660312]),
'cost': 1.845248848436588}
```

This should give you a basic insight into PSO.

Reinforcement Learning

You did perform a whole chapter of Reinforcement learning (RL), but I am covering these two specific algorithms as part of evolutionary programming as the technique is more evolutionary based.

RL Algorithm One

Open the Jupyter Notebook in the Chapter 10 directory called: Chapter-10-12-RL-01.ipynb

Result:

```
Optimized Policy:
[ 0.  1.  2.  3.  4.  5.  6.  7.  8.  9. 10. 11. 12. 12. 11. 15. 16. 17.
 18.  6. 20. 21.  3. 23. 24. 25.  1.  2.  3.  4.  5.  6.  7.  8.  9. 10.
 11. 12. 38. 11. 10.  9. 42.  7. 44.  5. 46. 47. 48. 49. 50.  1.  2.  3.
  4.  5.  6.  7.  8.  9. 10. 11. 12. 13. 11. 10.  9. 17.  7. 19.  5. 21.
 22. 23. 24. 25.  1.  2.  3.  4.  5.  6.  7.  8.  9. 10. 11. 12. 12. 11.
 10.  9.  8.  7.  6.  5.  4.  3.  2.  1.]

Optimized Value Function:
[0.00000000e+00 7.24792480e-05 2.89916992e-04 6.95257448e-04
 1.16010383e-03 1.76906586e-03 2.78102979e-03 4.03504074e-03
 4.66214120e-03 5.59997559e-03 7.08471239e-03 9.03964043e-03
 1.11241192e-02 1.56793594e-02 1.61464431e-02 1.69517994e-02
 1.86512806e-02 1.98249817e-02 2.24047303e-02 2.73845196e-02
 2.83388495e-02 3.04937363e-02 3.61633897e-02 3.84953022e-02
 4.44964767e-02 6.25000000e-02 6.27174377e-02 6.33700779e-02
 6.45857723e-02 6.59966059e-02 6.78135343e-02 7.08430894e-02
```

```
7.46098323e-02 7.64884604e-02 7.93035477e-02 8.37541372e-02
8.96225423e-02 9.58723575e-02 1.09538078e-01 1.10939329e-01
1.13360151e-01 1.18457374e-01 1.21977661e-01 1.29716907e-01
1.44653559e-01 1.47520113e-01 1.53983246e-01 1.70990169e-01
1.77987434e-01 1.95990576e-01 2.50000000e-01 2.50217438e-01
2.50870078e-01 2.52085772e-01 2.53496606e-01 2.55313534e-01
2.58343089e-01 2.62109832e-01 2.63988460e-01 2.66803548e-01
2.71254137e-01 2.77122542e-01 2.83372357e-01 2.97038078e-01
2.98439329e-01 3.00860151e-01 3.05957374e-01 3.09477661e-01
3.17216907e-01 3.32153559e-01 3.35020113e-01 3.41483246e-01
3.58490169e-01 3.65487434e-01 3.83490576e-01 4.37500000e-01
4.38152558e-01 4.40122454e-01 4.43757317e-01 4.47991345e-01
4.53440603e-01 4.62529268e-01 4.73829497e-01 4.79468031e-01
4.87912680e-01 5.01265085e-01 5.18867627e-01 5.37617932e-01
5.78614419e-01 5.82817988e-01 5.90080452e-01 6.05372123e-01
6.15934510e-01 6.39150720e-01 6.83960814e-01 6.92560339e-01
7.11950883e-01 7.62970611e-01 7.83963162e-01 8.37972371e-01
0.00000000e+00]
```

That is the first RL complete.

RL Algorithm Two

Here is the second RL algorithm.

Open the Jupyter Notebook in the Chapter 10 directory called: Chapter-10-13-RL-02.ipynb

The results are shown in Figure 10-13.

	1	2	3	4	5
1	3.31	8.79	4.43	5.32	1.49
2	1.52	2.99	2.25	1.91	0.55
3	0.05	0.74	0.67	0.36	-0.4
4	-0.97	-0.44	-0.35	-0.59	-1.18
5	-1.86	-1.35	-1.23	-1.42	-1.98

	1	2	3	4	5
1	21.98	24.42	21.98	19.42	17.48
2	19.78	21.98	19.78	17.8	16.02
3	17.8	19.78	17.8	16.02	14.42
4	16.02	17.8	16.02	14.42	12.98
5	14.42	16.02	14.42	12.98	11.68

***Figure 10-13.** Second RL algorithm*

Traveling Salesman Problem

I covered this problem previously, and I will now show you how to handle the same problem with evolutionary programming.

> Open the Jupyter Notebook in the Chapter 10 directory called: Chapter-10-14-TSP-01.ipynb

Your result is shown in Figure 10-14.

Gen: 499 | best fit: 53661.69

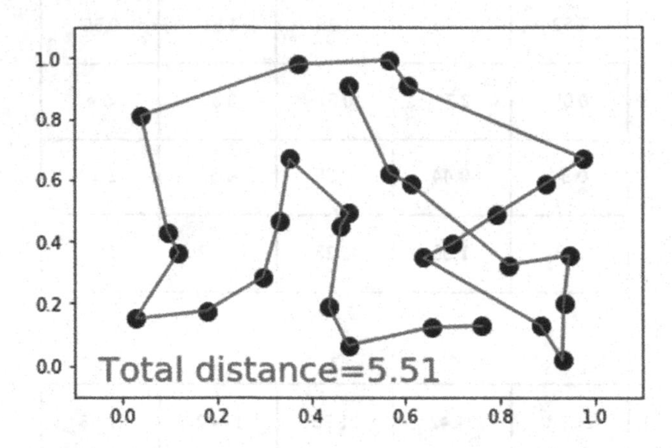

Figure 10-14. *Traveling Salesman*

Example 2:

With 25 cities, a brute force approach would have to test over ***300 sextillion routes***! The evolutionary programming works just as effectively but many factors are more efficient.

Open the Jupyter Notebook in the Chapter 10 directory called: Chapter-10-15-TSP-02.ipynb

The results are shown in Figure 10-15.

```
Initial distance: 2052.2221426003775
Final distance:    802.6666546711987
```

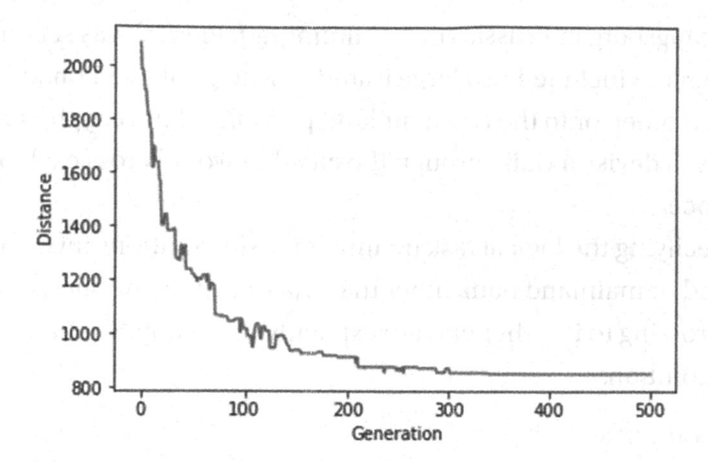

Figure 10-15. *TSP for 25 Cities*

Solve Ny Words, Please!

I have noted earlier that evolutionary programming is good at finding a data set that has high dimensionality as it can adapt to the process as it discovers the ecosystem.

I will take you through a simple test on a sentence I wrote, and your program must solve it.

Open the Jupyter Notebook in the Chapter 10 directory called: Chapter-10-16-Words-01.ipynb

The code should result in a text sentence:

```
'I love Denise very much!'
```

> EP is 240.8000 times faster than Brute Force on 14 letters.
>
> That shows how effective and efficient EP is against text strings discovery.

I will show you the next example to show how you use EP to find different routes on a map.

Seven Bridges of Konigsberg

The Seven Bridges of Konigsberg is a historically notable problem in mathematics. Its negative resolution by Leonhard Euler in 1736 laid the foundations of graph theory and prefigured the idea of topology.

The city of Konigsberg in Prussia (now Kaliningrad, Russia) was set on both sides of the Pregel River and included two large islands - Kneiphof and Lomse - which were connected to each other, or to the two mainland portions of the city, by seven bridges. The problem was to devise a walk through the city that would cross each of those bridges once and only once.

By way of specifying the logical task unambiguously, solutions involving either reaching an island or mainland bank other than via one of the bridges, or accessing any bridge without crossing to its other end are explicitly unacceptable. Euler proved that the problem has no solution.

See if you can solve it.

Open the Jupyter Notebook in the Chapter 10 directory called: Chapter-10-17-Konigsberg-01.ipynb

```
import matplotlib
import numpy as np
matplotlib.use('TkAgg')
%matplotlib inline

import networkx as nx
import matplotlib.pyplot as plt
# Create the graph
graph = nx.MultiGraph()
graph.add_nodes_from([0, 3])

edges = [
    (0, 1, {'key': 'Grüne Brücke', 'color': 'blue', 'width': 5}),
    (0, 1, {'key': 'Köttelbrücke', 'color': 'red', 'width': 2}),
    (1, 2, {'key': 'Krämerbrücke', 'color': 'blue', 'width': 5}),
    (1, 2, {'key': 'Schmiedebrücke', 'color': 'red', 'width': 2}),
    (2, 3, {'key': 'Holzbrücke', 'color': 'blue', 'width': 5}),
    (3, 1, {'key': 'Dombrücke', 'color': 'blue', 'width': 5}),
    (3, 0, {'key': 'Hohe Brücke', 'color': 'blue', 'width': 5})
]

graph.add_edges_from(edges);
```

```
# Plot the graph
plt.figure()

labels = {}
for i, j, label in graph.edges(data='key'):
    if (i, j) not in labels:
        labels[(i, j)] = label
    else:
        labels[(i, j)] += '\n' + label

colors = [x for _, _, x in graph.edges(data='color')]
widths = [x for _, _, x in graph.edges(data='width')]

pos = nx.spring_layout(graph)
nx.draw_networkx(graph, pos, edge_color=colors, with_labels=True,
width=widths)
nx.draw_networkx_edge_labels(graph, pos, edge_labels=labels);
```

The result is shown in Figure 10-16 - Seven Bridges of Konigsberg.

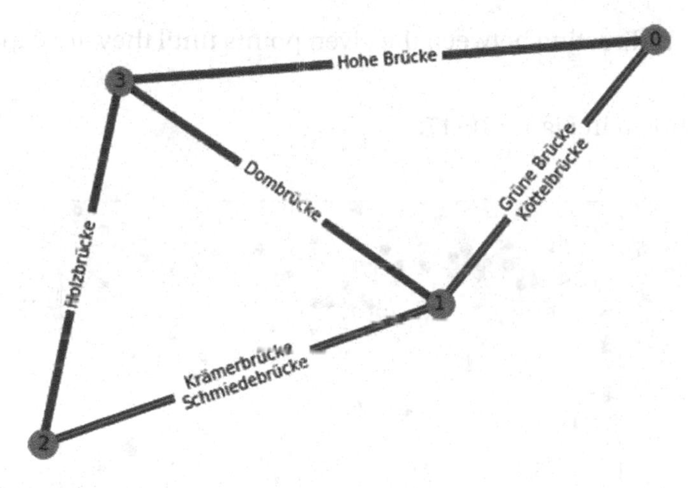

Figure 10-16. *Seven Bridges of Konigsberg*

Well done; you have modeled the problem, but this problem has no acknowledged solution as identified beforehand.

Tip Not all problems have solutions. The sign of knowledge is when you can determine that fact with a minimum amount of effort that will give you both an effective and efficient solution.

Next, we will look at a problem that has a solution.

Multi-depot Vehicle Scheduling Problem

Multiple Vehicle Routing Simulation using a naive greedy algorithm.

Red/Green Triangles: Vehicles

Blue circles: Targets that have not yet been reached

Pink circles: Targets that have already been reached

Example:

Open the Jupyter Notebook in the Chapter 10 directory called: Chapter-10-18-MVRS-01.ipynb

The trucks are redirecting between the given points until they are capable of visiting them all.

The result is shown in Figure 10-17.

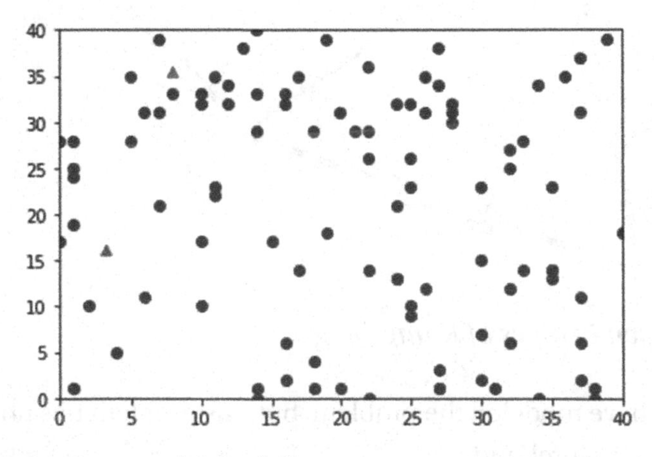

Figure 10-17. *Multi-depot vehicle delivery*

Simulation Using Schedules

I will introduce you to a Python library I found a while back that enables you to simulate a scheduling solution.

I suggest you open Jupyter Notebook in directory Chapter 10 called: Chapter-10-19-Basic-Scheduling-01.ipynb

pip install pyschedule

Load pyschedule and create a scenario with ten steps planning horizon

```
from pyschedule import Scenario, solvers, plotters
S = Scenario('dog_pyschedule',horizon=10)

# Create two resources
Angus, Jock = S.Resource('Angus'), S.Resource('Jock')

# Create three tasks with lengths 1,2 and 3
mail, food, hole = S.Task('steal_mail',1), S.Task('eat_food',2),
S.Task('dig_hole',3)

# Assign tasks to resources, either Angus or Jock,
# the %-operator connects tasks and resource
mail += Angus|Jock
food += Angus|Jock
hole += Angus|Jock

# Solve and print solution
S.use_makespan_objective()
solvers.mip.solve(S,msg=1)

# Print the solution
print(S.solution())
```

The result is shown in Figure 10-18.

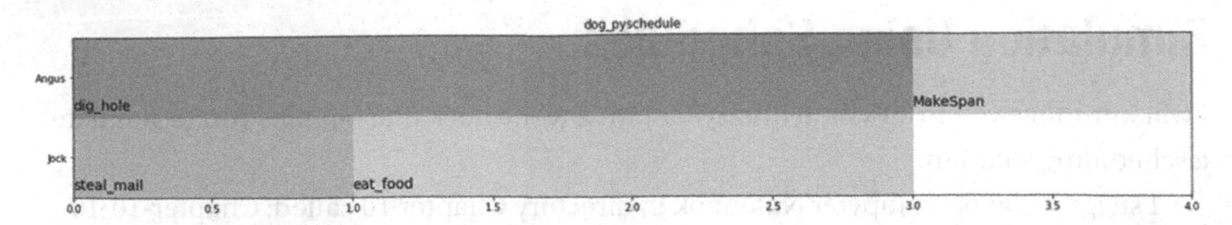

Figure 10-18. *Jock and Angus at Work*

Let's investigate a more complex simulation schedule.

Example:

I suggest you open the Jupyter Notebook in directory Chapter 10 called: Chapter-10-20-Bicycle-Shop.ipynb

> Andre and Laurence optimize their Bicycle Paint Shop using pyschedule

Andre and Laurence are running a paint shop for bicycles where they recycle old bicycles with fresh colors. Today they have to paint a green and a red bicycle. To get started they import pyschedule and create a new scenario. We use hours as granularity and expect a working day of at most 10 hours, so we set the planning horizon to 10. Some solvers do not need this parameter, but the default solver requires it:

```
from pyschedule import Scenario, solvers, plotters
S = Scenario('Bicycle_Paint_Shop', horizon=10)
```

Then they create themselves as resources:

```
Andre = S.Resource('Andre')
Laurence = S.Resource('Laurence')
```

Painting a bicycle takes two hours. Moreover, after the bicycle has been painted, it needs to get post-processed (e.g., tires pumped), which takes one hour (which is the default). This translates into four tasks in total:

```
Green_Paint, Red_Paint = S.Task('Green_Paint', length=2), S.Task('Red_
Paint', length=2)
Green_Prep, Red_Prep = S.Task('Green_Prep'), S.Task('Red_Prep')
```

Clearly, one can only do the post-processing after the painting with an arbitrary gap in between. For the red paint we are a little stricter; here we want to start the post-processing exactly one hour after the painting since this is the time the color takes to dry:

```
S += Green_Paint < Green_Prep, Red_Paint + 1 <= Red_Prep
```

Each task can be done by either Andre or Laurence:

```
Green_Paint += Andre|Laurence
Green_Prep += Andre|Laurence

Red_Paint += Andre|Laurence
Red_Prep += Andre|Laurence
```

So, let's have a look at the scenario:

```
S.clear_solution()
print(S)
```

We haven't defined an objective yet. We want to finish all tasks as early as possible, and so we use the MakeSpan and check the scenario again:

```
S.use_makespan_objective()
print(S)
```

Hence, we want to minimize the position of the MakeSpan task subject to the constraint that it is scheduled after all other tasks. Thus, the position of the MakeSpan is the length of our schedule. Now that we have the first version of our scenario, let's solve and plot it:

Set some colors for the tasks:

```
task_colors = { Green_Paint   : '#A1D372',
                Green_Prep    : '#A1D372',
                Red_Paint     : '#EB4845',
                Red_Prep      : '#EB4845',
                S['MakeSpan'] : '#7EA7D8'}
```

A small helper method to solve and plot a scenario:

```
def run(S) :
    if solvers.mip.solve(S):
        get_ipython().run_line_magic('matplotlib', 'inline')
        plotters.matplotlib.plot(S,task_colors=task_colors,fig_size=(15,2))
    else:
        print('no solution exists')
run(S)
```

See the result in Figure 10-19.

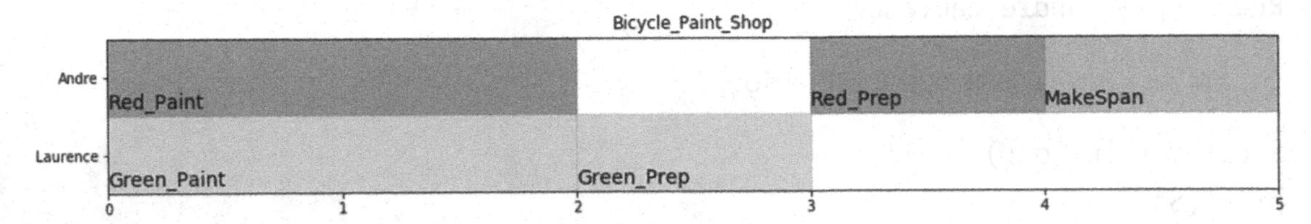

Figure 10-19. *Workshop (1)*

Note that it could happen that somebody needs to paints the red bicycle and then do the green post-processing. This would be annoying; switching bicycles takes too much time. We use the following constraints to ensure that the green/red painting and post-processing is always done by the same people:

```
#Green_Prep will use the same resources as Green_Paint if there is an
overlap in resource requirement
Green_Prep += Green_Paint*[Andre,Laurence]
# Same for Red_Prep and Red_Paint
Red_Prep += Red_Paint*[Andre,Laurence]
run(S)
```

The result is shown in Figure 10-20.

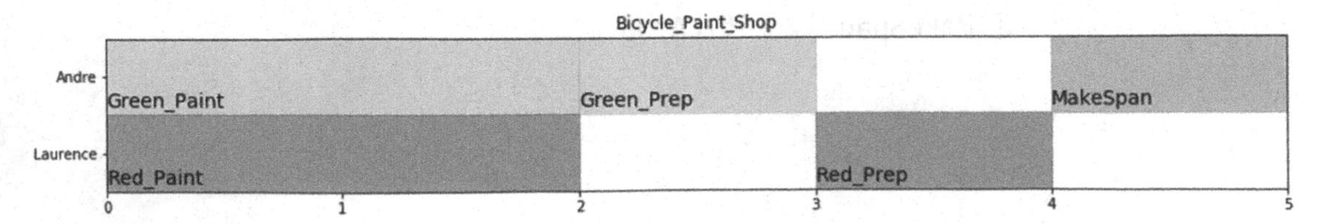

Figure 10-20. *Workshop (2)*

This schedule completes after four hours and suggests painting both bicycles at the same time. However, Andre and Laurence have only a single paint shop, which they need to share:

```
Paint_Shop = S.Resource('Paint_Shop')
Red_Paint += Paint_Shop
Green_Paint += Paint_Shop
run(S)
```

The results are shown in Figure 10-21.

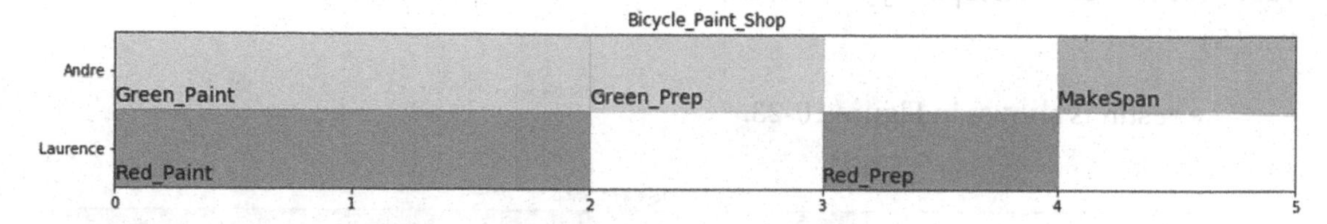

Figure 10-21. *Workshop (3)*

Great; everybody can still go home after five hours and have a late lunch! Unfortunately, Andre receives a call that the red bicycle will only arrive after two hours:

```
S += Red_Paint > 2
run(S)
```

The result is shown in Figure 10-22.

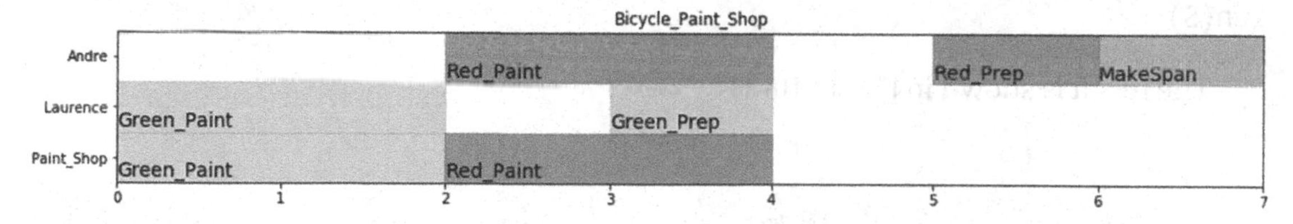

Figure 10-22. *Workshop (4)*

Too bad; everything takes now size hours to finish. Therefore, Andre and Laurence decide to schedule a lunch after the third hour and before the fifth hour:

```
Lunch = S.Task('Lunch')
Lunch += {Andre, Laurence}
S += Lunch > 3, Lunch < 5
task_colors[Lunch] = '#7EA7D8'
S.clear_objective() #we need to remove the objective and readd it because
of the new lunch task
S.use_makespan_objective()
task_colors[S['MakeSpan']] = '#7EA7D8'
run(S)
```

The result is shown in Figure 10-23.

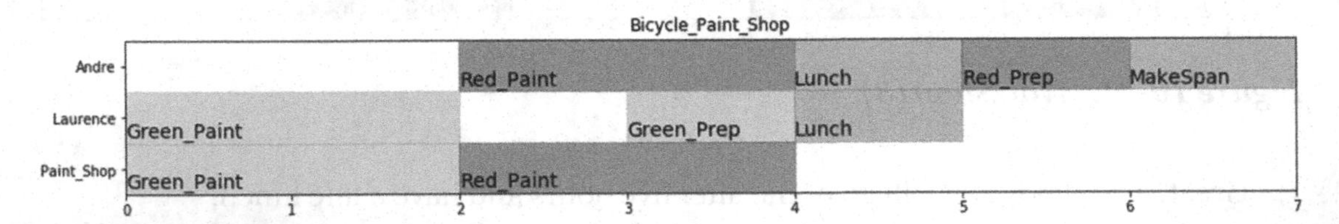

Figure 10-23. *Workshop (5)*

Andre is a morning person and wants to finish three hours of work before lunch, that is, before the third hour:

```
S += Andre['length'][0:3] >= 3
run(S)
```

The result is shown in Figure 10-24.

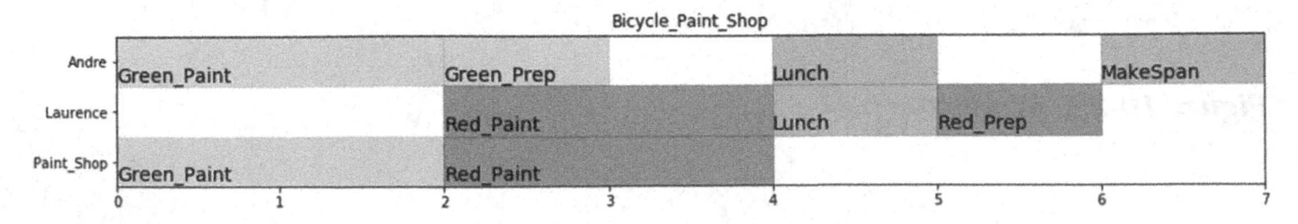

Figure 10-24. *Workshop (6)*

The weather forecast is really good for the afternoon, so Andre and Laurence decide to close the shop after lunch, that is, they fix the horizon to 5 hours. Unfortunately, the following happens:

```
S.horizon = 5
run(S)
```

Result:

> No solution exists. Andre and Laurence need 7 hours, they fix the horizon to 7 hours.

```
S.horizon = 7
run(S)
```

The result is shown in Figure 10-25.

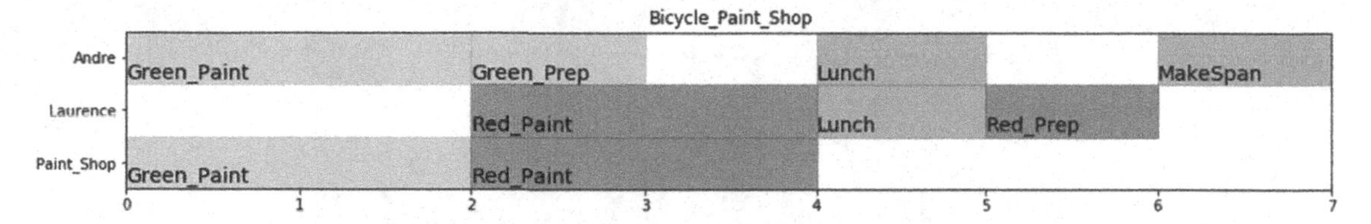

Figure 10-25. *Workshop (7)*

Well done; you can run the Paint Shop.

Bonus:

> I suggest you look at the two bonus examples. Both have three resources, and working this shows you how to get larger teams planned.

- Chapter-10-21-Basic-Scheduling-Three-Way.ipynb.
- Chapter-10-22-Bicycle-Shop-Three-way.ipynb.

Tip I have successfully used up to 20 resources and over 100 tasks. You can use the process for bigger processes also.

What Have You Achieved?

Well done; you can now create an investigative machine learning problem that requires Evolutionary Computing.

You can handle small-scale scheduling problems.

What's Next?

You have successfully completed the Evolutionary Computing chapter.

You can now use simulations and evolutionary techniques to discover and learn solutions for everyday problems in your environment.

The rest of this book will cover the application of the knowledge you have achieved by reaching the end of this chapter.

Next, I will introduce you to the interesting field of Mechatronics in Chapter 11.

CHAPTER 11

Mechatronics: Making Different Sciences Work as One

Mechatronics is a multidisciplinary field of engineering that includes a combination of mechanical engineering, electronics engineering, computer engineering, telecommunications engineering, systems engineering, and control engineering.

The most prominent, single physical example of this field is the motor vehicles we drive.

The next example is the increasing amount of robotic systems that are starting to become common in our daily lives.

I will explain the building blocks of a Mechatronics solution in this chapter and explain to you how these individual building blocks are enhanced by your machine learning skills.

Computer Engineering

Computer engineering is a discipline that integrates numerous fields of computer science and electronics engineering required to develop computer hardware and software.

Computer Systems

Computer systems require capabilities from us to design the architecture, parallel processing, and dependability of the overall ecosystem from the physical hardware to the communication networks required to enable them to interact everywhere in the world.

A. F. Vermeulen, *Industrial Machine Learning*, https://doi.org/10.1007/978-1-4842-5316-8_11

Engineers working in computer systems work on research projects that allow for reliable, secure, and high-performance computer systems. These projects are designing processors for multi-threading and parallel processing with an ever-increasing ability to process increasing processing capability and ML requirements.

I will guide you through a few basic discovery processes to understand your own computer systems.

Load the Jupyter notebook from example directory Chapter 11 called:

```
Chapter-011-01-Computer-System-01.ipynb
```

Let's get your system's name:

```
import platform
print(platform.node())
```

What operating system is on your system?

```
import platform
print(platform.system())
```

What CPU's do your system use?

```
import platform, subprocess

def get_processor_info():
    if platform.system() == "Windows":
        return platform.processor()
    elif platform.system() == "Darwin":
        return subprocess.check_output(['/usr/sbin/sysctl', "-n",
        "machdep.cpu.brand_string"]).strip()
    elif platform.system() == "Linux":
        command = "cat /proc/cpuinfo"
        return subprocess.check_output(command, shell=True).strip()
    return ""
cpustr = get_processor_info()
print(cpustr)
```

How busy is your CPU?

```
import psutil
print(psutil.cpu_percent())
```

```
How much diskspace is free?
```

```
import shutil
total, used, free = shutil.disk_usage('/')
print('Total:', total)
print('Used :', used)
print('Free :', free)
```

What is your IP address?

```
import socket
s = socket.socket(socket.AF_INET, socket.SOCK_DGRAM)
s.connect(('10.255.255.255', 1))
print('Your IP address is', s.getsockname()[0])
s.close()
```

How quick is your network connection?

```
from pythonping import ping
response_list = ping('8.8.8.8', count=20, size=100, timeout=1, verbose=False)
print('Minimum Return %5.2f ms' % (response_list.rtt_min_ms))
print('Average Return %5.2f ms' % (response_list.rtt_avg_ms))
print('Maximum Return %5.2f ms' % (response_list.rtt_max_ms))
```

Now that you have some basic techniques to investigate the ecosystem, I suggest we process data to demonstrate your ecosystem's capabilities to perform ML.

I will now show you how to make a process run in parallel.

Load the Jupyter notebook from example directory Chapter 11 called:

```
Chapter-011-02-Multi-Processing-01.ipynb
```

In this first processing technique, you will need a new library called "threading."

Tip I am adding the command %%time to my code. This will give you feedback on the time spent by the process while executing.

We will investigate using threads for parallel processing.
Definition of a Thread:

> A thread of execution is the smallest sequence of programmed instructions that can be managed independently by a computer scheduler.
>
> We will test a range of threads [2, 4, 8, 16, 32, 64, 128] for the same process.

- 1 - Processing complete: 2 threads for 10000000 calculations: 3.27 s
- 2 - Processing complete: 4 threads for 10000000 calculations: 3.17 s
- 3 - Processing complete: 8 threads for 10000000 calculations: 3.19 s
- 4 - Processing complete: 16 threads for 10000000 calculations: 3.11 s
- 5 - Processing complete: 32 threads for 10000000 calculations: 3.01 s
- 6 - Processing complete: 64 threads for 10000000 calculations: 2.81 s
- 7 - Processing complete: 128 threads for 10000000 calculations: 2.61 s

With parallel threaded procedures, the time decreases:

- 2 to 4 improves by 3.06%
- 2 to 8 improves by 2.45%
- 2 to 16 improves by 4.89%
- 2 to 32 improves by 7.95%
- 2 to 64 improves by 14.07%
- 2 to 128 improves by 20.18%

This shows clearly that the threads improve the processing capability of the programs.

Tip I normally perform a test on each of my new servers or after any major changes on my servers to work out what the "optimum" number of threads is for a given load.

I will now take you through another technique called parallel processes.

You basically create separate processes on the server that run independently from each other.

Warning In the next example, there is a counter totalsize = int(10e7); this will run for a long period of time.

If you want, you could change it to totalsize = int(10e3) to speed it up.

Load the Jupyter notebook from example directory Chapter 11 called:

```
Chapter-011-03-Multi-Processing-02.ipynb
Here we go.
Firstly you need a new library called "multiprocessing" that enables you to
create new processes on the system.
```

We will test a range of threads [2, 4, 8, 16, 32, 64, 128] for the same process.

- 1 - Processing complete: 2 threads for 10000000 calculations: 2.08 s
- 2 - Processing complete: 4 threads for 10000000 calculations: 1.78 s
- 3 - Processing complete: 8 threads for 10000000 calculations: 1.74 s
- 4 - Processing complete: 16 threads for 10000000 calculations: 1.72 s
- 5 - Processing complete: 32 threads for 10000000 calculations: 1.64 s
- 6 - Processing complete: 64 threads for 10000000 calculations: 1.61 s
- 7 - Processing complete: 128 threads for 10000000 calculations: 1.53 s

With parallel threads procedures, the time decreases:

- 2 to 4 improves by 2.25%
- 2 to 8 improves by 3.38%
- 2 to 16 improves by 7.61%
- 2 to 32 improves by 9.72%
- 2 to 64 improves by 14.08%
- 2 to 128 improves by 20.56%

Performance improvement is nearly the same, but the big difference is, however, the speed increases using parallel procedures against threads:

The procedures are 220% on average faster.

So, it takes the same amount of time +/- to run threads=4 than procs=8.

Tip With the right programming, the use of parallel procedures can save you massive amounts of time.

Just think what the difference between a two-hour running deep learning algorithm and a four-hour process means for you in terms of productivity.

Computer Vision for Robotics

Computer engineers focus on developing visual sensing technology to sense an environment, representation of an environment, and manipulation of the environment.

The ability for an IML ecosystem to identify objects via vision is a major development area in the ML ecosystem. The largest usage of these skills is robotics (Chapter 12) and the Fourth Industrial Revolution (Chapter 13) that necessitates your understanding of this complex but rewarding area of knowledge.

Load the Jupyter notebook from example directory Chapter 11 called: Chapter-011-04-Webcam-01.ipynb

---

Warning Make sure you have the OpenCV system installed.

Install with:

```
pip install opencv-python
```

Tip If your system's graphics card supports the latest OpenCV software, I suggest you install it. The newer graphics card has improved processing algorithms in the core drivers that make the processing more efficient.

I will take you through three steps of examples to show you how you add video processing capabilities to your machine learning knowledge base.

You start with a technique that process an image stored on the hard drive as you then can see the processing on a fixed image. I will later progress you to a full real-time webcam feed.

We will now interact with an image of a small dog.

Load the Jupyter notebook from example directory Chapter 11 called: Chapter-011-04-Webcam-01.ipynb

This start loads the libraries you need:

```
import cv2 as cv
```

Load this image of our dog:

```
imgfile = '../../data/dog.png'
img = cv.imread(imgfile, cv.IMREAD_UNCHANGED)
```

You now extract information on the image:

```
imgdim = img.shape
imgheight = img.shape[0]
imgwidth = img.shape[1]
imgchannels = img.shape[2]

print('Image Dimension    : ',imgdim)
print('Image Height       : ',imgheight)
print('Image Width        : ',imgwidth)
print('Number of Channels : ',imgchannels)
```

Your results are:

```
Image Dimension    :  (2421, 2513, 4)
Image Height       :  2421
Image Width        :  2513
Number of Channels :  4

This informs you that the image is 2421 x 2513 pixels with 4 channels.

The next steps will show you the image:
```

There is the dog ... He is feeling blue.

I will now show you how to rescale the image to 30% of its original size.

Tip Rescaling an image saves you massive amounts of storage space. The machine learning also runs more quickly on smaller images.

Here we go ... You will shrink the dog 30% ... 10% ... 5% ... Now you have a super miniature blue dog.

Warning When you store training data, I suggest you do not rescale more than 50% of the size because you lose 75% of the pixels. Remember you are rescaling on both axes.

If the machine learning trains on images that are 1000 by 1000 pixels, I suggest you rescale all the images to this scale. Machine learning works best with consistencies on features that do not help with the features of the final requirement. In this case, the size of the image carries no features to determine if it is a dog. Therefore, do not add the extra variance of image size if it does not enhance the process.

Warning Do not rescale only one axis because the distortion will render most machine learning void. Always keep the original perspective.

You can now close the example and stop the Jupyter notebook.

The next step is to get your dog back but perform a processing algorithm to expose a set of features from the image. In the example, we are going to investigate; we want to convert the image into a set of outlines that will later help with assisting the machine learning to discover enhanced features.

Load the Jupyter notebook from example directory Chapter 11 called: Chapter-011-05-Webcam-02.ipynb

The robot processes the image of 'Weener' but only needs the shape of 'Weener' to know it is the dog.

Tip Machine learning does not observe the world the same way we do.

The result is shown in Figure 11-1.

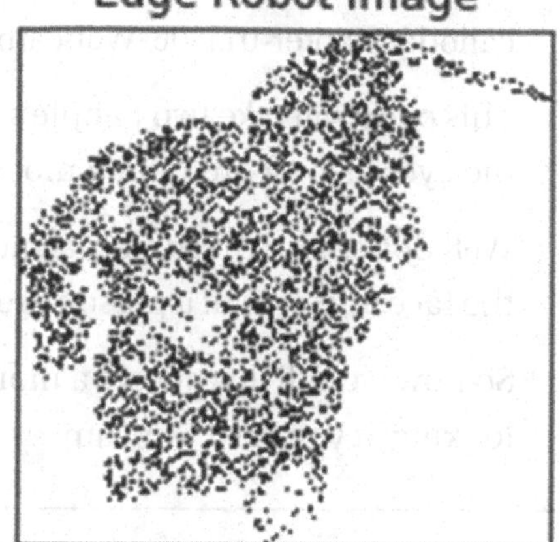

Figure 11-1. *Dog Vision (1)*

The image on the right is not clear on the screen, but if you download it from your Jupyter notebook, you will see what the robot will see (Figure 11-2).

Figure 11-2. *Dog Vision (2)*

The complex grayscale image is now reduced to a black and white outline image of the dog. The machine learning can more easily extract features from this image to determine that it is a dog, and you can observe that it is facing toward the robot.

So, your robot can see!

I will now take you through an example that finds the face and eyes on a picture of a person.

Load the Jupyter notebook from example directory Chapter 11 called: Chapter-011-06-Webcam-03.ipynb

This code will take two people's pictures and identify the face and the eyes using a pretrained model for each of these functions.

Well done; your machine learning model can now identify where the face and eyes of a person are.

So now we will investigate a more dynamic environment by looking at you via a webcam.

Warning This example will use your webcam. It, however, does not send the images to anybody else.

Load the Jupyter notebook from example directory Chapter 11 called: Chapter-011-07-Webcam-01.ipynb

The process will dynamically capture the webcam and then detect your face and eyes.

Well done; your machine learning model can now identify where the face and eyes are of a person that looks at your webcam.

If you used this on your robot as an input for your machine learning, you can collect information via the discovery of people looking at your webcam.

Tip This solution is basic but enough to act as a security system if you use more than one webcam.

Now that we have the robot seeing you, I will move on to getting your robot to talk to you.

Signal and Speech Processing

Computer engineers in this area develop improvements in human-computer interaction, including speech recognition and synthesis, medical and scientific imaging, or communications systems.

Signal Processing

Signal processing is useful when you are using sensors and require converting a continuous range of values into discrete values.

Load the Jupyter notebook from example directory Chapter 11 called: Chapter-011-08-Signal-01.ipynb

```
import matplotlib
matplotlib.use('TkAgg')
%matplotlib inline

import numpy as np
import scipy.signal as signal
import matplotlib.pyplot as plt

b1 = signal.firwin(1963, 0.5, pass_zero=False)
w1, h1 = signal.freqz(b1)
p1 = 20*np.log10(np.abs(h1))

plt.title('Digital filter frequency response - Low Pass')
line1, = plt.plot(w1, p1)

plt.setp(line1, color='b', linewidth=1.0)

plt.ylabel('Amplitude Response (dB)')
plt.xlabel('Frequency (rad/sample)')
plt.grid()
plt.show() (Figure 11-3)
```

Figure 11-3 shows the response graph for a low pass filter.

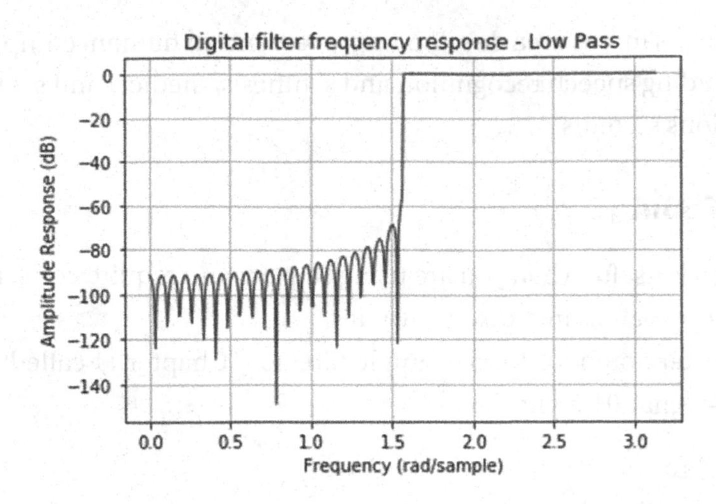

Figure 11-3. *Digital filter frequency response – Low Pass*

```
b2 = signal.firwin(1963, 0.5, pass_zero=True)
w2, h2 = signal.freqz(b2)
p2 = 20*np.log10(np.abs(h2))

plt.title('Digital filter frequency response - High Pass')
line2, = plt.plot(w2, p2)

plt.setp(line2, color='y', linewidth=1.0)

plt.ylabel('Amplitude Response (dB)')
plt.xlabel('Frequency (rad/sample)')
plt.grid()
plt.show() (Figure 11-4)
```

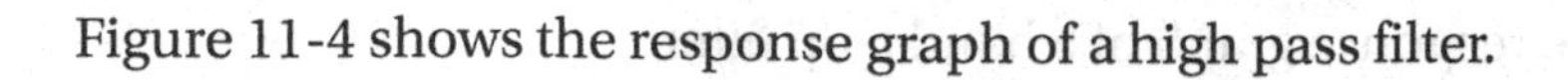

Figure 11-4 shows the response graph of a high pass filter.

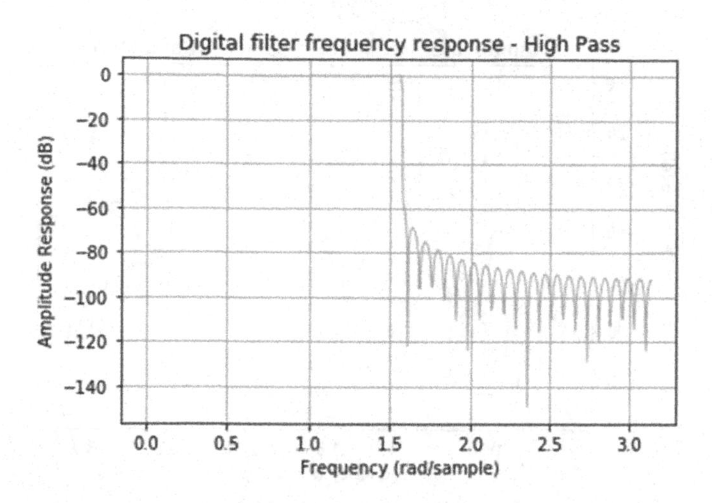

Figure 11-4. *Digital filter frequency response – High Pass*

```
b3 = signal.firwin(1963, [0.3, 0.8], pass_zero=False)
w3, h3 = signal.freqz(b3)
p3 = 20*np.log10(np.abs(h3))

plt.title('Digital filter frequency response - Band Stop')
line3, = plt.plot(w3, p3)

plt.setp(line3, color='r', linewidth=1.0)

plt.ylabel('Amplitude Response (dB)')
plt.xlabel('Frequency (rad/sample)')
plt.grid()
plt.show()  (Figure 11-5)
```

Figure 11-5 shows the response graph of a band stop filter.

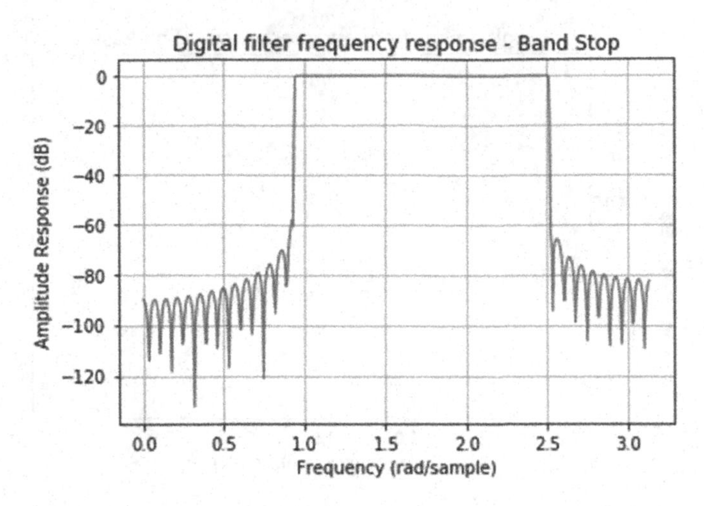

Figure 11-5. *Digital filter frequency response – Band Stop*

```
b4 = signal.firwin(1963, [0.3, 0.8], pass_zero=True)
w4, h4 = signal.freqz(b4)
p4 = 20*np.log10(np.abs(h4))

plt.title('Digital filter frequency response - Band Pass')
line4, = plt.plot(w4, p4)

plt.setp(line4, color='g', linewidth=1.0)

plt.ylabel('Amplitude Response (dB)')
plt.xlabel('Frequency (rad/sample)')
plt.grid()
plt.show()  (Figure 11-6)
```

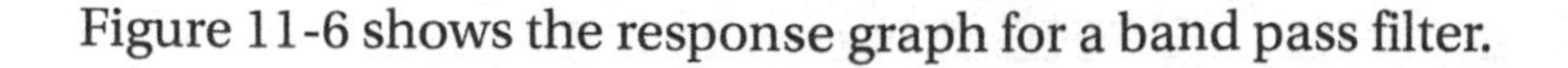

Figure 11-6 shows the response graph for a band pass filter.

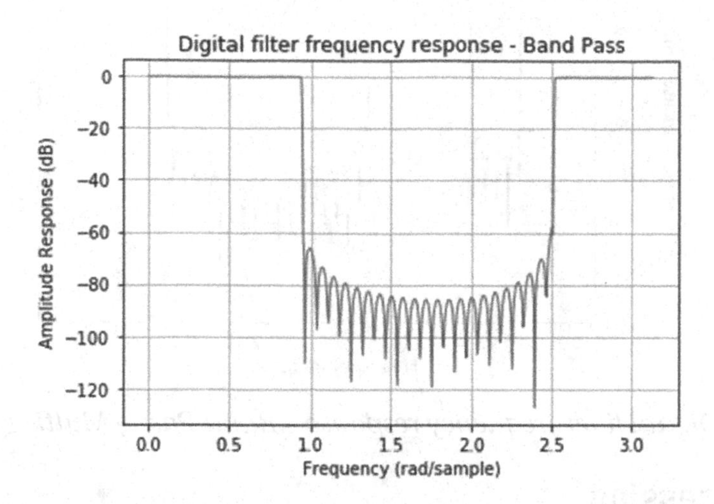

***Figure 11-6.** Digital filter frequency response – Band Pass*

```
b5 = signal.firwin(2019, [0.1, 0.2, 0.4, 0.8], pass_zero=True)
w5, h5 = signal.freqz(b5)
p5 = 20*np.log10(np.abs(h5))

plt.title('Digital filter frequency response - Band Pass')
line5, = plt.plot(w5, p5)

plt.setp(line5, color='k', linewidth=1.0)

plt.ylabel('Amplitude Response (dB)')
plt.xlabel('Frequency (rad/sample)')
plt.grid()
plt.show() (Figure 11-7)
```

Figure 11-7 shows the response graph for a multi-band pass filter.

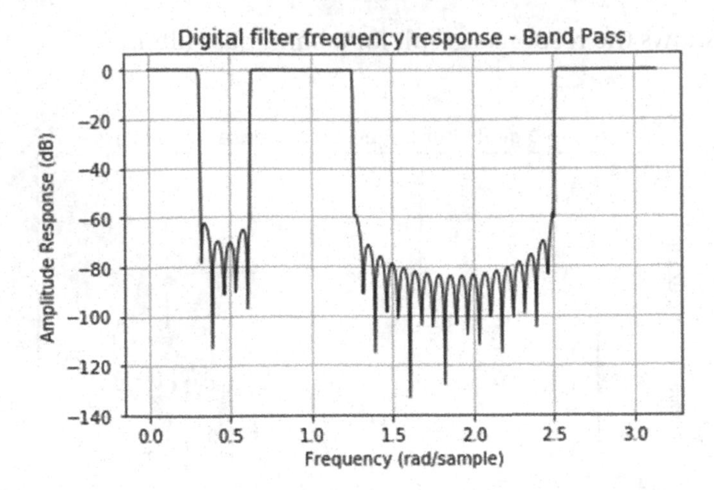

Figure 11-7. *Digital filter frequency response – Band Pass – Multi*

Speech Processing

I will now show you how to give your machine learning a voice. On Windows please add the pywin32 extensions.

This is done by execution this command: conda install -c anaconda pywin32

Load the Jupyter notebook from example directory Chapter 11 called: Chapter-011-09-Speak-01.ipynb

The main command to learn is: `engine.say('Hello Friend.')`

Tip The ability to speak back to humans is a widespread application in the smart cities we are building. These voices are already responding to actions from humans by warning us about dangers or giving directions.

Mechanical Engineering

Mechanical engineering is the discipline that applies engineering, physics, engineering mathematics, and materials science principles to design, analyze, manufacture, and maintain mechanical systems. It is one of the oldest and broadest of the engineering disciplines.

The mechanical engineering field requires an understanding of core areas including mechanics, dynamics, thermodynamics, materials science, structural analysis, and electricity.

The field has continually evolved to incorporate advancements; today mechanical engineers are pursuing developments in such areas as composites, mechatronics, and nanotechnology.

Robotics is the application of mechatronics to create robots, which are often used in industry to perform tasks that are dangerous, unpleasant, or repetitive. These robots may be of any shape and size, but all are preprogrammed and interact physically with the world. To create a robot, an engineer typically employs kinematics (to determine the robot's range of motion) and mechanics (to determine the stresses within the robot).

Robots are used extensively in industrial engineering. They allow businesses to save money on labor, perform tasks that are either too dangerous or too precise for humans to perform economically, and to ensure better quality. Many companies employ assembly lines of robots, especially in Automotive Industries and some factories are so robotized that they can run by themselves. Outside the factory, robots have been employed in bomb disposal, space exploration, and many other fields. Robots are also sold for various residential applications, from recreation to domestic applications.

Pulley System

Pulleys have been used by humans for thousands of years to move heavy loads around. Why would I show you pulleys?

In Chapters 12, 13, 14 and 15, I will discuss in detail how pulleys are an important part of enabling machine learning to manipulate the real world via industrialized machine learning.

In this chapter I will cover some basic introductory concepts on the pulleys, but it is a highly specific setup I will describe. If you are interested in this field, I suggest you get a book on pulleys and rigging principles.

This field is a specialized field in its own right.

Tip I suggest you as a machine learning engineer understand the real-world parameters of the system you are modeling, to ensure your models are implementable once you have your outcome.

The Internet-of-Things sensor and machines not currently being run by data decisions are exposing the data science and machine learning engineers to have subject matter insights these fields not as a normal part of their areas of expertise.

I will cover a few introductory concepts when dealing with dragging loads; there are a few terminologies you will need to know.

Dead Weight

The weight of the load you want to move in kilograms.

Tow Weight

When you drag an object across any surface, it resists the movement. This resistance is called Tow Resistance and is noted as a Tow Resistance Coefficient that is given as a % of extra effort required to tow the load.

Calculation:

$$TowWeight = DeadWeight\ (1 + Tow\ ResistanceCoefficient)$$

Mechanical Advantage

The Mechanical Advantage is the factor of enhancement the pulley system provides and is determined by the principle:

$$MechanicalAdvantage = \frac{OutputForce}{InputForce}$$

Directional Advantage

Directional Advantage is the ability to guide a force from a specific direction to move objects.

I will show you how to use a grid principal to deploy specific pivot points to enable your machine learning to control these pivot point locations to enable your machine learning to alter its own environment and learn how to discover and manage the real world.

Cable Length

The pulley system entails a length of cable to maneuver. It is typically coiled on a type of drum to store it out of harm's way. Cable length is the distance between force and the first pulley plus the amount of cable lengths between the pulley system and the load.

It gets determined mostly by the locations of the following four points:

- Force engine's X Y position
- Pulley pivot point's X Y position

- Load's X Y position
- Pulley system's design (i.e., number of pulleys)

Basic Pulley System

I will now show you a few basic calculations to explain what the pulley system gives you.

Load the Jupyter notebook from example directory Chapter 11 called: Chapter-011-10-Pulleys-01.ipynb

The first pulley system we look at is a mechanical advantage of 1 (Figure 11-8).

The pulley system gives you no mechanical advantage, but it does give you a change in direction. So, you can drag a load toward your anchor point by pulling the cable toward your pulling engine.

Tip Change in direction is important when you want to drag loads around other objects.

I will discuss this concept in more detail in Chapter 15 when we design our full-scale industrial solution.

Figure 11-8 shows a basic pulley system with one force, one pulley and one load.

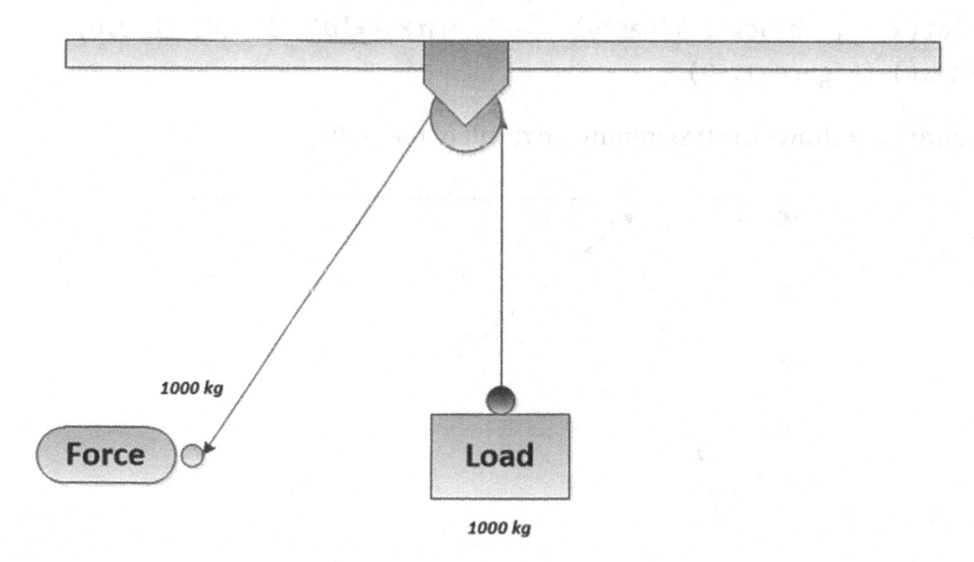

***Figure 11-8.** Basic Pulley System*

Let's investigate the three calculations we need.

```
import math
import numpy as np
from matplotlib import pyplot as plt

Pulleys=1

Force_Loc = [0,0]
Load_Loc = [100,0]
Pulley_Loc = [20,100]

dead_weight = 1000
tow_resistance = 0.01

data = np.array([
    Force_Loc,
    Pulley_Loc,
    Load_Loc
    ])

x, y = data.T

plt.plot(x, y, 'black', linestyle='--', marker='D', linewidth=2.0)
plt.show() (Figure 11-9)
```

Figure 11-9 shows the programmatic result of the pulley.

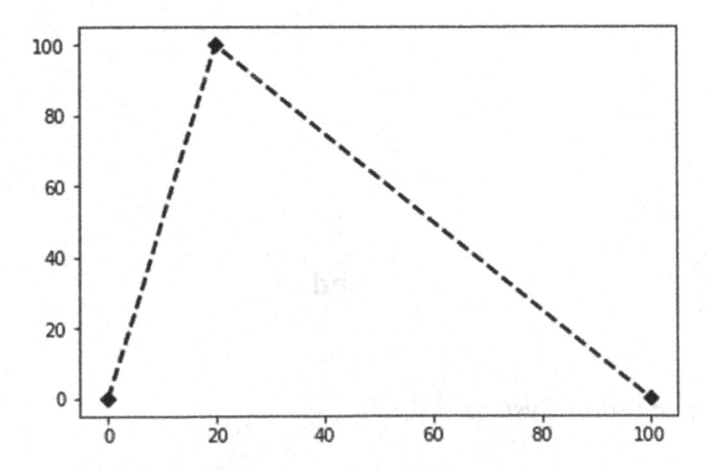

Figure 11-9. *Basic Pulley System Plot*

```
x1=Force_Loc[0]
y1=Force_Loc[1]
x2=Pulley_Loc[0]
y2=Pulley_Loc[1]
dist_force_pulley = math.hypot(x2 - x1, y2 - y1)
print('You are %0.5f meters from pulley system' % dist_force_pulley)
```

```
You are 101.98039 meters from pulley system
```

```
x1=Pulley_Loc[0]
y1=Pulley_Loc[1]
x2=Load_Loc[0]
y2=Load_Loc[1]
dist_pulley_load = math.hypot(x2 - x1, y2 - y1)
dist_pulleys_load = dist_pulley_load * Pulleys
print('You are %0.5f meters from load and need %0.5f meters of cable in pulleys
when using %0d pulleys' % (dist_pulley_load, dist_pulleys_load, Pulleys))
```

```
You are 128.06248 meters from load and need 128.06248 meters of cable in
pulleys when using 1 pulleys.
```

```
minimum_cable_length = (dist_force_pulley + dist_pulleys_load)
print('You need minimum %0.5f meters of cable in system when using %0d
pulleys' % (minimum_cable_length, Pulleys))
```

You need minimum 230.04288 meters of cable in system when using 1 pulley

```
tow_weight = dead_weight  * (1 + tow_resistance)
Load_Needed = (tow_weight * 9.80665)
mechanical_advantage = Pulleys
Force_Needed = Load_Needed/mechanical_advantage
print('System needs a %0.5f newton force when using %0d pulleys for a %0.5f
newton load (%0.3f kg of dead weight with tow resistance of %0.3f %%)' %
(Force_Needed, Pulleys, Load_Needed, dead_weight, tow_resistance*100))
```

The system needs a 9904.71650 newton force when using 1 pulley for a 9904.71650 newton load (1000.000 kg of dead weight with tow resistance of 1.000%).

Well done; you can now calculate how to use a pulley system.

You will use this again in Chapters 12, 13, 14, and 15.

The second pulley system we look at is a mechanical advantage of two (Figure 11-10).

To handle the new requirements of two pulleys, you could simply change the variable in your current notebook:

Pulleys=2

Or load the Jupyter notebook from example directory Chapter 11 called: Chapter-011-11-Pulleys-02.ipynb

Figure 11-10 shows a double pulley system with one force, one load but two moving pulleys. Note that the third pulley does not move due to the fix anchor point.

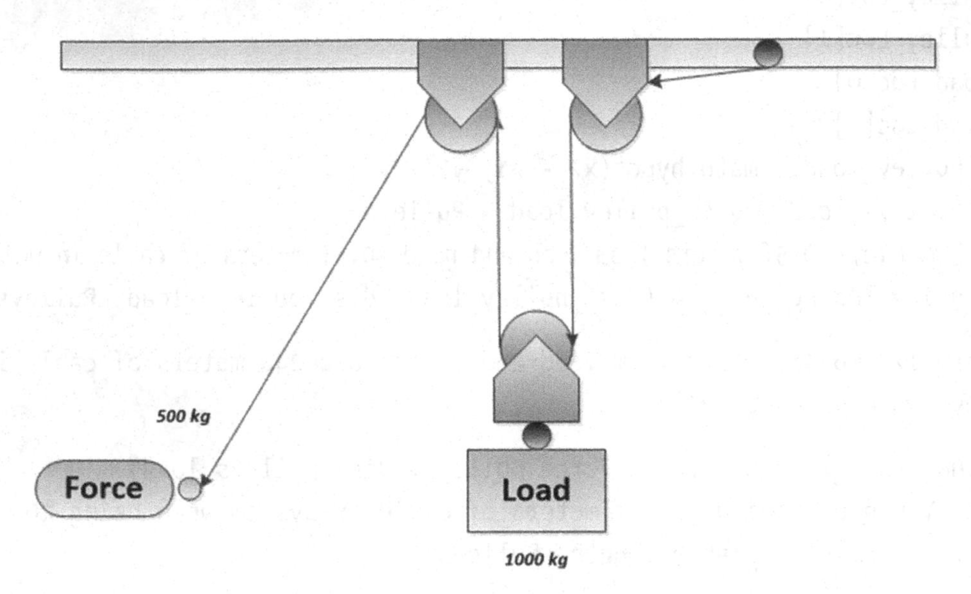

Figure 11-10. *Double Pulley System*

The results are:

You are 101.98039 meters from the pulley system.

You are 128.06248 meters from the load and need 256.12497 meters of cable in pulleys when using two pulleys.

You need a minimum of 358.10536 meters of cable in a system when using two pulleys.

System needs a 4952.35825 newton force when using two pulleys for a 9904.71650 newton load (1000.000 kg of dead weight with tow resistance of 1.000%).

So, one pulley was 9904.71650 newton, but two pulleys are 4952.35825 newton.

You can now move double the load with the same force.

The third pulley system we look at is a mechanical advantage of 4 (Figure 11-11).

To handle the new requirement of four pulleys, you could simple change the variable in your current notebook:

Pulleys=4

Or load the Jupyter notebook from example directory Chapter 11 called: Chapter-011-12-Pulleys-03.ipynb

Figure 11-11 shows a more complex four pulley system with a fix anchor point.

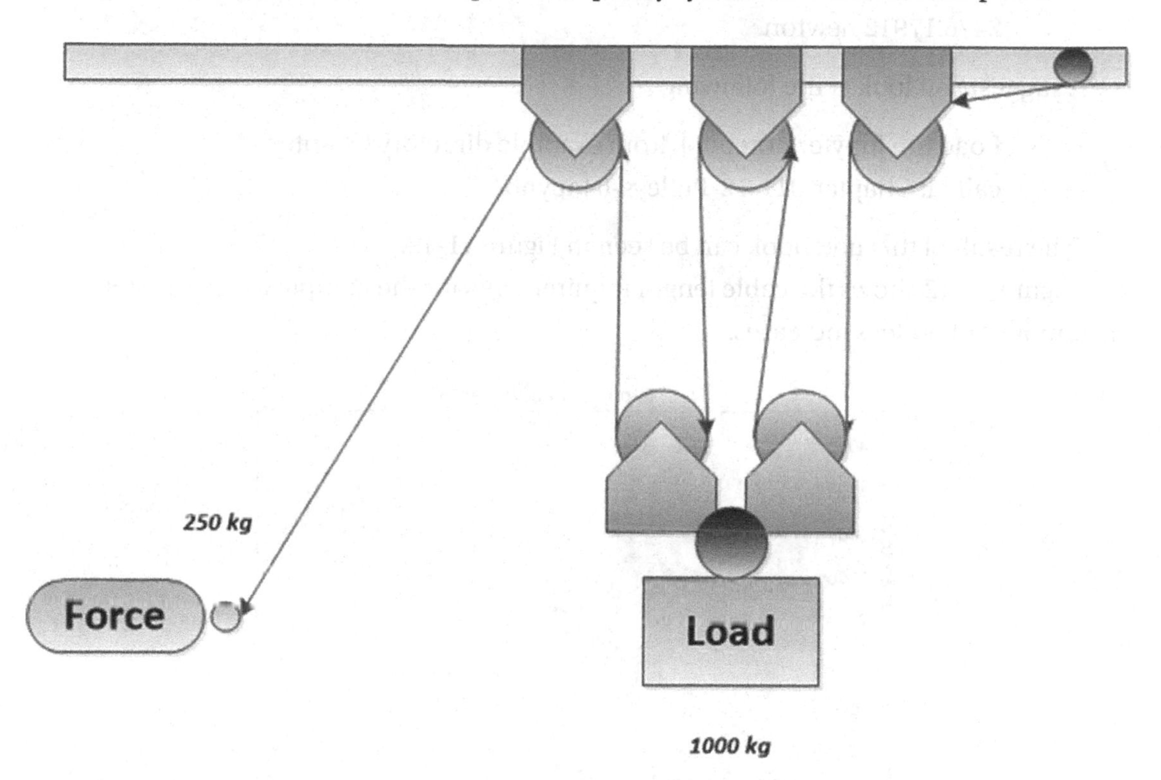

Figure 11-11. *Complex Pulley System*

Results:

> You are 101.98039 meters from the pulley system.
>
> You are 128.06248 meters from the load and need 512.24994 meters of cable in pulleys when using four pulleys.
>
> You need a minimum 614.23033 meters of cable in a system when using four pulleys.
>
> System needs a 2476.17912 newton force when using four pulleys for a 9904.71650 newton load (1000.000 kg of dead weight with tow resistance of 1.000%).
>
> So, one pulley was 9904.71650 newton, but four pulleys are 2476.17912 newton.

I suggest you look at the following:

> Load the Jupyter notebook from example directory Chapter 11 called: Chapter-011-12-Pulleys-04.ipynb

The result of this notebook can be seen in Figure 11-12.

Figure 11-12 shows the cable length requirements for the complex pulley system as the amount of pulleys increases.

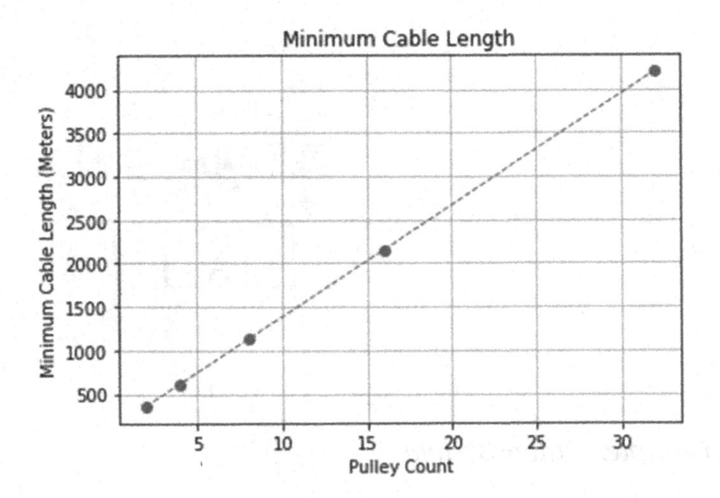

Figure 11-12. *Cable Length Plot*

The force results can be seen in Figure 11-13.

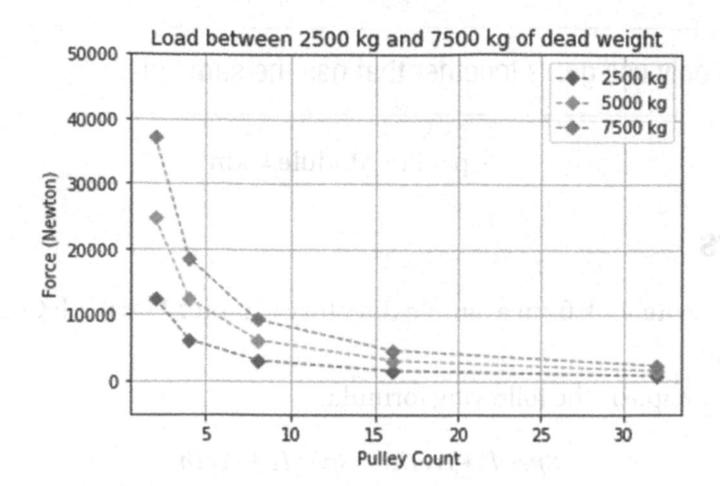

Figure 11-13. *Force vs. Pully Count*

Well done; you have the basic concepts of pulleys. I will discuss this subject in Chapter 14 and then finally in Chapter 15 to show how your knowledge assists with a full-scale industrialized machine learning problem.

Gears

A gear is a toothed machine wheel or cylinder that meshes with another toothed component to transmit motion or to change speed or direction. This basic principal gives your machine learning the mechanical capabilities to manipulate the real world.

I will give you basic insight into this process.

Using ISO (International Organization for Standardization) guidelines, Module Size is designated as the unit representing gear tooth sizes.

Warning Other sizing methods are also used! Please check to ensure the gears match properly.

```
Module (m)
m = 1 (p = 3.1416)
m = 2 (p = 6.2832)
m = 4 (p = 12.566)
```

If you multiply Module by Pi, you can obtain Pitch (p).

Pitch is the distance between corresponding points on adjacent teeth.

Tip You can only use gears together that has the same pitch.

$$p = \text{Pi x Module} = \pi m$$

Single Gears

Load the Jupyter notebook from example directory Chapter 11 called: Chapter-011-14-Gears-01A.ipynb

The basic principle is the following formula:

$$Speed1 * Teeth1 = Speed2 * Teeth2$$

Figure 11-14 shows a simple two gear system.

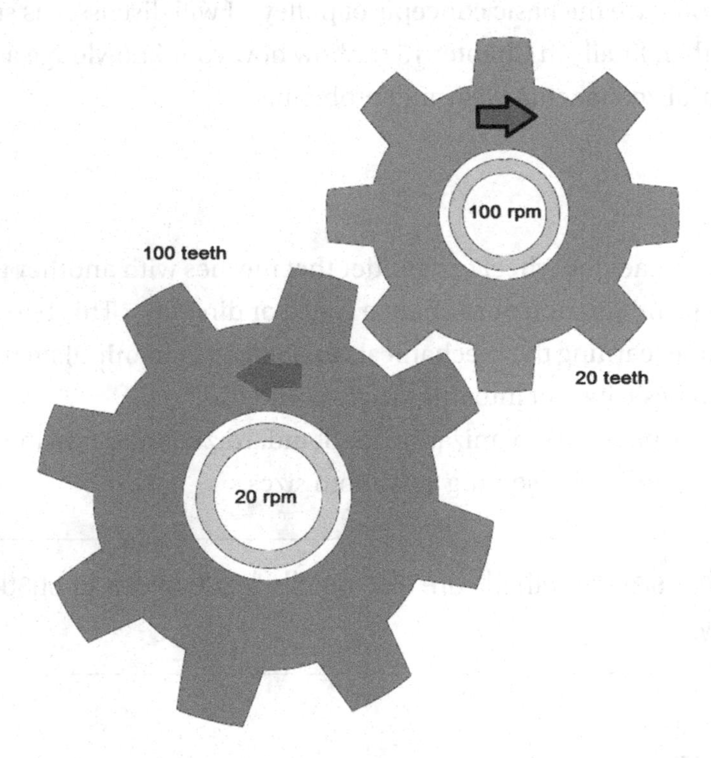

Figure 11-14. *Single Gears*

The example demonstrates how to calculate the gear, speed, and power ratios for the gears shown (Figure 11-14).

Result:

```
Apply Green  1.500 kW ( 716.197 newton meter) on 100 Teeth turning
at  20.000 rpm
Supply Amber 1.500 kW ( 143.239 newton meter) on  20 Teeth turning at
100.000 rpm
Power: 0.200 factor
Speed: 5.000 factor
```

Load the Jupyter notebook from example directory Chapter 11 called: Chapter-011-15-Gears-01B.ipynb

Change:

```
n1 = 20   # Teeth Gear Green
n2 = 100  # Teeth Gear Amber
```

Result:

```
Apply Green 1.500 kW ( 716.197 newton meter) on 20 Teeth turning at
20.000 rpm Supply Amber 1.500 kW (3580.986 newton meter) on 100 Teeth
turning at 4.000 rpm
Power: 5.000 factor
Speed: 0.200 factor
```

See: `https://www.engineersedge.com/gear_formula.htm`

Composite Gears

This gear system consists of four gears (Green, Amber, Yellow, and Red). The Amber and Yellow are connected via a single drive rod (i.e., if Amber turns at 10 rpm, Yellow also turns at 10 rpm and all power applied to Amber is 100% transferred to Yellow). See Figure 11-15.

Note These composite gears systems are the ones most commonly used in any application.

Load the Jupyter notebook from example directory Chapter 11 called: Chapter-011-16-Gears-02A.ipynb

Figure 11-15 shows a composite gear system that improves the effectiveness of the gears.

Figure 11-15. *Composite Gears*

Note that the star gears are connecting via a shaft. This creates a composite gears system that is common in real-world applications.

Results:

Apply Green 1.500 kW (716.197 newton meter) on 100 Teeth turning at 20.000 rpm.

Supply Amber 1.500 kW (143.239 newton meter) on 20 Teeth turning at 100.000 rpm.

Apply Yellow 1.500 kW (143.239 newton meter) on 200 Teeth turning at 100.000 rpm.

Supply Red 1.500 kW (10.743 newton meter) on 15 Teeth turning at 1333.333 rpm.

Power: 0.015 ratio.

Speed: 66.667 ratio.

You change 20 rpm into over 1300 rpm.

Well done; you have completed your first gears example.

Load the Jupyter notebook from example directory Chapter 11 called: Chapter-011-17-Gears-02B.ipynb.

Your simply change the following to change the gears from a speed increase to a power increase.

```
n1 = 20  # Teeth Gear Green
n2 = 100 # Teeth Gear Amber
n3 = 15  # Teeth Gear Yellow
n4 = 200 # Teeth Gear Red
```

```
Apply Green 1.500 kW ( 716.197 newton meter) on 20 Teeth turning at 20.000 rpm
Supply Amber 1.500 kW (3580.986 newton meter) on 100 Teeth turning at 4.000 rpm

Apply Yellow 1.500 kW (3580.986 newton meter) on 15 Teeth turning at 4.000 rpm
Supply Red 1.500 kW (47746.483 newton meter) on 200 Teeth turning at 0.300 rpm

Power: 66.667 factor Speed: 0.015 factor
```

Well done; you now have a basic knowledge base on the use of gears. I will cover more practical implementations of this knowledge in Chapters 12, 13, 14, and 15 when I will explain how to use this knowledge.

Lift

The lift principal is widely used to enable the application of a small amount of force to generate a required larger force.

A lever (Figure 11-16) is a simple machine that consists of two material components and two work components:

- A beam or solid rod
- A fulcrum or pivot point
- An input force (effort)
- An output force (load or resistance)

Load the Jupyter notebook from example directory Chapter 11 called: Chapter-011-18-Lift-01.ipynb

$$M1\,a = M2\,b$$

```
M1 - The mass on one end of the fulcrum (the input force)
a  - The distance from the fulcrum to M1
M2 - The mass on the other end of the fulcrum (the output force)
b  - The distance from the fulcrum to M2
```

Figure 11-16 is a simple lever.

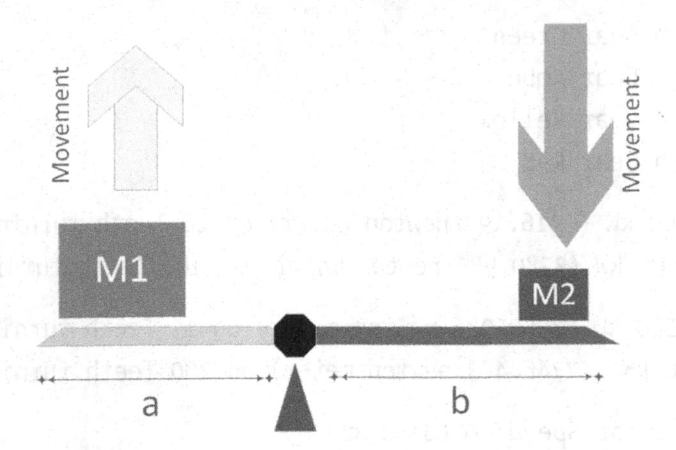

Figure 11-16. *Basic Lever*

I will quickly show you how to draw a lever like the one in the example (Figure 11-17).

```
from PIL import ImageFont, Image, ImageDraw
im = Image.new('RGBA', (600, 200), (255, 255, 255, 255))
draw = ImageDraw.Draw(im)
draw.line((10,100, 200,100), fill=128, width=10)
draw.line((210,100, 550,100), fill=128, width=10)

draw.ellipse((195, 105, 215, 170), fill=(255,0,0,255))
draw.ellipse((155, 155, 255, 170), fill=(0,0,0,255))

f = ImageFont.load_default()
draw.text( (5, 80), "a x M1",  font=f, fill=255)
draw.text( (535, 80), "b x M2",  font=f, fill=255)

im.show() (Figure 11-17)
```

Figure 11-17 is the result for the simple lever.

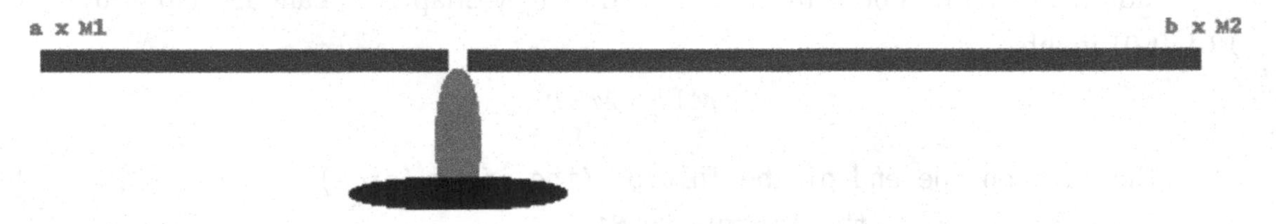

Figure 11-17. *Lever plot*

The four components of a lever can be combined in three basic ways, resulting in three classes of levers.

Class 1 Levers

This is a common configuration where the fulcrum is in between the input and output forces. Real-world use would be scales or tools like a crowbar. See Figure 11-18.

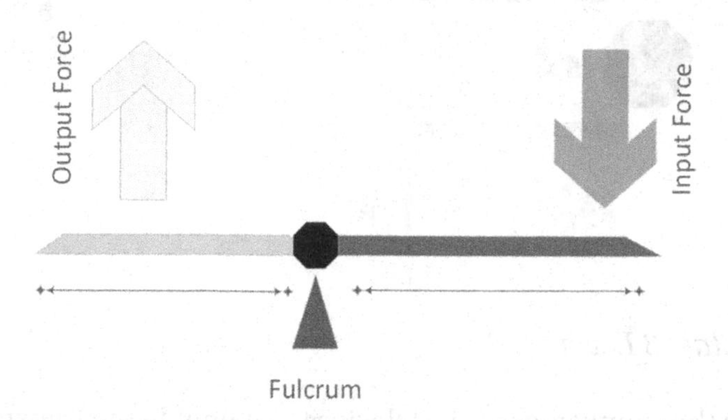

Figure 11-18. *Class 1 Lever*

Class 2 Levers

The resistance comes between the input force and the fulcrum; a real-world use is a wheelbarrow. See Figure 11-19.

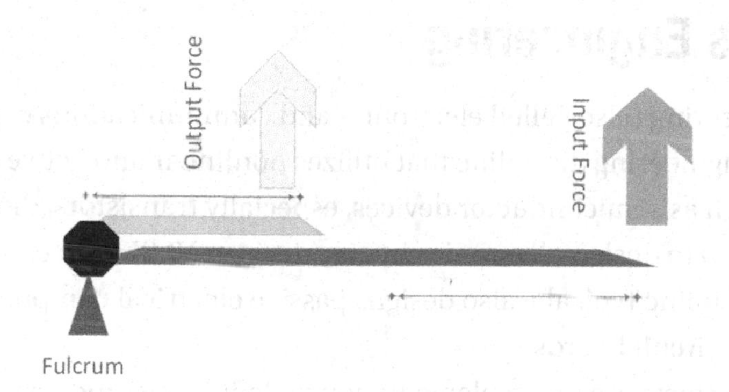

Figure 11-19. *Class 2 Lever*

Class 3 Levers

The fulcrum is on one end and the resistance is on the other end, with the effort in between the two, such as with a pair of tweezers or BBQ tongs. See Figure 11-20.

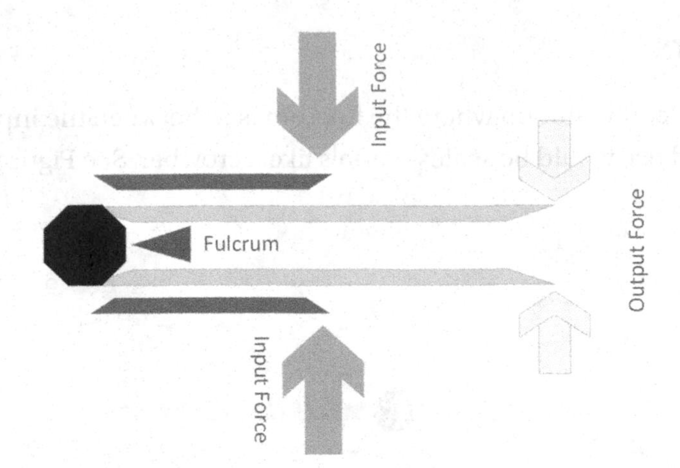

Figure 11-20. *Class 3 Lever*

The levers are the simply mechanical devices that enable the transfer of power and movement to force the actions the machine learning requires.

The most common use for you would be robotics and industrial manufacturing that I will discuss in detail in Chapters 12, 13, 14, and 15 on how to use this knowledge to enable your machine learning to change the real world.

Electronics Engineering

Electronic engineering (also called electronics and communications engineering) is an electrical engineering discipline that utilizes nonlinear and active electrical components (such as semiconductor devices, especially transistors, diodes, and integrated circuits) to design electronic circuits, devices, VLSI devices, and their systems. The discipline typically also designs passive electrical components, usually based on printed circuit boards.

Signals and systems are the implementation of definitions, and properties of Laplace transform continuous-time and discrete-time Fourier series, continuous-time and discrete-time Fourier Transform, z-transform.

Load the Jupyter notebook from example directory Chapter 11 called: Chapter-011-19-Electronics-01.ipynb

I want to introduce you to a simple set of electronic devices. The electronic engineering field has many more complex systems, but for the purpose of this book I will only explain a few basics.

Logical Gates

Logical electronic gates are the building blocks of the electronic integrated circuit world. These basic circuits are combined to form more complex logical circuit that can be programmed to perform decisions or monitor statuses for a machine learning process.

I will introduce you to some of the basic circuits to give you an introduction to these tools to enable you to have insight into how you could use these to assist your IML earning processes.

AND Gate

The AND gate is a logical decision process embedded within an electronic circuit. The gate reacts against two inputs a and b with a single output x. See Figure 11-21.

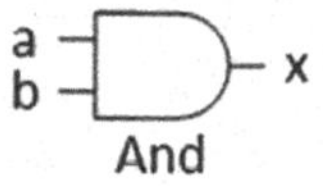

Figure 11-21. *AND Gate*

There are a basic four options:

- a=0, b=0 then x=0
- a=0, b=1 then x=0
- a=1, b=0 then x=0
- a=1, b=1 then x=1

```
for a in range (2):
    for b in range (2):
        x= a and b
        print(' %0d and %0d => %0d' % (a, b, x))
```

Results:

```
0 and 0 => 0
0 and 1 => 0
1 and 0 => 0
1 and 1 => 1
```

OR Gate

There are the following options for OR gates (Figure 11-22).

- a=0, b=0 then x=0
- a=0, b=1 then x=1
- a=1, b=0 then x=1
- a=1, b=1 then x=1

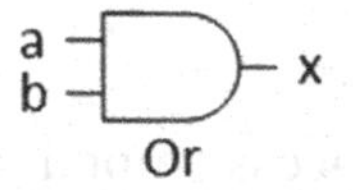

Figure 11-22. *OR Gate*

```
for a in range (2):
    for b in range (2):
        x= a or b
        print(' %0d or %0d => %0d' % (a, b, x))
```

Results:

The following four results are for OR gate:

```
0 or 0 => 0
0 or 1 => 1
1 or 0 => 1
1 or 1 => 1
```

Invertor

There are the following options for Invertors (Figure 11-23):

- a=0 then x=1
- a=1 then x=0

Figure 11-23. *Invertor*

```
for a in range (2):
        x= not a
        print(' %0d => %0d' % (a, x))
```

Basic Logic System

This circuit is a combination of two AND gates, an OR gate, and a NOT gate (Figure 11-24).

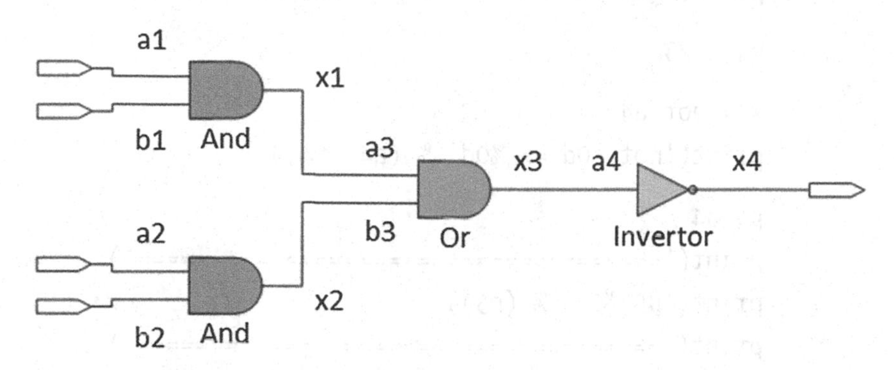

Figure 11-24. Basic Logic System

This is a simple but common logic circuit to check for four (a1, b1, a2, b2) inputs to produce a single output (x4).

Let's take this process one step at a time to explain what is happening.

```
t=0
for p1 in range (2):
    for p2 in range (2):
        for p3 in range (2):
            for p4 in range (2):
                a1=p1
                b1=p2
                a2=p3
                b2=p4
                t+=1
                print('===================================')
                print('c= %0d:p1=%0d p2=%0d p3=%0d p4=%0d' % (t, p1, p2,
                p3, p4))
                print('===================================')
                x1=a1 and b1
                print('%0d and %0d => %0d' % (a1,b1,x1))
```

```
        x2=a2 and b2
        print('%0d and %0d => %0d' % (a2,b2,x2))

        a3 = x1
        b3 = x2

        x3=a3 or b3
        print('%0d or %0d => %0d' % (a3,b3,x3))

        a4 = x3

        x4= not a4
        print('not %0d => %0d' % (a4, x4))

        p5=a4
        print('====================================')
        print('p5=%0d' % (p5))
        print('====================================')
```

```
====================================
c= 1:p1=0 p2=0 p3=0 p4=0
====================================
0 and 0 => 0
0 and 0 => 0
0 or 0 => 0
not 0 => 1
====================================
p5=0
====================================
```

Telecommunications Engineering

Telecommunications engineering is an engineering discipline centered on electrical and computer engineering that seeks to support and enhance telecommunication systems.

A transmitter (information source) takes information and converts it to a signal for transmission.

The transmission medium over which the signal is transmitted:

- Wired communication
- Wireless communication

The receiver (information sink) receives and converts the signal back into required information.

Load the Jupyter notebook from example directory Chapter 11 called: Chapter-011-20-Telecomms-01.ipynb

Run the notebook to get insights into how the simulations of the following telecommunication networks are working:

- Wired Telecommunication

  This network has only two nodes: one transmitting and another receiving. This is a typical IoT ecosystem. Simulations result in Point-to-Point Chapter-011-20-Telecomms-01-01.txt

- Central Wireless Telecommunication

  This network has one central node transmitting and two nodes receiving. This is a typical robot control ecosystem. Simulations result in Point-to-Point Chapter-011-20-Telecomms-01-02.txt

- Single-Channel Wireless Telecommunication

  This network has three nodes transmitting and three nodes receiving on the same bandwidth. This is a typical unmanaged ecosystem. Simulations result in Point-to-Point Chapter-011-20-Telecomms-01-03.txt

---

Warning This style network is not effective or efficient at communicating. I suggest you avoid this model as it causes chaotic communication and wastes bandwidth.

- Tri-Mesh Wireless Telecommunication

  This network has three nodes transmitting and three nodes receiving on a mesh bandwidth. Simulations result in Point-to-Point Chapter-011-20-Telecomms-01-04.txt

---

Tip Mesh networks are managed and highly effective as well as efficient.

Systems Engineering

Systems engineering is an interdisciplinary field of engineering and engineering management that focuses on how to design and manage complex systems over their life cycles. See Figure 11-25.

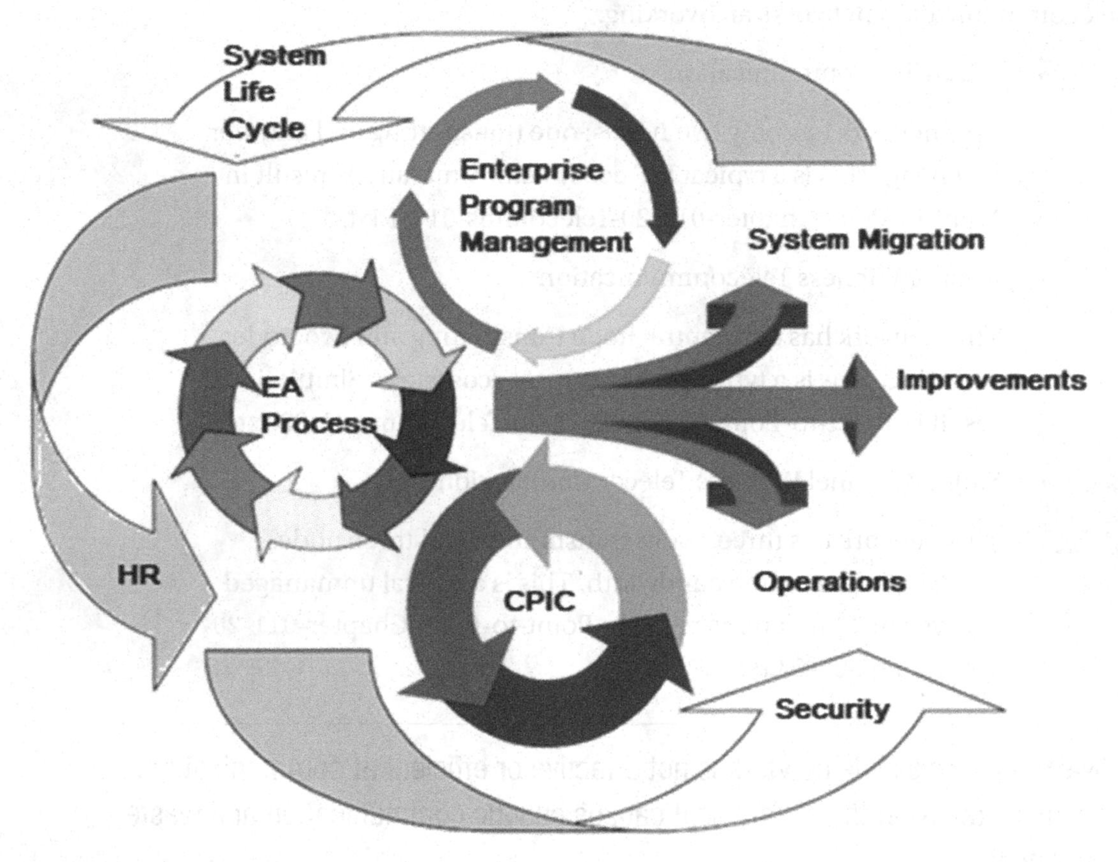

Figure 11-25. *Enterprise Life-Cycle Ecosystem*

Systems engineering is a complex field in its own. I will only show this quick introduction as it is not a field that I use regularly. I suggest you read about ecosystem management processes as they are important for use in highly industrialized ecosystems.

The basic components are as follows.

Enterprise Program Management

This unit ensures the programs of work are completed on time and within budget. The main contact that you as a machine learning engineer would have with this unit is via a technical project manager.

They handle all the functional and nonfunctional requirements for the work you will undertake.

Enterprise Architecture Process

The enterprise architecture unit handles everything that is related to the hardware and the Internet-of-Things devices. They will supply you with the physical equipment to perform you industrialized machine learning. They also own the robots and communication equipment.

Human Resources

They keep your human-in-the-loop resources in perfect condition and ensure the right person is available at the right time. Without soft skills and the ability to manage people, you will not achieve success.

System Life Cycle

This is a process that ensures that the old systems get removed as per the decommissioning agreements and ensures new systems are brought online without interruption of the daily services of the machine learning environment.

Central Processing Information Center (CPIC)

The CPIC is where the machine learning lives within the ecosystem. All the processing happens in this unit after you have handed your new machine learning over via your DevOps procedures. They are the custodians of all the active machine learning models in their systems and ensure its processed against the enterprise data sources as per your instructions as a data scientist and machine learning engineer.

Security

The cybersecurity and physical security of the machine learning ecosystem are an important unit in the ecosystem because with the current requirement to place Internet-of-Things devices in unprotected environments, unwelcome access or misuse of these devices becomes a certainty. Your security staff are the people that will ensure that your machine learning is not used for unfair or damaging purposes.

Control Engineering

Control engineering or control systems engineering is an engineering discipline that applies automatic control theory to design systems with desired behaviors in control environments.

It uses sensors and detectors to measure the output performance of the process being controlled; these measurements are used to provide corrective feedback helping to achieve the desired performance. Systems designed to perform without requiring human input are called automatic control systems.

A time series is a series of data points indexed (or listed or graphed) in time order. Most commonly, a time series is a sequence taken at successive equally spaced points in time. Thus, it is a sequence of discrete-time data.

Control Simulation

Install:

```
conda install -c conda-forge control
Install a fortan compiler.
```

Load the Jupyter notebook from example directory Chapter 11 called: Chapter-011-21-Control-01.ipynb

```
import matplotlib
matplotlib.use('TkAgg')
%matplotlib inline
```

```
from scipy import *
from pylab import *
a = zeros(500)
a[:100]=1
b = fft(a)
plot(abs(b))
show()  (Figure 11-26)
```

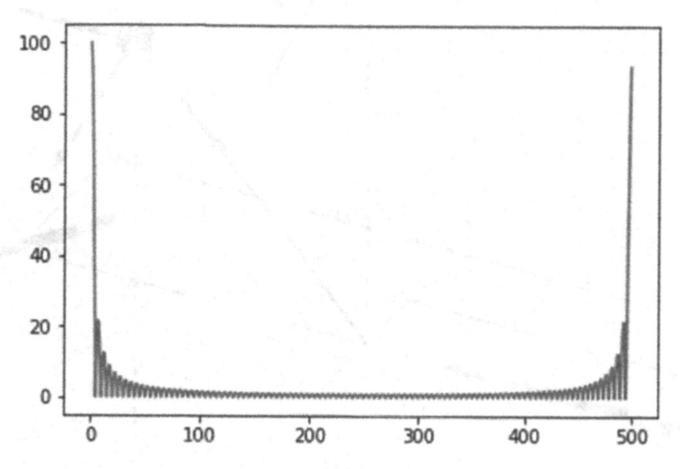

Figure 11-26. *Fourier Transformation Plot*

This notebook has several control-style configurations that you can investigate.

Dual-Loop Controller for a Bicycle

The next example will take you through a sequential dual-loop controller for a bicycle to simulate how machine learning could control the bicycle by simply controlling balance and direction of travel.

This is achieved in the same manner a human would ride the bicycle.

Load the Jupyter notebook from example directory Chapter 11 called: Chapter-011-22-Control-02.ipynb

This notebook simulates a bicycle balancing.

Here are the control vectors for the process. The code will supply insights into how the process is working. See Figure 11-27.

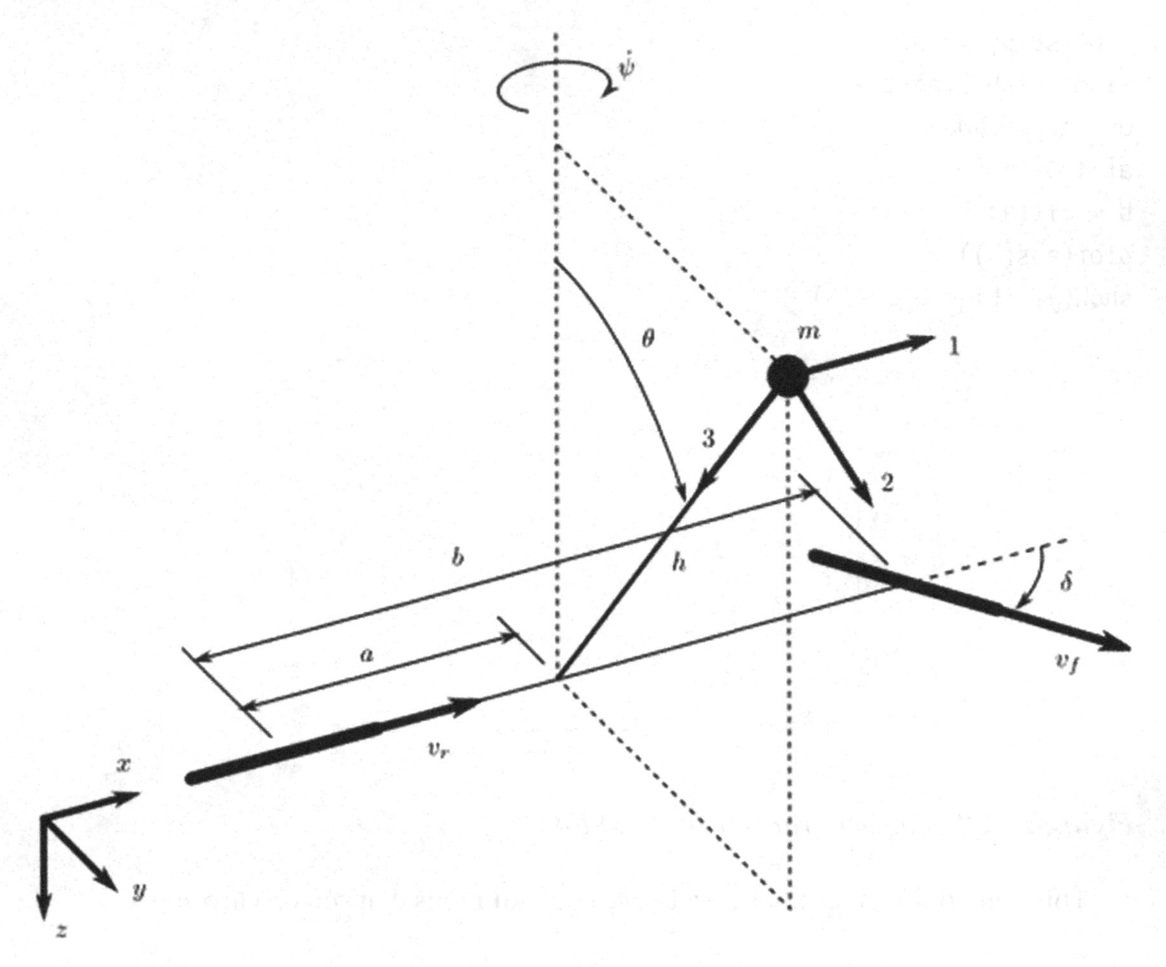

***Figure 11-27.** Bicycle Control Design*

The structure of the controller is shown in the following block diagram (Figure 11-28).

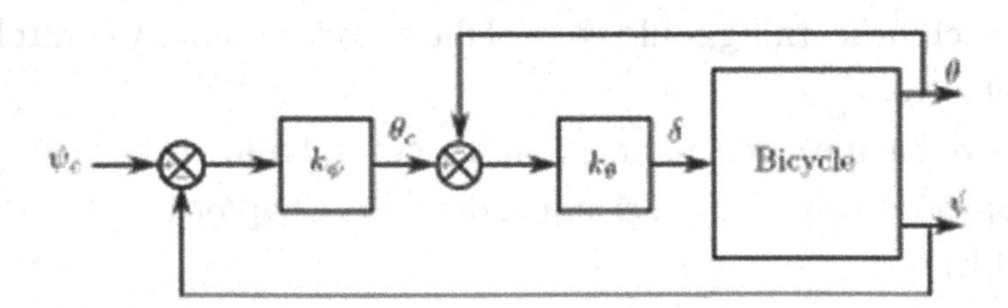

***Figure 11-28.** Bicycle Control Concept*

Load the Jupyter notebook from example directory Chapter 11 called: Chapter-011-23-Control-03.ipynb

```
Install: pip install harold
```

The notebook creates models via Harold that defines two classes to represent the dynamic models: Transfer and State classes.

These are the transfer states of the process:

$$G_1 = \frac{z-1}{z^3 - zs^2 + z},\ G_2 = \begin{bmatrix} \frac{1}{s+2} & \frac{s+3}{s^2+s-4} \\ 0 & \frac{s+2}{s-3} \end{bmatrix},\ G_3 = \left[\begin{array}{cc|c} 0 & 1 & 0 \\ -2 & -0.5 & 1 \\ 0 & 3.5 & 1 \end{array}\right],\ G_4 \left[\begin{array}{ccc|cc} -4 & -2 & 2 & 1 & 1 \\ 0 & -5 & 2 & 0 & 1 \\ 4 & -3 & -4 & 2 & -1 \\ -1 & 5 & 2 & 0 & 0 \end{array}\right]$$

Your results are:

```
Discrete-Time Transfer function with sampling time: 0.100 (10.000 Hz.)
 1 input and 1 output

  Poles(real)    Poles(imag)    Zeros(real)    Zeros(imag)
-------------  -------------  -------------  -------------
            0              0              1              0
            1              0
            1              0
```

You can directly identify the pole zero structure with the process.

The MIMO state representations are preferred exclusively, and MIMO transfer models explicitly require a list of lists structure.

$$G_5 = \frac{1}{s^2+5s+1}\begin{bmatrix} s+1 & 2 \\ s+3 & 4 \end{bmatrix}$$

Results:

```
[[-4. -2.  2.]
 [ 0. -5.  2.]
 [ 4. -3. -4.]]
```

Finally, you can create random State models via the random_stable_model.

You should force the random model to have more chance to have oscillatory modes by changing the probability distribution of the selected poles.

Assume that you want a discrete-time model with majority poles on the imaginary axis and occasional integrators.

Results are:

```
Continuous-time state representation
5 states, 2 inputs, and 3 outputs
  Poles(real)     Poles(imag)  Zeros(real)    Zeros(imag)
------------ ------------ ------------ ------------
    -7.36961       6.61705
    -7.36961      -6.61705
    -3.03534       0
    -7.30188       0
    -5.08623       0

Discrete-Time state representation with sampling time: 0.010 (100.000 Hz.)
20 states, 2 inputs, and 3 outputs
  Poles(real)     Poles(imag)  Zeros(real)    Zeros(imag)
------------ ------------ ------------ ------------
    0.991056      0.133444
    0.991056     -0.133444
    0.995934      0.0900819
    0.995934     -0.0900819
    1.06971      0
    0.99681       0.079816
    0.99681      -0.079816
    0.998778      0.0494188
    0.998778     -0.0494188
    1             0
    0.999386      0.0350363
    0.999386     -0.0350363
    0.999515      0.0311479
    0.999515     -0.0311479
    0.999561      0.0296121
    0.999561     -0.0296121
    0.999602      0.0281987
    0.999602     -0.0281987
    0.999592      0.0285679
    0.999592     -0.0285679
```

Conversions of Models are used in order to convert one model representation to another, and the relevant functions are transfer_to_state() and state_to_transfer().

Note Typically, the conversion from a Transfer model to a State model leads to a non-minimal representation.

```
Continuous-time state representation
4 states, 2 inputs, and 2 outputs
  Poles(real)     Poles(imag)     Zeros(real)     Zeros(imag)
-------------   -------------   -------------   -------------
           -2               0              -2    9.55463e-08
            3               0              -2   -9.55463e-08
            2               0               2    0
           -2               0

Discrete-Time Transfer function with sampling time: 0.100 (10.000 Hz.)
 1 input and 1 output

  Poles(real)     Poles(imag)     Zeros(real)     Zeros(imag)
-------------   -------------   -------------   -------------
        -0.25        0.193649      -0.0251582               0
        -0.25       -0.193649        -3.97484               0
```

Minimal Realization

Minimalism is an essential property of any representation for reliable computations and hence we can use the minimal_realization() function. This function uses a distance metric to the closest rank-deficient matrix pencil for State models and a straightforward walk over the poles and zeros for Transfer representations.

Note The tolerance for a decision can be adjusted.

Warning MIMO poles and zeros don't necessarily cancel each other even though the values are identical. This is because for MIMO representations, pole and zero directionalities can be different.

You can see this for the representation H, which you obtained before.

```
Continuous-time state representation
3 states, 2 inputs, and 2 outputs
  Poles(real)  Poles(imag)  Zeros(real)  Zeros(imag)
------------- ------------- ------------- -------------
          -2             0             2             0
           2             0            -2             0
           3             0
```

Discretization and Non-discretization

Discretization of models is mostly an art or result of experience rather than pure science. Though the methods are sound, selecting the right method and the relevant sampling period is a machine learner's choice.

```
Discrete-Time Transfer function with sampling time: 0.300 (3.333 Hz.)
 1 input and 1 output

  Poles(real)  Poles(imag)  Zeros(real)  Zeros(imag)
------------- ------------- ------------- -------------
     0.792052      0.172336     -0.869152             0
     0.792052     -0.172336
```

```
Here are the current known discretization methods:

Discretization methods
==============================
bilinear
tustin
zoh
foh
forward difference
forward euler
forward rectangular
backward difference
backward euler
backward rectangular
```

```
lft
>>
<<
```

You can now convert back these models; G3 will be converted using the default method `Tustin`; however, G6d will be converted via a zero-order hold method.

Note Had this information not been present, the resulting continuous model would be slightly different than what you started with.

```
Continuous-Time Transfer function
 1 input and 1 output

  Poles(real)    Poles(imag)  Zeros(real)    Zeros(imag)
-------------  -------------  -------------  -------------
         -0.7       0.714143
         -0.7      -0.714143

Continuous-Time Transfer function
 1 input and 1 output

  Poles(real)    Poles(imag)    Zeros(real)    Zeros(imag)
-------------  -------------  -------------  -------------
    -0.705414       0.708951        6.66667              0
    -0.705414      -0.708951       -95.2327              0
```

Model Algebra

Basic algebraic operations and feedback operations are implemented via typical *,+,-,@ and feedback() functions. The shape and sampling time compatibility are checked before the operations and errors are raised in case of mismatches. One exception is the operation with an SISO model over an MIMO model. For example, G1 is an SISO model and G3 is a sampling time matching MIMO model; hence following is allowed, which multiplies each entry of G3 with G1. For multiplication of MIMO models, the matrix multiplication rules and hence matrix multiplication operator @ are followed.

```
print(G1 * G3)

Discrete-Time state representation with sampling time: 0.100 (10.000 Hz.)
5 states, 1 input, and 1 output
  Poles(real)    Poles(imag)    Zeros(real)    Zeros(imag)
------------  ------------  ------------  ------------
     -0.25       0.193649     -0.0251582            0
     -0.25      -0.193649     -3.97484              0
      0          0             1                    0
      1          0
      1          0

print(G4 @ G2)

Continuous-time state representation
7 states, 2 inputs, and 1 output
  Poles(real)    Poles(imag)  Zeros(real)    Zeros(imag)
------------  ------------  ------------  ------------
    -6.76735       0
    -3.11633       1.45258
    -3.11633      -1.45258
    -2             0
     3             0
     2             0
    -2             0

CL = feedback(G3, G1)
CL_min = minimal_realization(CL)
print(CL)

Discrete-Time state representation with sampling time: 0.100 (10.000 Hz.)
5 states, 1 input, and 1 output
  Poles(real)    Poles(imag)    Zeros(real)    Zeros(imag)
------------  ------------  ------------  ------------
   -1.30016        0            -0.0251582            0
    0.912951       1.46766      -3.97484              0
    0.912951      -1.46766      -0                    0
    1              0             1                    0
   -0.0257453      0             1                    0
```

You should observe that there is a pole/zero cancellation at 1, which is removed afterward in the process.

```
Discrete-Time state representation with sampling time: 0.100 (10.000 Hz.)
4 states, 1 input, and 1 output
  Poles(real)      Poles(imag)      Zeros(real)      Zeros(imag)
-------------    -------------    -------------    -------------
   -1.30016            0             5.76839e-15               0
   -0.0257453          0            -0.0251582                 0
    0.912951         1.46766       1                           0
    0.912951        -1.46766      -3.97484                     0
```

Basic Plotting Functionality

The common plots that are needed are already available.

For frequency domain plotting, the default units are Hz and powers of ten for amplitudes.

```
impulse_response_plot(G4)
```

See Figure 11-29.

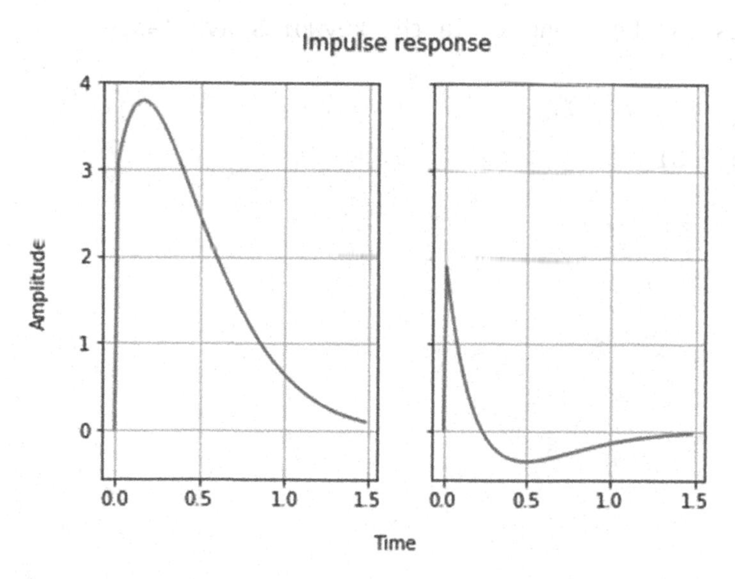

Figure 11-29. *Impulse Response (1)*

The discrete-time plant plots are automatically drawn as stairs.

```
G4_d = discretize(G4, 0.1, method='zoh')
impulse_response_plot(G4_d)
(See Figure 11-30)
```

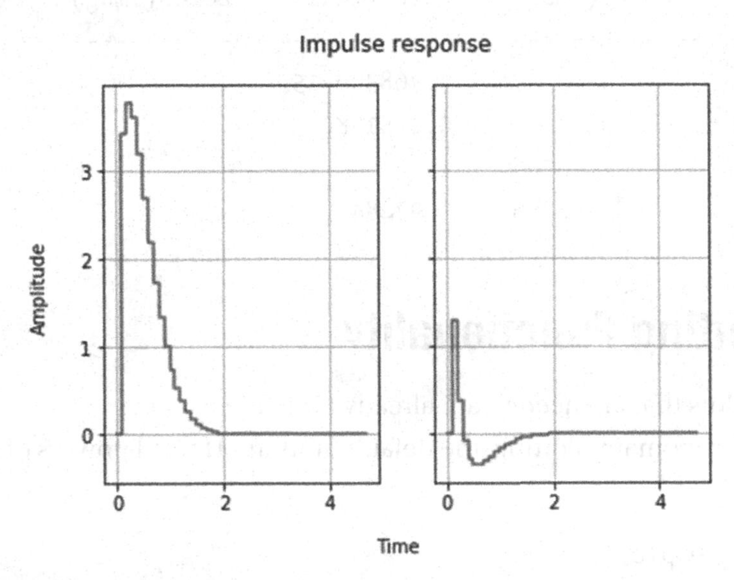

Figure 11-30. *Impulse Response (2)*

```
The plot units can be changed via the keywords available.
bode_plot(G4)
```

(See Figure 11-31.)

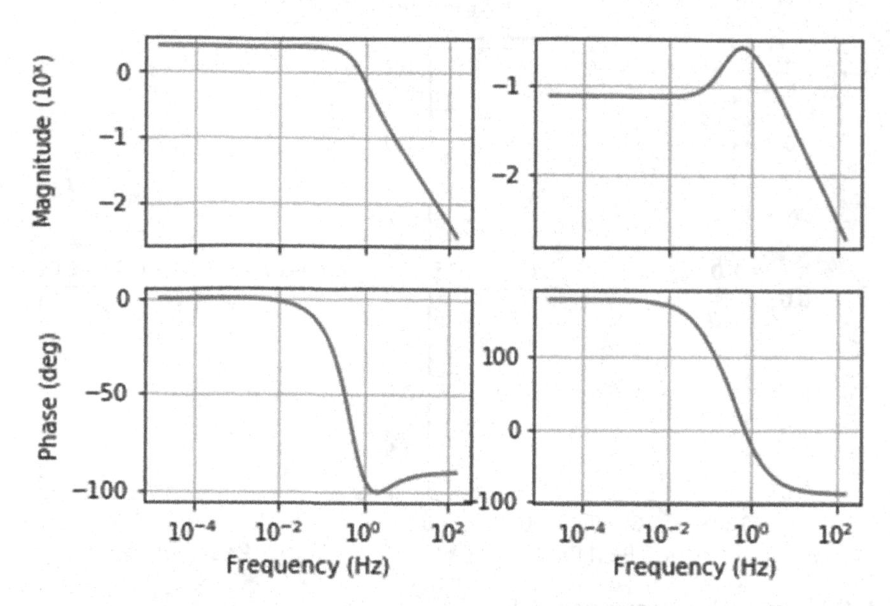

***Figure 11-31.** Impulse Response (3)*

```
bode_plot(G4, use_db=True, use_hz=False, use_degree=False)
```

(See Figure 11-32.)

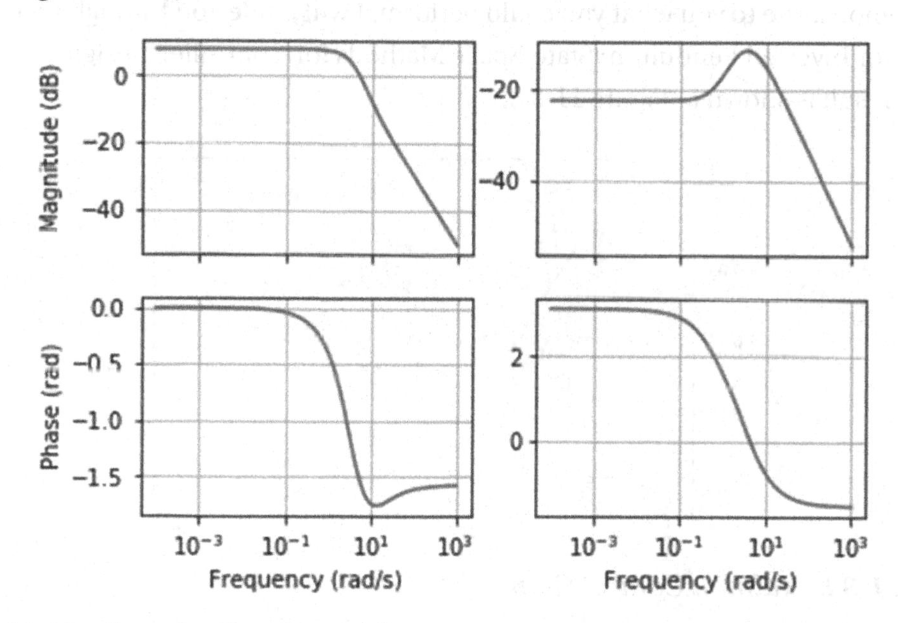

***Figure 11-32.** Impulse Response (4)*

```
nyquist_plot(G2);
```

(See Figure 11-33.)

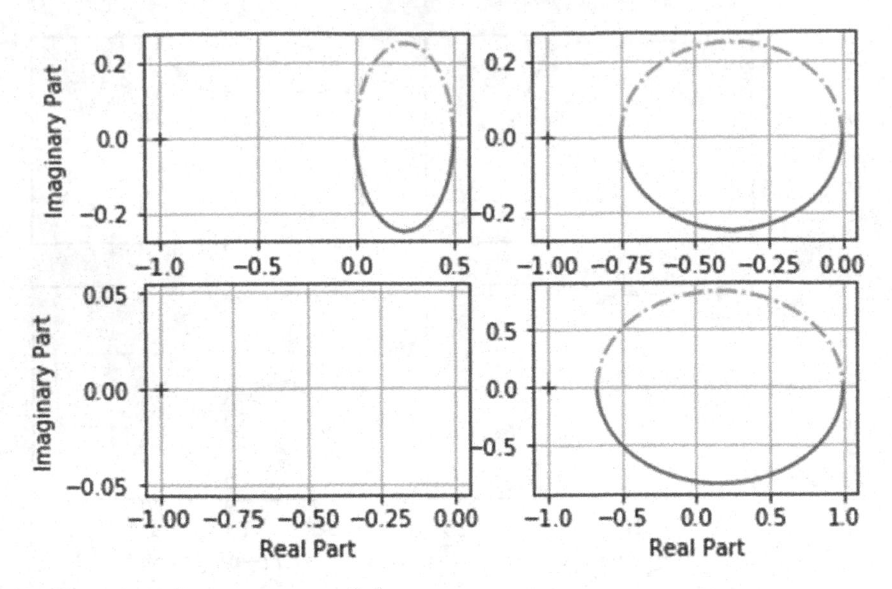

Figure 11-33. *Impulse Response (5)*

An LQR Example

Just to demonstrate to you what you could perform, I will guide you through an LQR example for Inverted Pendulum: State-Space Methods for Controller Design.

Your result is shown in Figure 11-34.

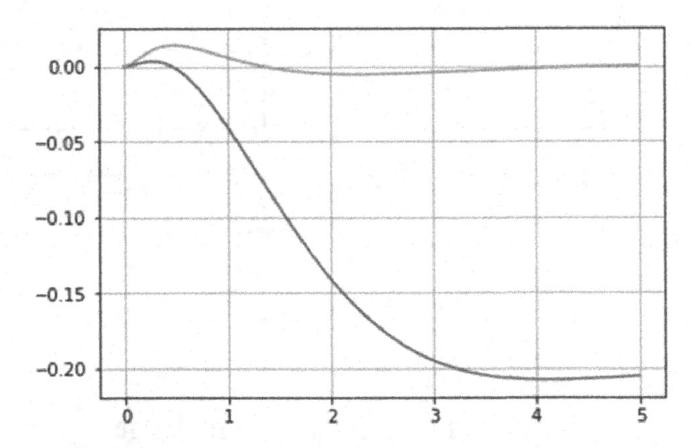

Figure 11-34. *Kalman Control Gain*

Now, you increase the weights on the position states.

Your results are shown in Figure 11-35.

New Controller gains: [[-70.71067812 -37.83445399 105.52978192 20.92384375]]

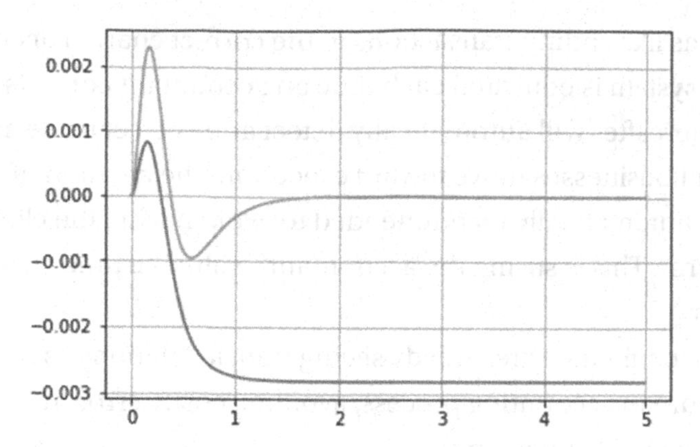

***Figure 11-35.** New Controller Gain*

Modern Control Theory

Modern control theory is carried out in the state space and can deal with multiple-input and multiple-output (MIMO) systems.

Resilient control systems extend the customary focus of addressing only scheduled instabilities to frameworks and attempt to address numerous types of unanticipated disruption; into specific, adapting, and metamorphosing activities of the control system in response to malicious actors, uncharacteristic failure modes, and undesirable human acts.

Evolutionary computation is a family of algorithms for global optimization inspired by genetic evolution, and the subfield of AI and computing learning on these systems. They are a family of population-based trial-and-error problem solvers with a metaheuristic or stochastic optimization character.

Active Disruptor

The evolutionary nature of modern-day machine learning is silently but highly effectively disrupting specific unwanted human activities.

Accounting Services

In cloud accounting platforms, machine learning systems are already working in the background. Accounting technology companies are applying machine learning principles to defeat the challenge many that non-accounting-savvy small businesses owners face in areas like coding transactions to the correct chart of accounts codes. The machine learning system is educated each time an accountant corrects errors in their clients' files and thereafter will automatically detect and correct these accounting errors.

When the small businesses move toward execution, the creation of their next invoice, account codes are automatically recommended to avoid having the client mistakenly create the same error. This is saving the accountant's valuable time cleaning up these undesirable errors.

Many jobs in accountancy are already seeing transformations as machine learning is overruling the human accounting process, avoiding costly error in reporting, double payments, or unacceptable invoicing.

AI-based systems are reducing the hours used by lawyers and loan officers for assessments of agreements in the finance industry. Major companies have installed natural language processing software to screen through tens of thousands commercial loan agreements, decreasing the time needed to sort out these agreements from hundreds of thousands of hours to seconds.

I have seen start-ups demanding that machine learning deliver innovative ways to battle cybercrime and malicious hackers as they no longer hire people to keep them safe from these attacks. This is creating a domain where small, highly niched players can compete with bigger companies on equal footing by simply using these AI systems. The other advantage of these systems is that the distributed nature of these AI means that once the AI works out how to disrupt the unwanted attack, it immediately can spread this knowledge to all the attached customers.

User-Based Insurance

Drivers shopping for car insurance with Progressive can opt to install a company-provided telematics device that measures how much risk for an accident they pose, based on how often they speed and engage in harsh braking (a sign of distraction or tailgating). Safe drivers are offered premium discounts.

Accident Reduction

Fleets are beginning to turn an ML-assisted way of identifying high-risk drivers called predictive or prescriptive analytics. These algorithms use records of driver behavior and various demographic and industry data to predict the probability that they'll be involved in a collision over the following 12 months. Ones those that pose the greatest risk are identified, and fleets can remediate their poor driving behavior with additional training and coaching.

Real-time repair estimates are close to becoming a reality. Many insurance carriers are experimenting with ML-based technologies to help drivers who've had an accident receive real-time repair estimates by taking photos of their vehicle damage with a smartphone camera. The app is being built based on thousands of images of damaged vehicles and their repair costs.

Predictive Maintenance

Based on historical data, programs have been developed that can predict when auto parts or systems are likely to fail and recommend replacements before those failures occur.

An AI mortgage chatbot that takes borrowers from application to closing is now performing the work done by 200+ call center people.

AI-enabled wearable devices can help health professionals monitor patients either on the hospital ward or at home. The wearable device captures oxygen saturation, respiration rate, pulse rate, temperature, movement, and posture. The machine learning can also interact with other devices in the environment: BP cuff, weighing scales, and a glucose monitor via a needle-free smart patch enables 24/7 monitoring of patients.

A mathematical formula developed with the help of AI had correctly determined the correct dose of immunosuppressant drugs to administer to organ patients.

The health-care industry manages costly back-office problems and inefficiencies. Activities that have nothing to do with patient care consume over half (51%) of a nurse's workload and nearly a fifth (16%) of physician activities. AI-based technologies, such as voice-to-text transcription, can improve administrative workflows and eliminate time-consuming non-patient-care activities, such as writing chart notes, filling prescriptions, and ordering tests. We estimate that these applications could save the industry $18 billion annually.

There is an AI algorithm proven against 21 dermatologists on its superior ability to identify skin cancers.

Poor data management however remains a key obstacle to the clinical application of AI. Much heavy lifting remains to be done to improve the data on which the future of AI relies. We need to bust data out of silos, so it can be easily accessed, queried, and analyzed. Using AI to correctly identify, categorize, and share information will lay the groundwork for future, breakthrough analyses.

3D Printing

The new techniques in 3D printing are creating major advances in the ability to produce custom machines and tools to assist the industrial machine learning to disrupt the world around itself with creating its own custom tools.

It is also disrupting the traditional manufacturing processes used by industry.

A prototype can now be created within hours and tested within days to enable the feedback to the 3D printing process to improve the tool to fit its purpose better for a more effective and efficient solution.

The introduction of 3D printing into machine learning solutions enables a machine learning algorithm to empower the human-in-the-loop to change the ecosystem with augmentations via robotics created by prediction from machine learning algorithms.

You could, in principle, manufacture any commercial item on a just-in-time basis that will result in a near to zero stock and investment in the product before it gets sold. Your machine learning could, in parallel, advise consumers to buy products that are only scheduled for release once they are paid for via the system.

This type of business model is now enabling small businesses to complete with the help of industrialized machine learning with traditional leaders in the field they are now disrupting.

Robotics

AI-assisted robotic techniques created by industrialized machine learning with robotics resulted in a five-fold reduction in surgical complications compared to when surgeons operated alone. When applied properly to orthopedic surgery, our analysis found AI-assisted robotic surgery could also generate a 21% reduction in a patient's length of stay in the hospital following surgery, as a result of fewer complications and errors, and it create $40 billion in annual savings.

I will discuss this area in detail further along in this book.

Practice Mathematics

I want to cover the following practical mathematics with you as I have found several business solutions where the simplest formula can change the speed that a machine learning algorithm performs.

You have already used a great library called SymPy (`www.sympy.org`) that we used at the end of Chapter 4.

The core reason is that SymPy is a Python library for symbolic mathematics. Its purposes are to be a full-featured computer algebra system (CAS). That means it will perform mathematics if you program the processes correctly.

I personally find this CAS system as a great source of simplifications of mathematics and formulas. It also creates the ecosystem in that you can proof the formula is mathematically and statistically sound.

Load the Jupyter notebook from example directory Chapter 11 called: Chapter-011-24-Solve-Math-01.ipynb

The notebook solves the following equation.

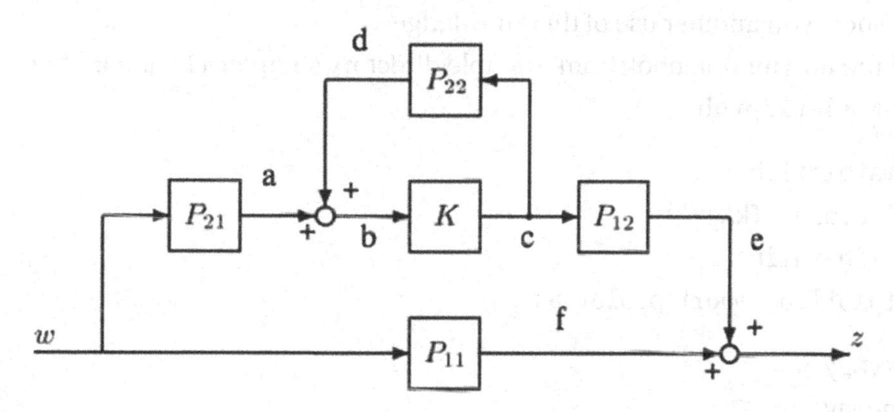

Here is the basic process:

$$
\begin{aligned}
b &= a+P_{22}c && (1)\\
c &= Kb && (2)\\
c &= Ka + KP_{22}c && (3)\\
c - KP_{22}c &= Ka && (4)\\
(I - KP_{22})c &= Ka && (5)\\
(I - KP_{22})^{-1}(I - KP_{22})c &= (I - KP_{22})^{-1}Ka && (6)\\
c &= (I - KP_{22})^{-1}Ka && (7)\\
&= K(I - P_{22}K)^{-1}a\text{~(push-through)} && (8)\\
z &= f+e && (9)\\
&= P_{21}w + P_{12}c && (10)\\
&= (P_{21} + P_{12}\underbrace{K(I - P_{22}K)^{-1}P_{12})w}_{c} && (11)
\end{aligned}
$$

I will show you another use of this knowledge:

Load the Jupyter notebook from example directory Chapter 11 called: Chapter-011-25-Solve-Math-02.ipynb

```
import matplotlib
matplotlib.use('TkAgg')
%matplotlib inline
from matplotlib import pyplot as plt

import sympy
import numpy
import mpmath
sympy.init_printing()

x = sympy.var('x')

expression = sympy.expand((x + 1)*(x + 2)*(x + 3))

print(sympy.solve(expression))
[-3, -2, -1]
s, t = sympy.var('s, t')
```

```
G = 1/(s + 1)**2
u = 1/s

gt = sympy.inverse_laplace_transform(G*u, s, t)
print(gt)
gt

-(t - exp(t) + 1)*exp(-t)*Heaviside(t)
```

$$-\left(t-e^{t}+1\right)e^{-t}\theta(t)$$

```
sympy.plot(gt, (t, 0, 10))
plt.show() (Figure 11-36)
```

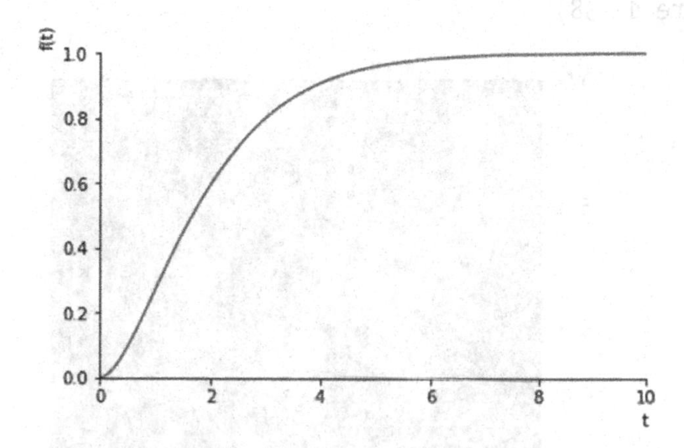

Figure 11-36. *Inverse Laplace Transform*

```
Gfunc = sympy.lambdify(s, G)
omega = numpy.logspace(-1, 2)
Gomega = Gfunc(1j*omega)

plt.loglog(omega, abs(Gomega))
plt.show() (Figure 11-37)
```

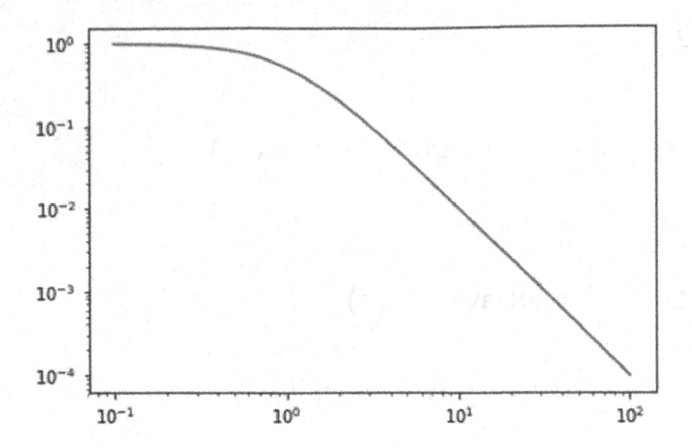

Figure 11-37. *G-Omega Plot*

```
mpmath.cplot(Gfunc, [-10, 10], [-10, 10], points=10000)
plt.show()(Figure 11-38)
```

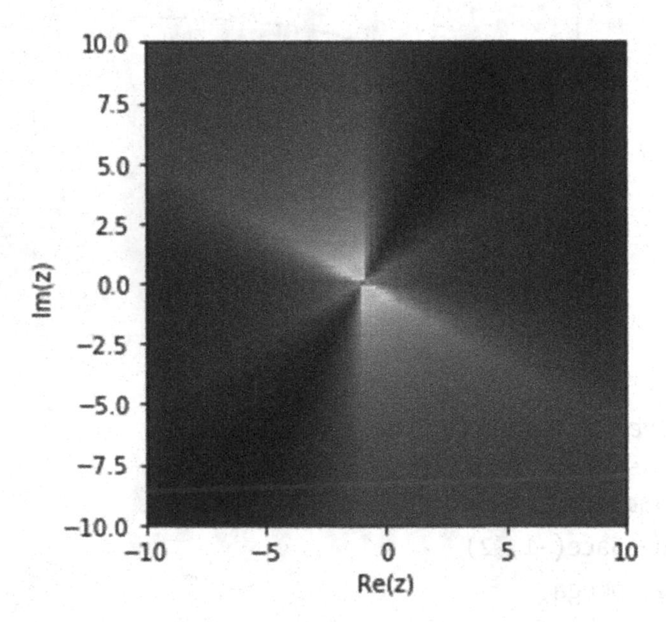

Figure 11-38. *G Plot (1)*

```
mpmath.cplot(sympy.lambdify(s, 1/(s**2 + s + 1)), [-5, 5], [-5, 5],
points=20000)
plt.show() (Figure 11-39)
```

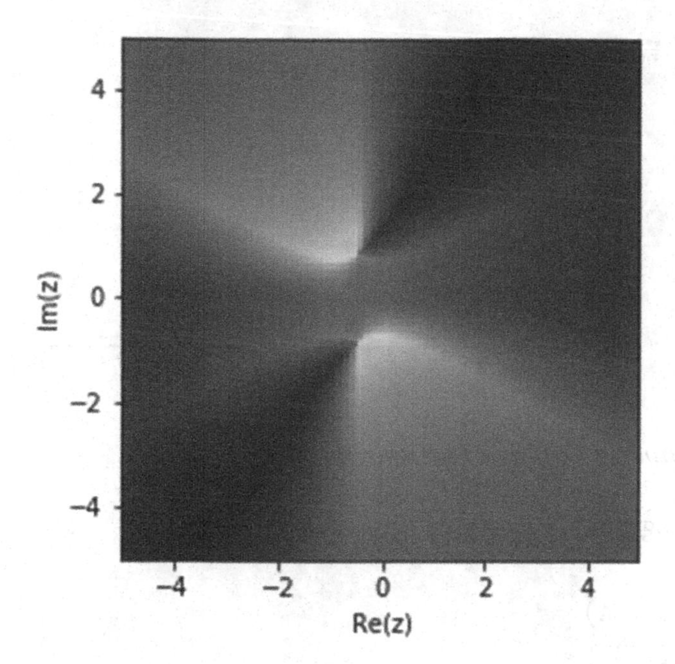

Figure 11-39. *G Plot (2)*

```
a = 2
s = 1 + a*1j
s = complex(1, 2)
```

```
print(numpy.exp(s))
```

```
(-1.1312043837568135+2.4717266720048188j)
```

```
print(numpy.exp(1)*complex(numpy.cos(2), numpy.sin(2)))
```

```
( 1.1312043837568135+2.4717266720048188j)
```

```
s = sympy.var('s')
t = sympy.var('t')
```

```
s, t = sympy.var('s t')
```

```
f = sympy.inverse_laplace_transform(1/(s**2 + s + 1), s, t)
```

```
sympy.plot(f, (t, 0, 10))
plt.show() (Figure 11-40)
```

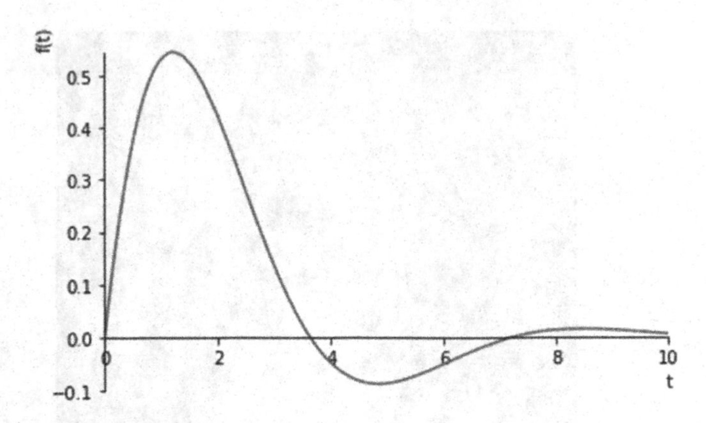

Figure 11-40. *Inverse Laplace Transform*

```
omega = numpy.logspace(-2, 2)
s = 1j*omega
Gw = 1/(s**2 + s + 1)
plt.loglog(abs(Gw))
plt.show() (Figure 11-41)
```

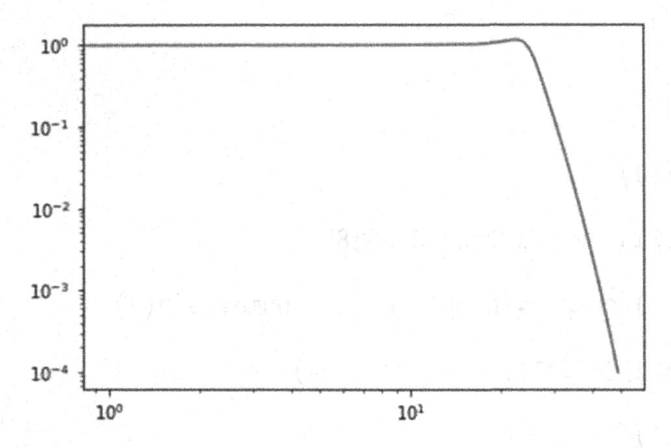

Figure 11-41. *Log Space Plot*

```
plt.semilogx(numpy.angle(Gw))
plt.show() (Figure 11-42)
```

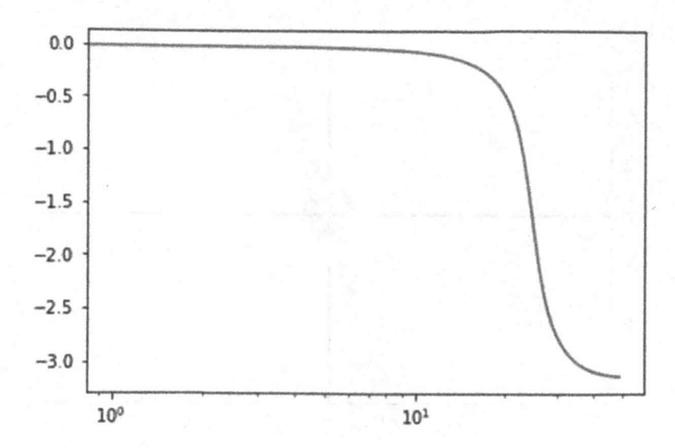

Figure 11-42. *Semi-Log Plot*

```
def plotvector(x, color='blue'):
    plt.plot([0, x[0,0]], [0, x[1,0]], color=color)

A = numpy.matrix([[1, 2], [4, 5]])

def vectdemo(theta):
    x = numpy.matrix([[numpy.cos(theta)],
                      [numpy.sin(theta)]])
    y = A*x
    #print(x)
    plotvector(x)
    plotvector(y, color='red')
    plt.axis([-10, 10, -10, 10])
    plt.axhline(0, color='black')
    plt.axvline(0, color='black')

for theta in numpy.linspace(0, 2*numpy.pi):
    vectdemo(theta);
plt.axis('equal');
plt.show() (Figure 11-43)
```

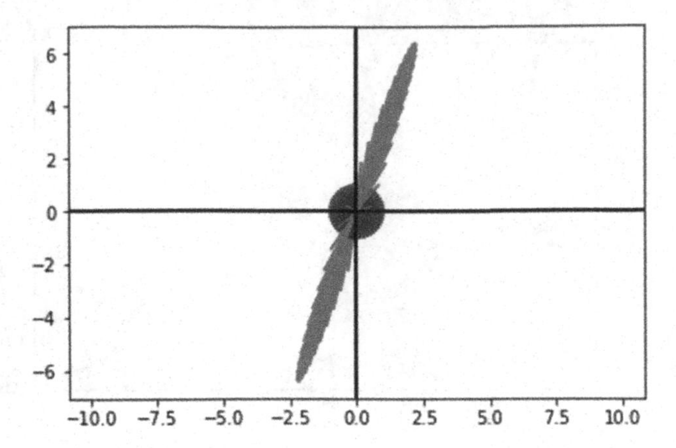

Figure 11-43. *Theta Plot*

I will guide you through a practical solution that processes sun spots to show how you apply the knowledge you have gained for the theory and practice in this chapter.

Load the Jupyter notebook from example directory Chapter 11 called: Chapter-011-26-Sunspot-01.ipynb

```
import matplotlib
matplotlib.use('TkAgg')
%matplotlib inline

import pandas as pd
import matplotlib.pyplot as plt
import numpy as np

def movingaverage(interval, window_size):
    window = np.ones(int(window_size))/float(window_size)
    return np.convolve(interval, window, 'same')

data = pd.read_csv('../../data/sunspots.csv', header=None)
#print(data.shape)
data.columns = ['year', 'spots']

x = data['year']
xm = x.min()
y = data['spots']

mn = int(y.min()) - 10
mx = int(y.max()) + 10
```

```
plt.plot(y, "k.")

xlabelstr="Months since " + str(xm)

y_av = movingaverage(y, 10)
plt.plot(y_av,"r")
plt.xlim(mn,mx)
plt.xlabel(xlabelstr)
plt.ylabel("No. of Sun spots")
plt.show() (Figure 11-44)
```

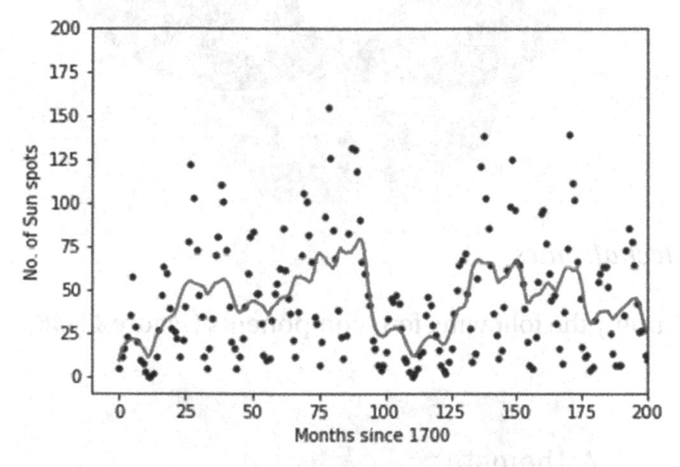

Figure 11-44. *Number of Sun Spots per Year*

What Should You Know?

Mechatronics

Understand that Mechatronics is a massive field (Figure 11-45).

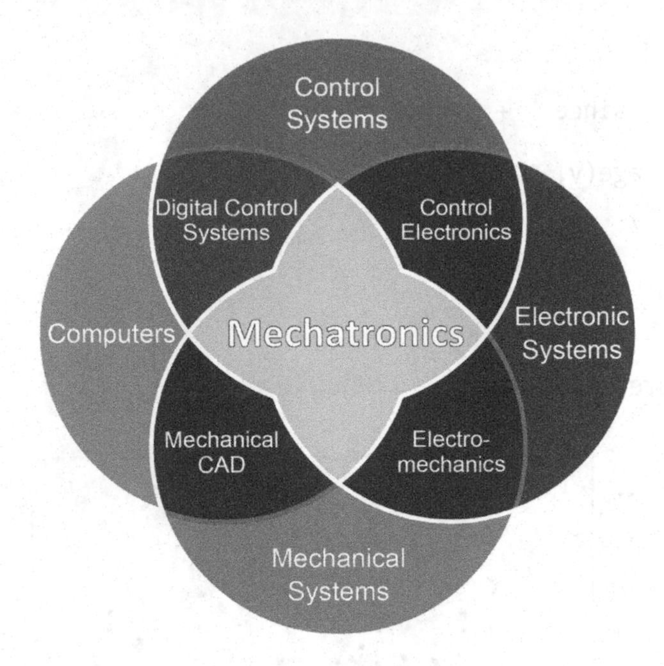

Figure 11-45. *Mechatronics*

I personally simplify the following four components (Figure 11-46).

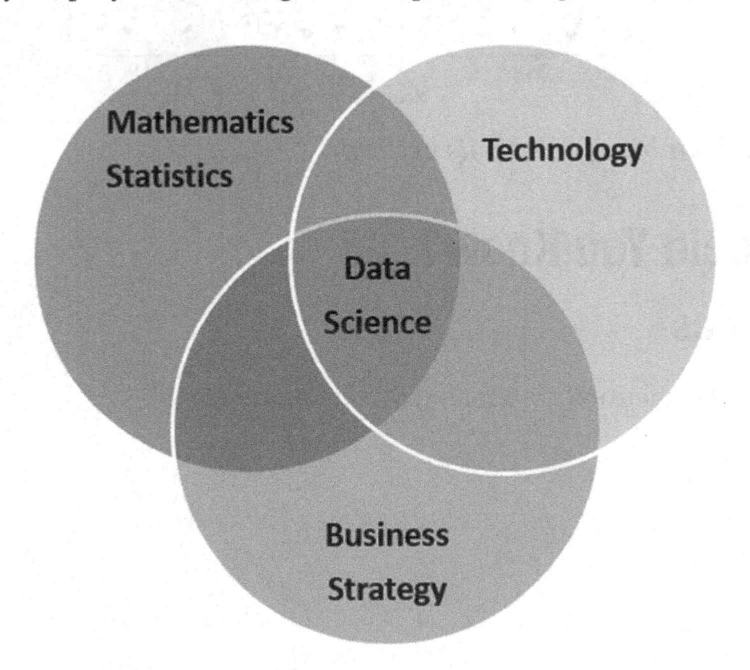

Figure 11-46. *Simplified components*

Data Science

This integrates the skills and knowledge you need to perform industrialized machine learning.

The overall skills you need is taken from these following three zones.

Mathematics and Statistics

You need the knowledge to calculate in your machine learning. This skill is a core requirement.

Technology

You need the ability to work with diverse technology from embedded machine learning in the Internet-of-Things devices to massive parallel processing in the cloud to prepare your machine learning. You need to understand how robots work and know which communication channels are needed.

Business Strategy

You need to understand the business of your customers. Without this insight, you are not performing industrialized machine learning. You need to sell your machine learning.

You need to make an impact!

What's Next?

Chapter 12 is all about Robotics. I will take all the knowledge you have acquired over the last eleven chapters and start to apply this knowledge to how to assist soft and hard robots to enforce industrialized machine learning onto the real world around us.

The next chapter will give you the ability to use machine learning as a disruptor of the world around you.

CHAPTER 12

Robotics Revolution

In this chapter you will learn the basic workings of a robot. And I will introduce you to a few commercial products that use robot toolkits to achieve industrialized machine learning (IML) via a direct action model using soft or hard robots.

Robots

A robot is a real or virtual machine that is equipped from the principal objective that it performs the work of a person, and that work is by design or controlled by a computer method.

General Machine Learning

General Machine Learning, also known as human-level Machine Learning or strong Machine Learning, is a type of Machine Learning that can understand and reason in its environment like a human would.

General Machine Learning has always been elusive.

I predict that the reservations that these robots could accomplish better performance than humans, will hold this research back until we create an ***Artificial Super Intelligence***.

Artificial Super Intelligence Machine Learning

When ***Artificial Super Intelligence*** type robots become significantly smarter than the best human brains in practically every field, including scientific creativity, general wisdom, and social skills, they will have achieved Artificial Super Intelligence Machine Learning.

By most experts' accounts, the gap between general and super artificial intelligence (AI) will be produced by a short jump in capability of the Machine Learning techniques. The resulting machine learning will then change the real world to improve its own learning curve to overtake humans.

A. F. Vermeulen, *Industrial Machine Learning*, https://doi.org/10.1007/978-1-4842-5316-8_12

Narrow Machine Learning

Narrow Machine Learning is the only form of Machine Learning that humanity has truly achieved so far. This is machine learning that is good at performing a single task, such as playing games, making purchase suggestions, sales predictions, and weather forecasts. Computer vision, natural language processing, is still at the current stage only at the narrow machine learning maturity level.

The Machine Learning that I will discuss in this chapter is only narrow. Therefore, I will only cover robots with a single purpose to perform single activities. There are two types of Machine Learning robots ... Soft and Hard.

Soft Robots

The concept of a soft robotics is the formation of an agent to perform a specific task on behalf of a human in a consistent and repeatable method.

Institutions are under pressure to be extra industrious and accomplish improved outcomes with increasingly scarce resources. Every day, personnel struggle to complete countless time-consuming professional processes that demand accuracy and speed. Soft Robots are a better way. Humans are error prone and responsibility for identical work over and over is a source of frustration among the workforce.

The blending of machine learning and robotics creates the perfect ecosystem to help humans achieve more with their time. You should use them only for the exception or the human-in-the loop moments in the process.

To illustrate what I organize daily, I will discuss a basic scanned email sorting process robot.

> You need to first load the following Jupyter Notebook from example directory Chapter 07: Chapter-012-01-Soft-Robot-01A.ipynb
>
> The processor creates a virtual mailing system on disk by creating simulation post offices.
>
> This notebook creates a virtual post office simulation that will act as a base for the next notebook.

Tip If you want, you can enlarge the simulation by increasing the parameters at the beginning of the notebook.

Currently:

```
postcodecnt=2
subpostcodecnt=3
```

That supports: 6 post offices with 36 mail items

Try:

```
postcodecnt=5
subpostcodecnt=5
```

That supports: 25 post offices with 625 mail items

I will now show you how to create and execute the soft robots that sorts the mail between the virtual post offices.

You need to second load the Jupyter notebook from example directory Chapter 07 called: Chapter-012-01-Soft-Robot-01B.ipynb

This notebook creates and executes six soft robots, one per post office, to sort the mail.

You have successfully built your first soft robots.

Well done; you can now perform simple processes. You can use the knowledge you gained in earlier chapters to generate Machine Learning methods that can use training sets to train the robot to predict what it should perform on any future data or actions it must take.

Industry Soft Robots

I use an industrial strength platform to perform the tasks I need by preparing machine learning solutions with supervised learning, unsupervised learning, or reinforcement learning algorithms. The results then get ported to a tool from UIPath (`www.uipath.com/`) (Figure 12-1).

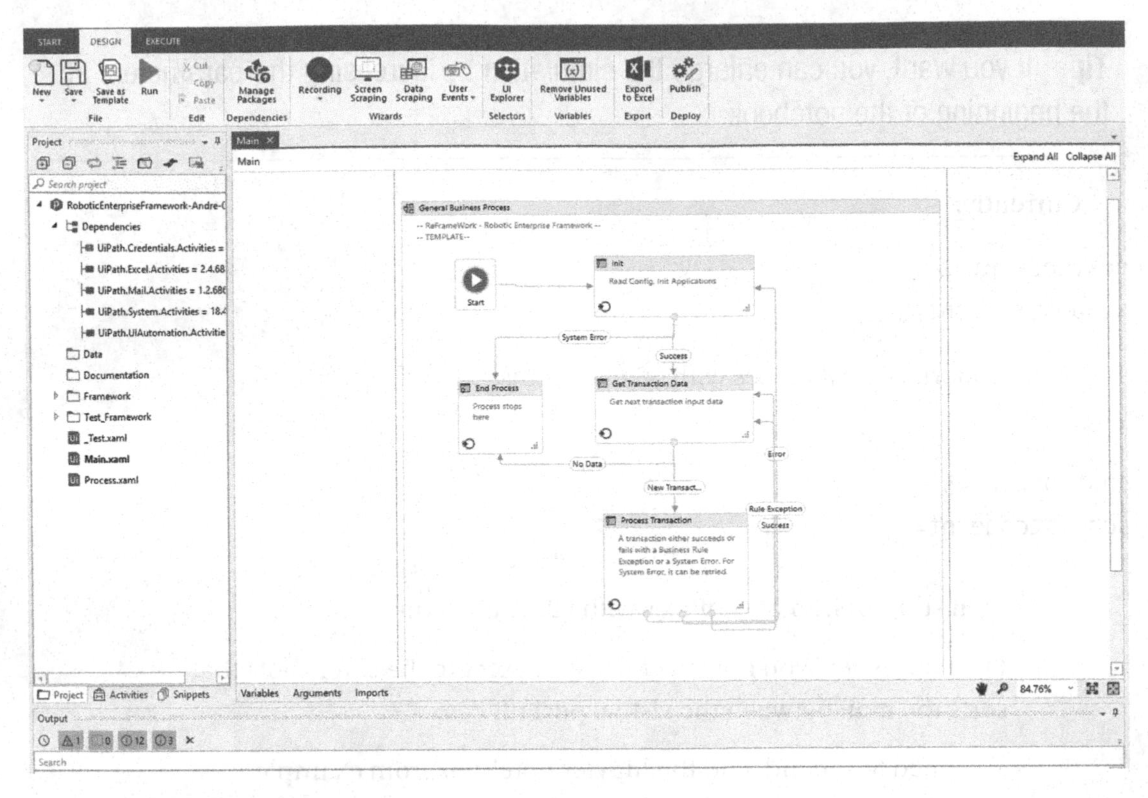

Figure 12-1. *Basic UIPath tool Layout*

This tool orchestrates an ecosystem with numerous soft robots that then perform work on my behalf.

My robots handle every aspect of my life that involves the repetitive application of previously identified procedures and methods to the matching incoming data.

Hard Robots

The latest developments are creating super-strength robots based around a precise ancient concept like origami. In addition, machine learning processes are evolving with the capability to interact with the real world. While the skeletons of these robots are made with traditional metals or plastics, the breakthrough comes from their being folded into a specific pattern. An origami-style internal structure gives the robot a phenomenal amount of strength - making it capable of lifting up to 1000 times its own body weight. Artificial nanofiber muscles power these skeletons to perform great work tasks.

Robots are evolving on a daily basis and they are here to stay.

I will now introduce a few basic mathematical concepts that will help with the understanding of robotics.

Basic Trigonometry

The movements of robots are driven by basic trigonometry mathematics (Figure 12-2).

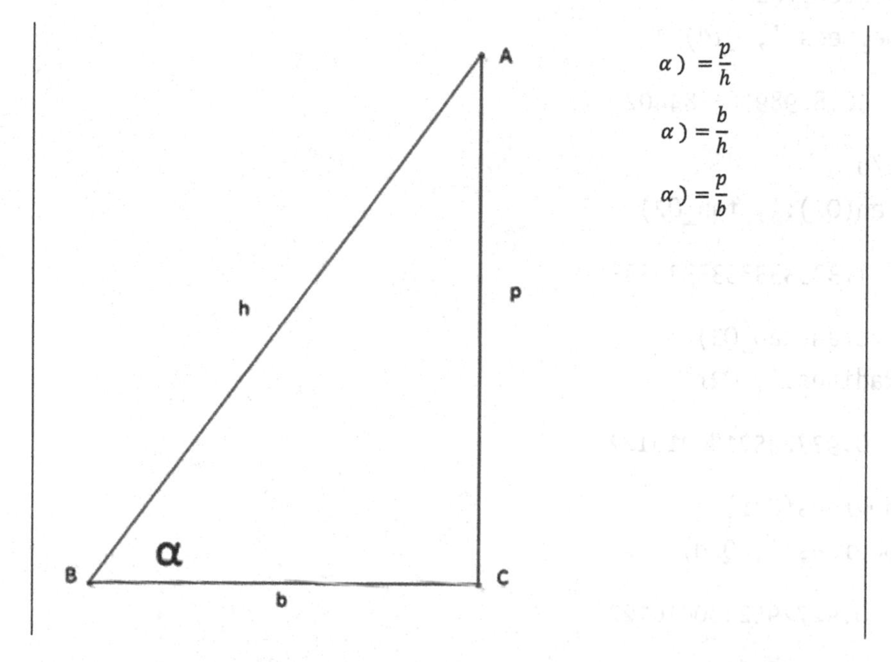

***Figure 12-2.** Basic Trigonometry*

Load the Jupyter Notebook from example directory Chapter 07 called: Chapter-012-03-Hard-Robot-01.ipynb

I will quickly introduce you to the basic mathematics against the trigonometry formulas:

```
import numpy as np
b=4
p=3
```

```
tan_Q1=p/b
print('Tan(Q):', tan_Q1)

Tan(Q): 0.75

Q1r=np.arctan(tan_Q1)
print('Radians:', Q1r)

Radians: 0.6435011087932844

Q1d=np.degrees(Q1r)
print('Degrees:', Q1d)

Degrees: 36.86989764584402

tan_Q2=b/p
print('Tan(Q2):', tan_Q2)

Tan(Q2): 1.3333333333333333

Q2r=np.arctan(tan_Q2)
print('Radians:', Q2r)

Radians: 0.9272952180016122

Q2d=np.degrees(Q2r)
print('Degrees:', Q2d)

Radians: 0.9272952180016122

Q2d=np.degrees(Q2r)
print('Degrees:', Q2d)

Degrees: 53.13010235415598

Q = Q1d + Q2d + 90
print('Test Sum Angles:', Q)

Test Sum Angles: 180.0

h = p / np.sin(Q1r)
print('Length h from length p and angle Q:', h)

Length h from length p and angle Q: 5.0
```

```
h= b / np.cos(Q1r)
print('Length h from length b and angle Q:', h)
```

```
Length h from length b and angle Q: 5.0
```

```
p = h * np.sin(Q1r)
print('Length p from length h and angle Q:', p)
```

```
Length p from length h and angle Q: 3.0
```

```
b = h * np.cos(Q1r)
print('Length b from length h and angle Q:', b)
```

```
Length b from length h and angle Q: 4.0
```

Now that you understand basic trigonometry mathematics, let's expand these concepts into controlling hard robots.

Rigid-Body robotics is the most common type of robotics in use.

Basic Robot

Figure 12-3 is an example of a basic robot.

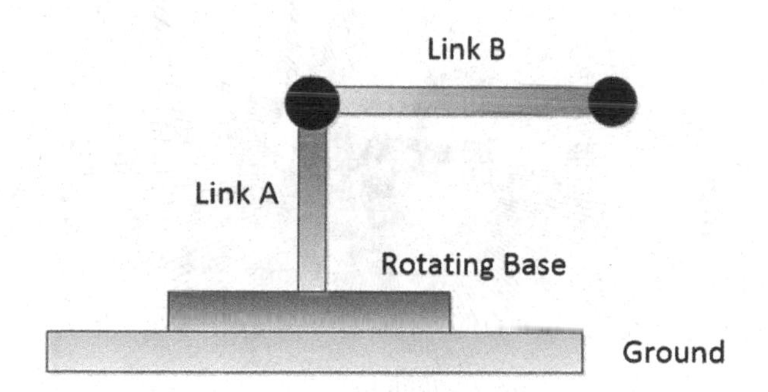

Figure 12-3. *Basic Robot*

Load the Jupyter Notebook from example directory Chapter 07 called: Chapter-012-04-Hard-Robot-1DOF-01A.ipynb

Figure 12-4 shows a basic robot grid.

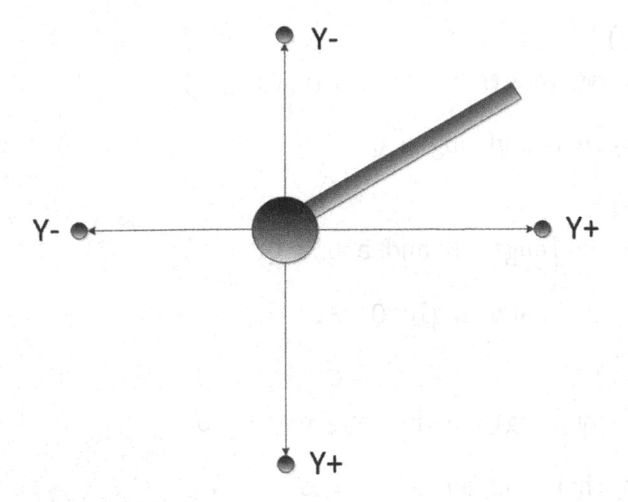

Figure 12-4. *Basic Robot Grid*

Your results are shown (Figure 12-5).

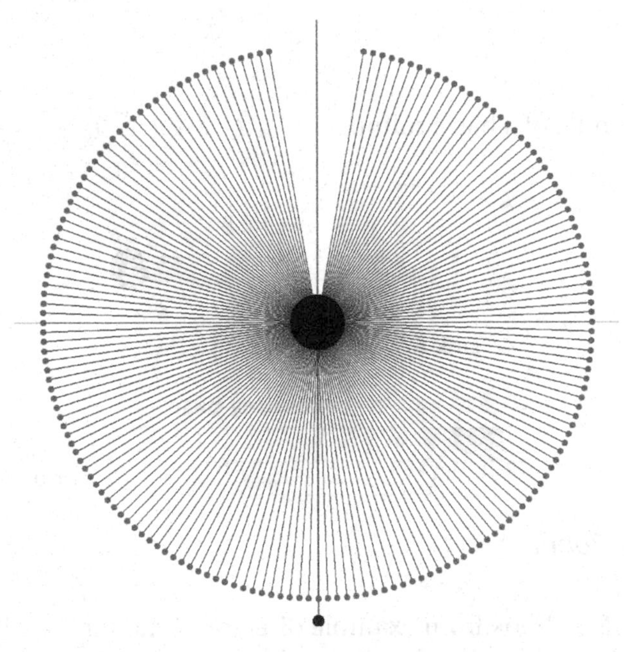

Figure 12-5. *Basic Robot Reach Diagram (170 degrees)*

This diagram shows clearly what the reach of the robot is using the given parameters.

Load the Jupyter Notebook from example directory Chapter 07 called: Chapter-012-04-Hard-Robot-1DOF-01B.ipynb

The only difference with the previous example is the change from 170 to 45 degrees of reach (Figure 12-6). The reach profile of the robot has been reduced as the robot has lost some of its ability to move more freely in the ecosystem.

Tip Always investigate the reach profile of your robot as this will save you hours of refitting once you have installed the robot if you get this phase wrong!

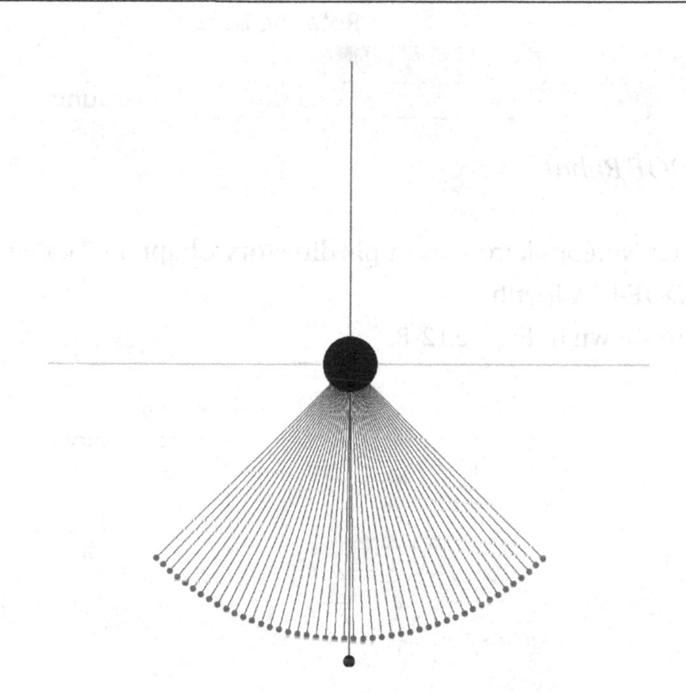

Figure 12-6. *Basic Robot Reach Diagram (45 degrees)*

Figure 12-7 shows a simple 2 DOF robot with rotating base that is the most common robot currently used.

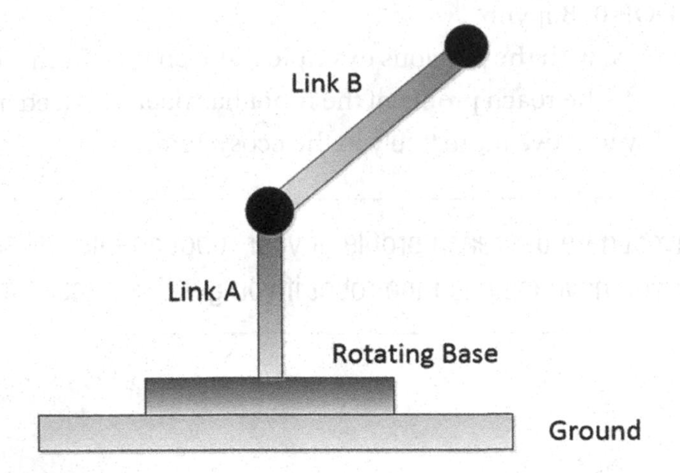

Figure 12-7. *2 DOF Robot*

Load the Jupyter Notebook from example directory Chapter 07 called: Chapter-012-05-Hard-Robot-2DOF-01A.ipynb

Your results are shown in Figure 12-8.

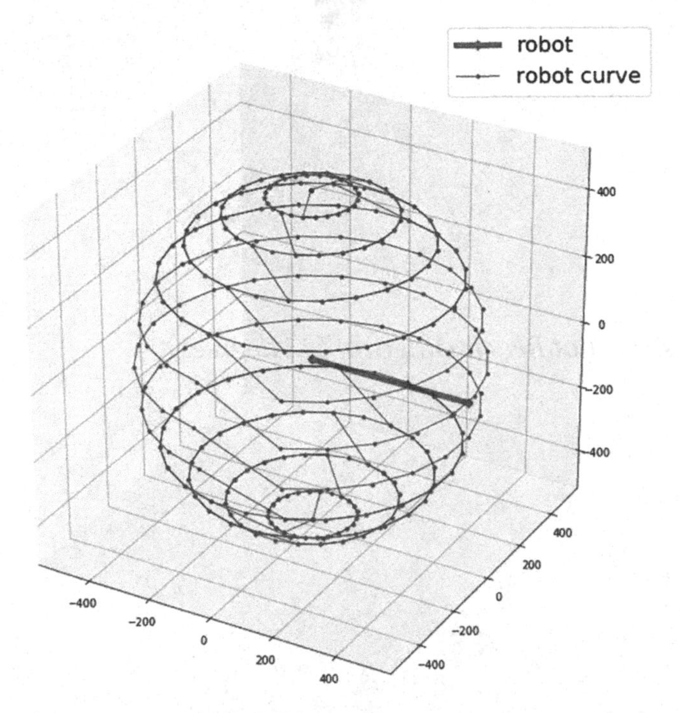

Figure 12-8. *2 DOF Robot Reach Diagram*

Load the Jupyter Notebook from example directory Chapter 07 called: Chapter-012-05-Hard-Robot-2DOF-01B.ipynb

Once again, the reach profile reduces! (See Figure 12-9.)

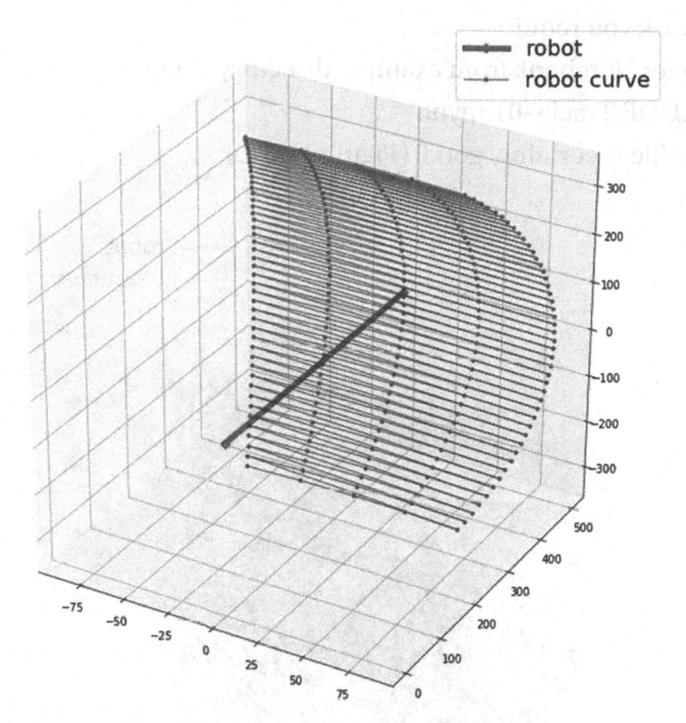

Figure 12-9. *Limited 2 DOF Robot Reach Diagram*

Path of Robot

The path of a robot is the route or area that is reachable by the robot. A stationary robot typically will only have a spherical reach path, but a mobile robot could have a better path as it can move the spherical reach capability closer to objects that are outside its current reach. I will show you how to simulate this capability to move the spherical reach across a linear path to increase the reach of the robot.

Load the Jupyter Notebook from example directory Chapter 07 called: Chapter-012-06-Hard-Robot-2DOF-Path-01.ipynb

This notebook also demonstrates where the reach profile includes a path. These robots can move their reach profile along a given fixed path. This makes the robot more effective and efficient to adapt to the ecosystem variances, but note the path still limits the robot's total freedom to move anywhere.

Robot with Tracks

The addition of tracks to a robot opens the world to a robot as its reach now becomes near unlimited. I will show you how to simulate that capability to investigate if your robot will be fit for the task you require.

Load the Jupyter Notebook from example directory Chapter 07 called: Chapter-012-07-Hard-Robot-2DOF-Tracks-01.ipynb

This reach profile is certainly good! (Figure 12-10.)

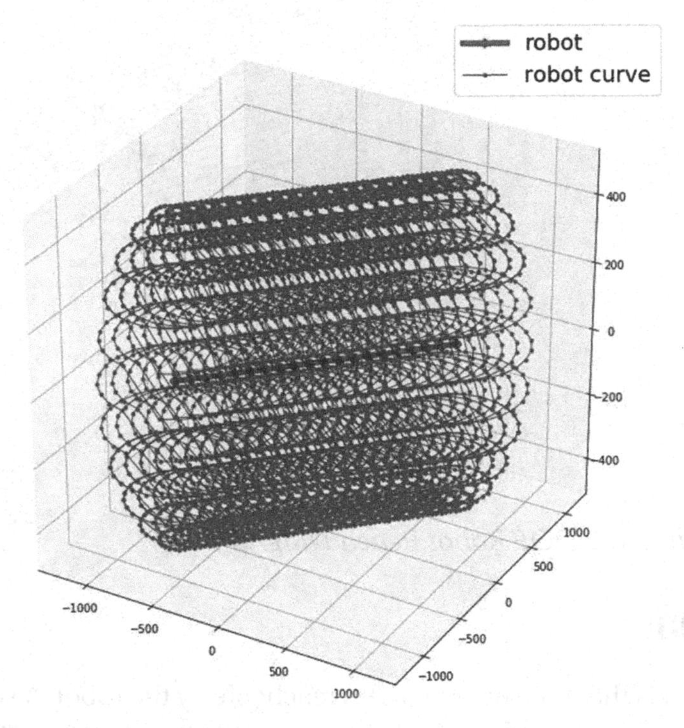

Figure 12-10. *2 DOF Robot Reach Diagram with linear movement*

As you now can calculate the basic reach profile of a robot arm, I suggest we investigate more complex robot configurations.

Anatomy of a Hard Robot

The anatomy of each robot is unique, and this creates an environment that is unique to each robot. It is important that you compile a detailed understanding of the capabilities of each of your robots.

The changes to soft robots are mostly very easy as they can be changed and redeployed within a short development cycle.

Hard robotics, however, have limitations as a physical robot or set of robots are used that does not easily change their shape or performance characteristics. See Figure 12-11.

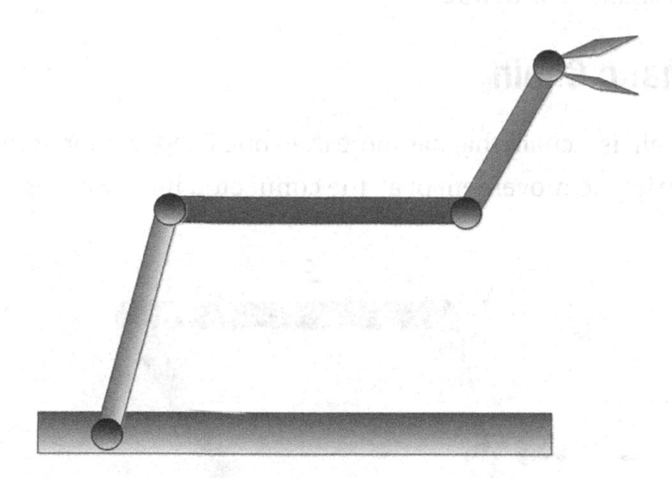

Figure 12-11. *Anatomy of a Hard Robot*

The field of Kinematics is the tool that helps you understand the physical environment and how you interact with it using your IML.

Kinematics

Kinematics is a branch of established mechanics that defines the motion of points, bodies (objects), and systems of bodies (groups of objects) without making an allowance for the forces that caused the motion.

The concept of reach of a robot is a basic but essential requirement for the successful deployment of machine learning in the real world. A simple capability of a robot having enough reach to perform a task enables the machine learning result to be implemented with success. The reach is accomplished through the preapproved path formed by the kinematic chains designed by the components of the robot.

Kinematic Chains

A kinematic chain is a known fixed amount of links either joined together or arranged in a manner that allows each other to move relative to one another along pre-agreed pathways. If the links are connected in a specific way so that no motion is possible, it results in a locked chain or structure.

Closed Kinematic Chain

A closed kinetic chain is a chain that has more than one fixed anchor point. These multiple anchor points restrict the movement of all the connected links. See Figure 12-12.

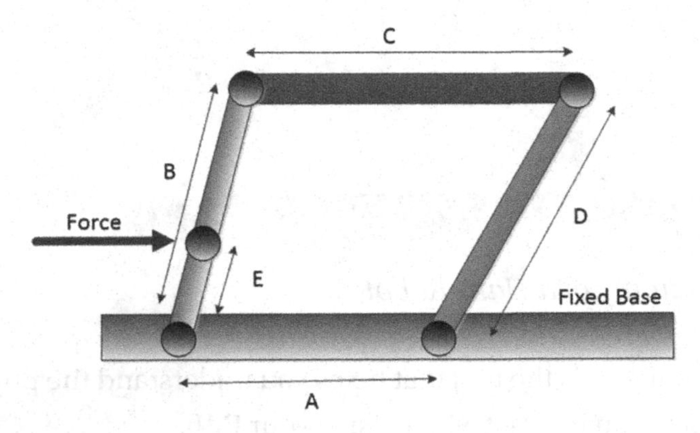

Figure 12-12. *Closed Kinematic Chain*

The application of the force will result in the complete chain responding to a fix preconfigured path.

This is useful when you want to ensure the preapproved movement is guaranteed.

Locked Chain or Structure

A locked chain or structure is a Kinetic Chain with only one possible position. See Figure 12-13.

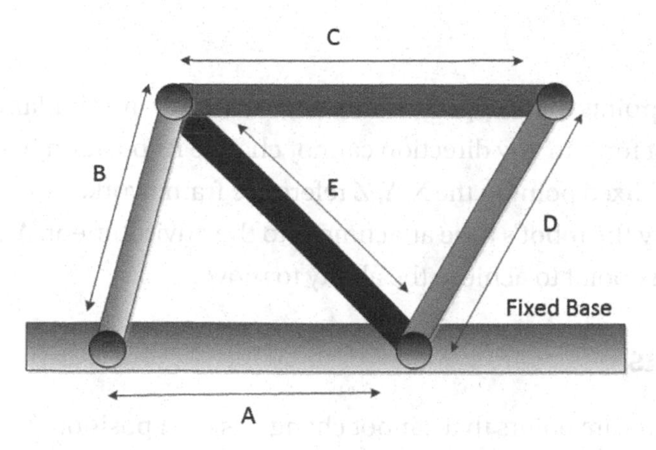

Figure 12-13. Locked chain

The structure is useful where you want to have a guaranteed point of reverence for the rest of the solution.

I have found that the creation from machine learning models via 3D printing ensures that your model's solution is turned into a fixed solution.

Open Kinematic Chain

The open kinematic chain is common in the robot's environment as most robots are a form of open kinetic chain. See Figure 12-14.

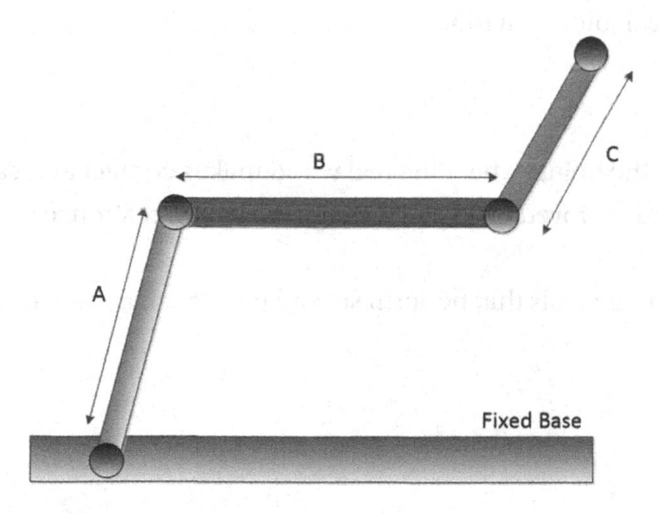

Figure 12-14. Open Kinematic Chain

The open kinematic chain typically consists of the following.

Fix Points

The fix points are points that support no movement as they are similar to a structure.

Any amount of force in any direction cannot change its position in the real world. It has a preapproved fixed point in the X, Y, Z reference framework.

This is typically the robot's base attachment to the environment. Any robot has to have at least one fix point to achieve the ability to move.

Fix Swivel Points

The fix swivel points are points that cannot change its own position from a preapproved fixed point in the X, Y, Z reference framework.

It can enable any link attached to it to move through a preapproved X, Y, Z path through the reference framework.

These are typical joints in a robot.

Non-fix Swivel Points

The non-fix swivel points are points that can change their own position along a preapproved fixed path in the X, Y, Z reference framework.

They can enable any link attached to them to move through a preapproved fixed X, Y, Z path through the reference framework.

These are typical joints in a robot.

Endpoints

The endpoints are the points where the real world makes contact and can change its own position along a preapproved fixed path though the X, Y, Z reference framework. See Figure 12-15.

These are normally tools that perform several tasks such as picking and placing objects.

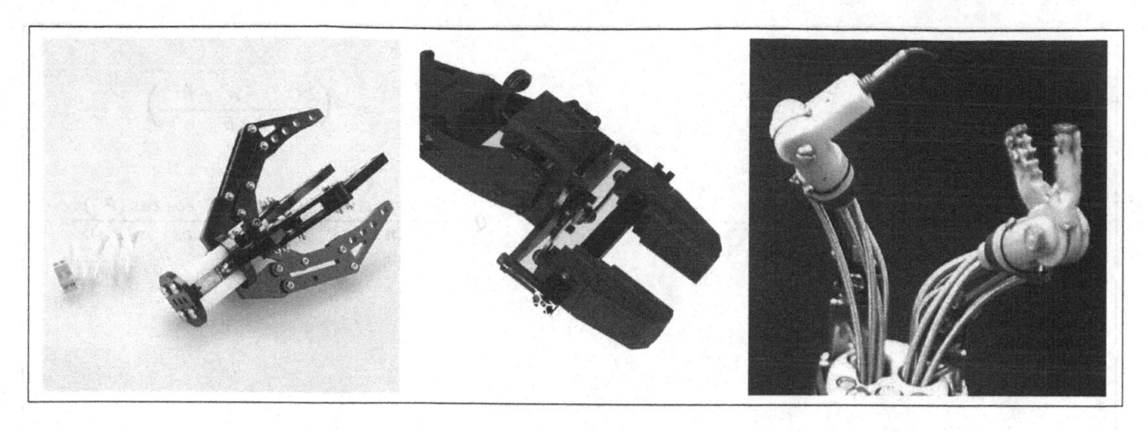

Figure 12-15. Endpoints

These endpoints can be humble pinchers but frequently are highly sophisticated tools that need complex machine learning to achieve effective and efficient results.

Degrees of Freedom (DOF)

The degree of freedom of a kinematic chain is computed from the number of links and the number and type of joints using the mobility formula of the chain.

Inverse Kinematics

The use of inverse kinematics enables you to calculate the settings of your robot to achieve a given goal or position in the real-world grid system. See Figure 12-16.

I will start with a simple two-arm robot to explain the concept.

Load the Jupyter Notebook from example directory Chapter 07 called: Chapter-012-08-Inverse-kinematics -2DOF-01.ipynb

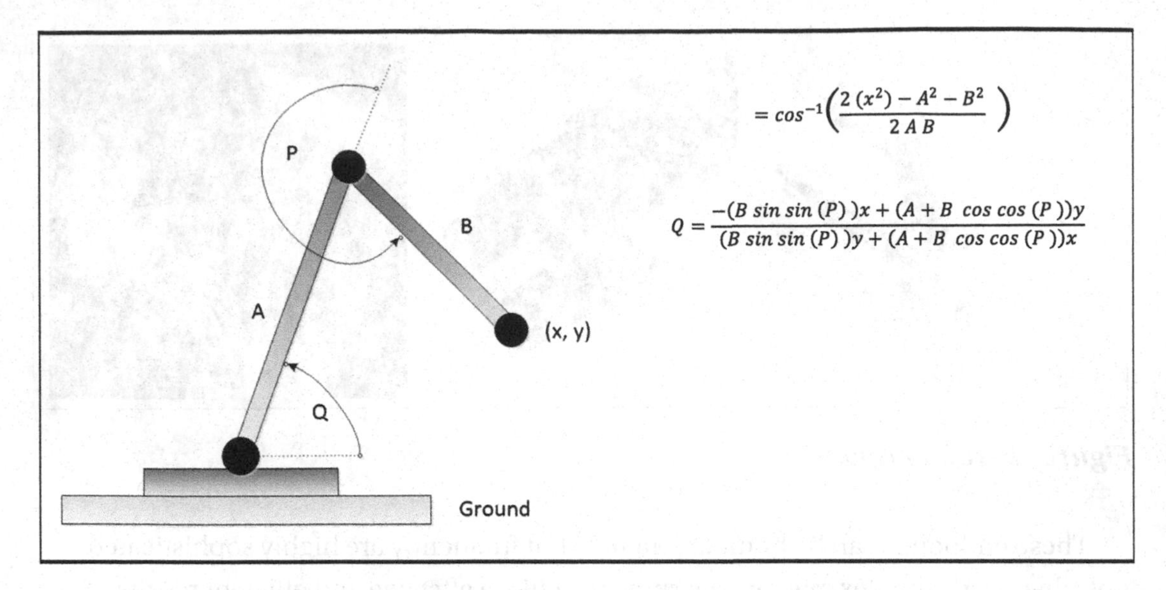

Figure 12-16. *Inverse Kinematics*

This notebook shows you how to calculate the settings needed to reach a given point in the reach profile (Figure 12-17).

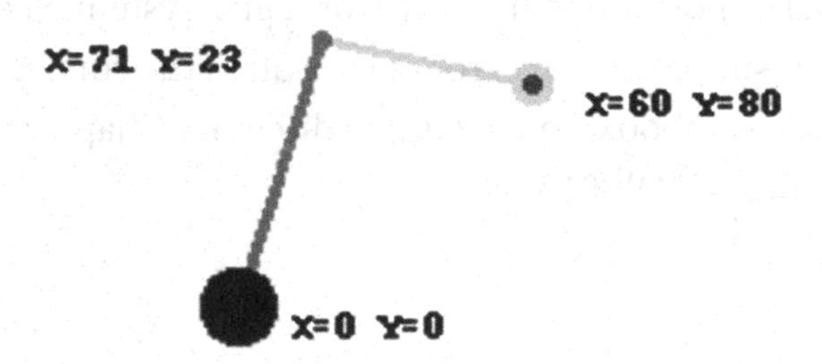

Figure 12-17. *Inverse Kinematics Plot (1)*

Load the Jupyter Notebook from example directory Chapter 07 called: Chapter-012-09-Inverse-kinematics -2DOF-02.ipynb

```
x=float(130)
```

See Figure 12-18.

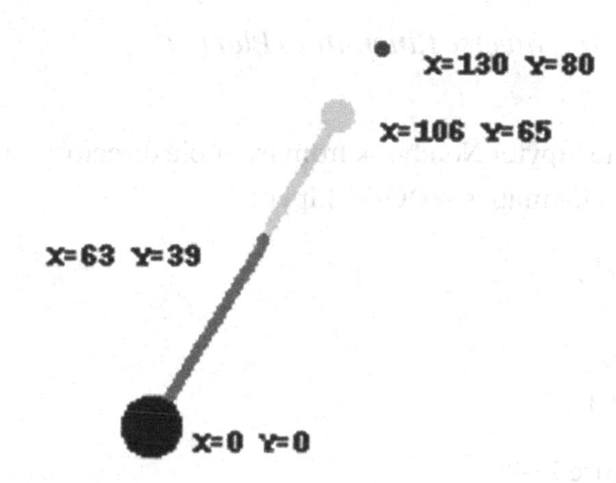

Figure 12-18. *Inverse Kinematics Plot (2)*

Load the Jupyter Notebook from example directory Chapter 07 called: Chapter-012-10-Inverse-kinematics -2DOF-03.ipynb

```
x=float(60)
y=float(20)
```

See Figure 12-19.

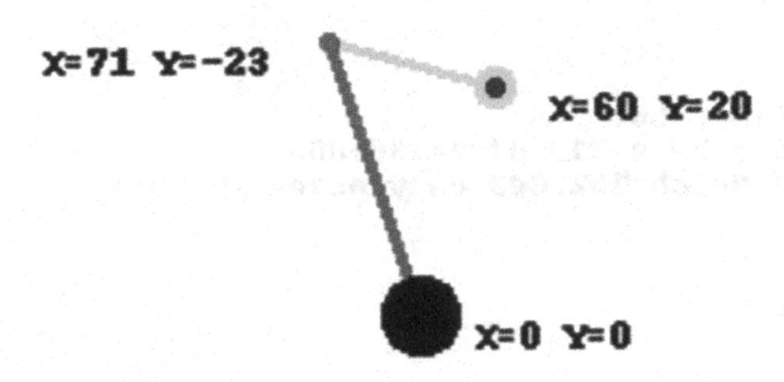

Figure 12-19. *Inverse Kinematics Plot (3)*

Load the Jupyter Notebook from example directory Chapter 07 called: Chapter-012-11-Inverse-kinematics -2DOF-04.ipynb

```
x=float(-20)
y=float(50)
A=float(50)
B=float(50)
```

See Figure 12-20.

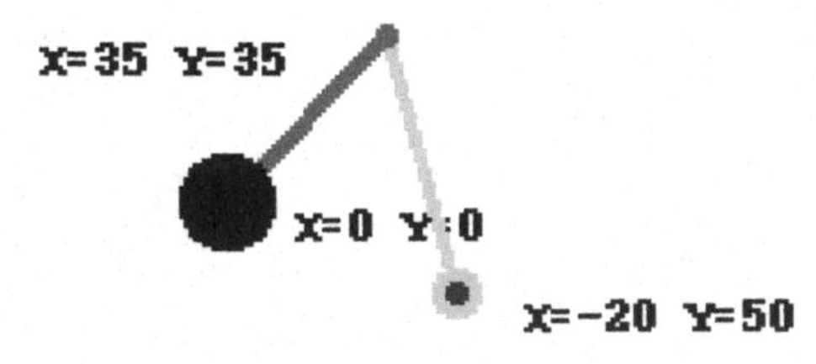

Figure 12-20. *Inverse Kinematics Plot (4)*

Differential Kinematics

See Figure 12-21 for an example of differential kinematics.

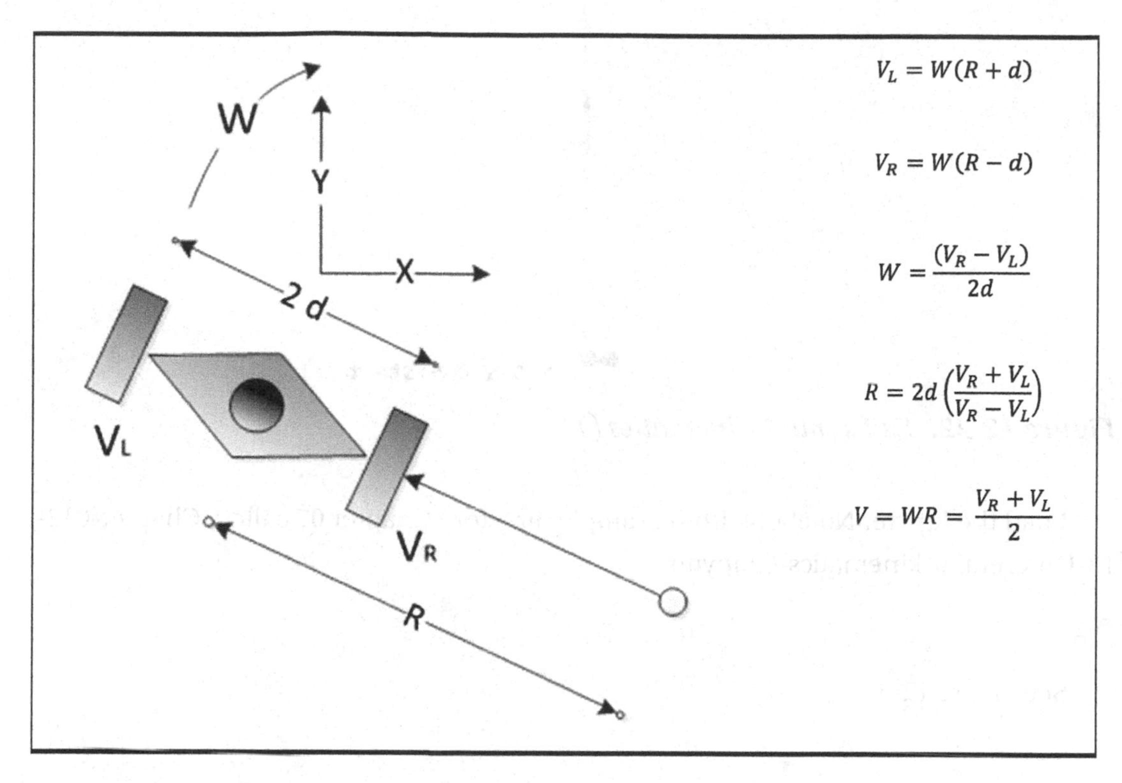

W
Y
X
2 d
V_L
V_R
R
$V_L = W(R + d)$
$V_R = W(R - d)$
$W = \frac{(V_R - V_L)}{2d}$
$R = 2d\left(\frac{V_R + V_L}{V_R - V_L}\right)$
$V = WR = \frac{V_R + V_L}{2}$

Figure 12-21. *Differential Kinematics*

Load the Jupyter Notebook from example directory Chapter 07 called: Chapter-012-12-Differential-kinematics-01.ipynb

This notebook demonstrates how to calculate the differential kinematics of a robot. See Figure 12-22.

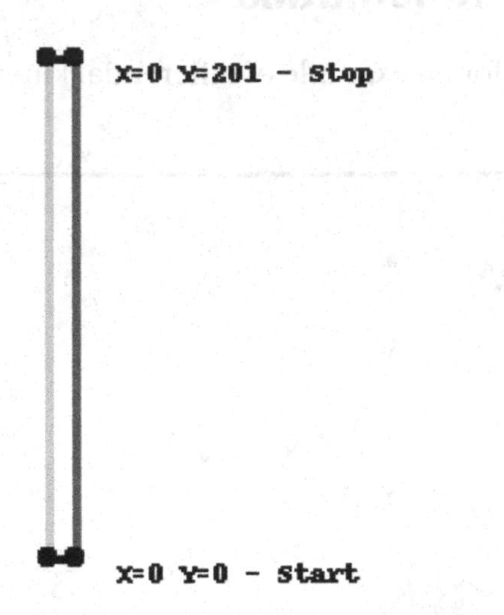

Figure 12-22. *Differential Kinematics (1)*

Load the Jupyter Notebook from example directory Chapter 07 called: Chapter-012-13-Differential-kinematics-02.ipynb

T=1

See Figure 12-23.

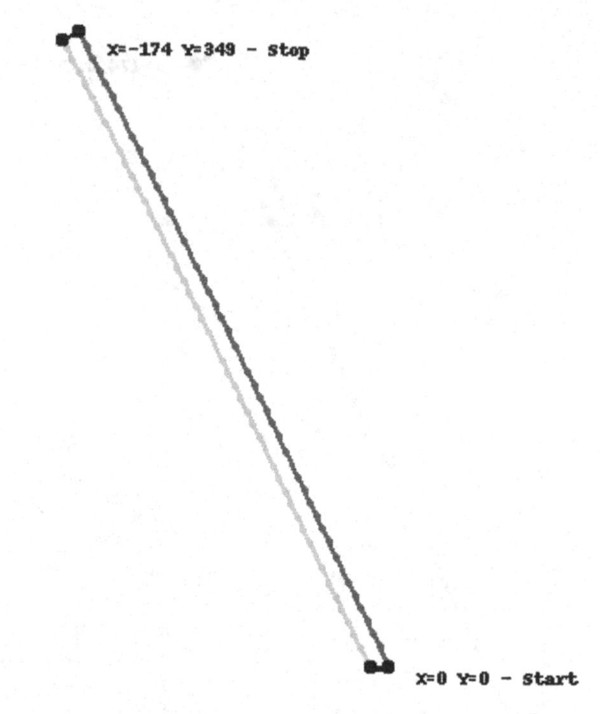

***Figure 12-23.** Differential Kinematics (2)*

Load the Jupyter Notebook from example directory Chapter 07 called: Chapter-012-14-Differential-kinematics-03.ipynb

T=2

See Figure 12-24.

Turn Right Robot - VL=200.000 VR=190.000 - Forward-Right

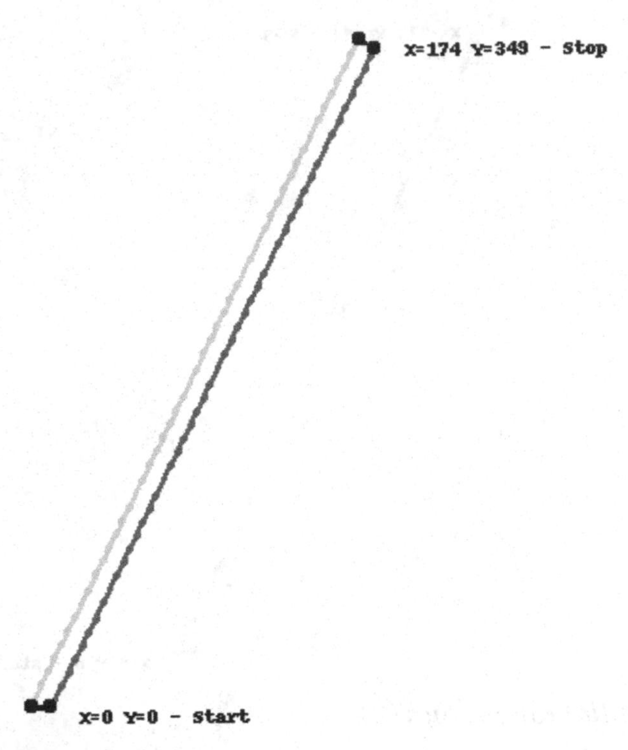

Figure 12-24. *Differential Kinematics (3)*

Load the Jupyter Notebook from example directory Chapter 07 called: Chapter-012-15-Differential-kinematics-04.ipynb

T=3

See Figure 12-25.

Turn left Robot - VL=-190.000 VR=190.000 - Forward-Pivot-Left

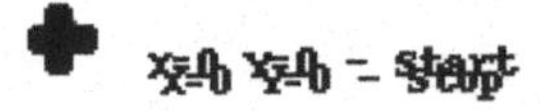

Figure 12-25. *Differential Kinematics (4)*

Load the Jupyter Notebook from example directory Chapter 07 called: Chapter-012-16-Differential-kinematics-05.ipynb

```
T=4
```

See Figure 12-26.

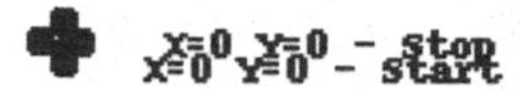

Figure 12-26. *Differential Kinematics (5)*

Load the Jupyter Notebook from example directory Chapter 07 called: Chapter-012-17-Differential-kinematics-06.ipynb

```
T=1
B=0.75
```

See Figure 12-27.

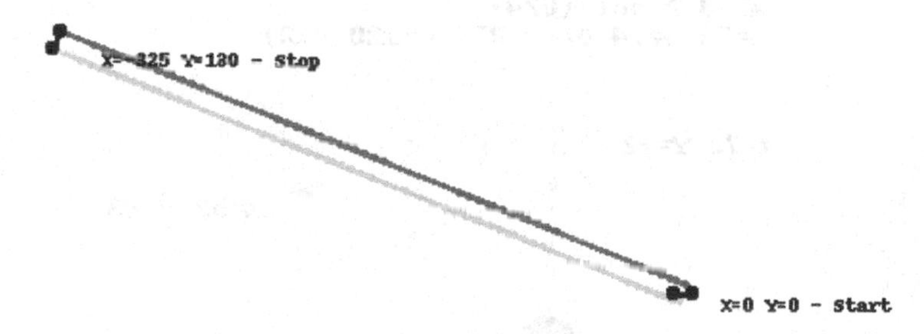

Figure 12-27. *Differential Kinematics (6)*

Load the Jupyter Notebook from example directory Chapter 07 called: Chapter-012-18-Differential-kinematics-07.ipynb

```
B=0.55
```

See Figure 12-28.

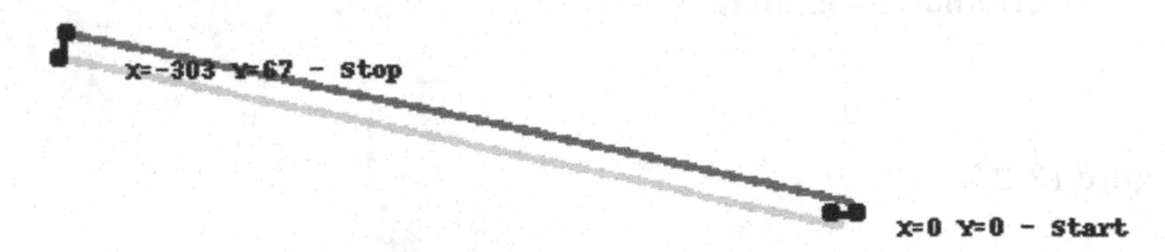

Figure 12-28. *Differential Kinematics (7)*

Evolutionary Robotics

Evolutionary Robotics is the concept where robots build robots that can be adapted to the current ecosystem by replacing or modifying a part of the robot because it has a modular structure.

I will demonstrate this trend by showing you how to modify a robot to have better reach or more precisely adaptive reach capabilities.

Load the Jupyter Notebook from example directory Chapter 07 called: Chapter-012-19-Evolutionary-01.ipynb

This notebook creates an evolutionary robot configuration that adapts to the environment (Figure 12-29).

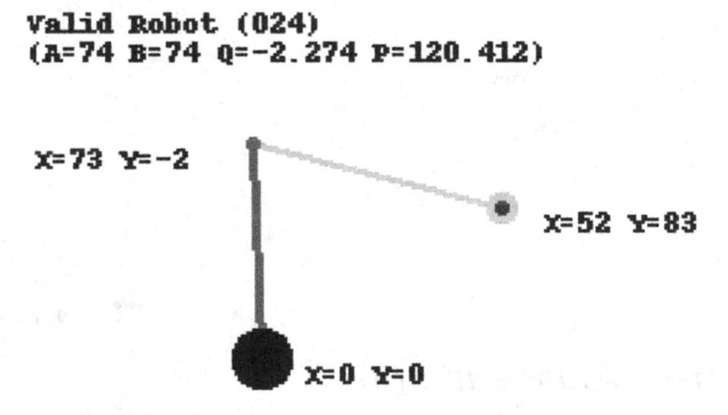

Figure 12-29. *Adapted Robot*

Multi-agent system

Multi-agent robots mean the use of more than one robot working together to achieve one specific task. These robots are normally controlled by one control system and are also called a cluster of robots or a work cell.

There is a good tool at: `https://robodk.com/`

I normally use this to simulate a multi-agent environment.

The library (`https://robodk.com/library`) is useful to get industrial robot information.

This tool enables you to create a simulation for your complete multi-agent environment.

Using the offline environment (`https://robodk.com/offline-programming`), you can use your Python knowledge to simulate a fully working industrial solution (Figure 12-30).

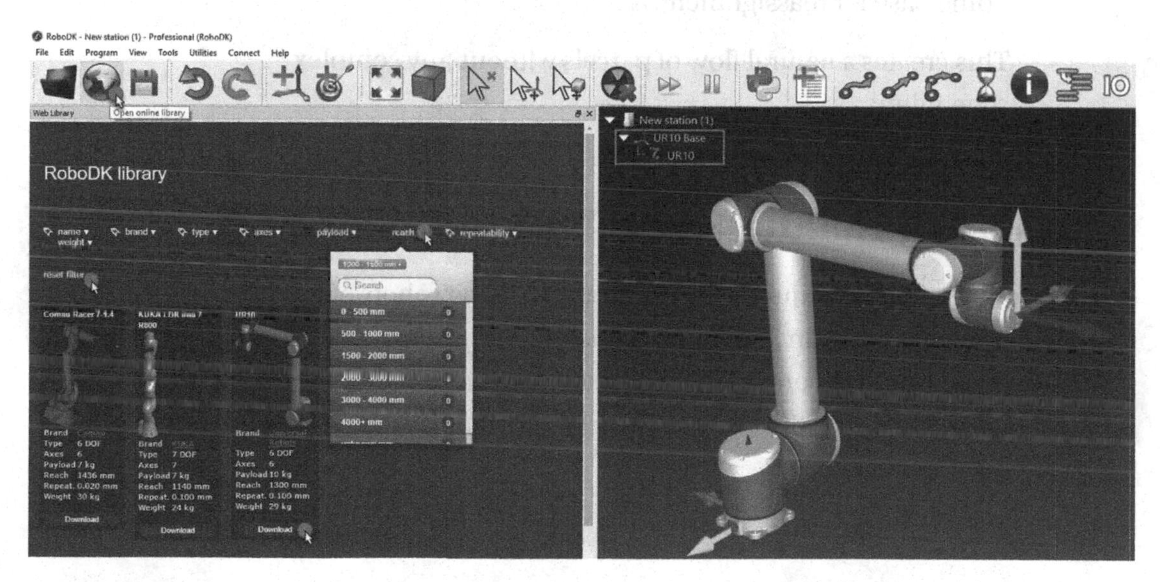

Figure 12-30. *RoboDK Tool*

Swarm Robotics

Swarm robotics is an approach to the coordination of multiple robots as a system that consists of large numbers of typically simple physical robots. It is designed so that a preferred cooperative behavior develops from the interactions between these robots and interactions of robots with the allocated environment.

The advantage of swarm robots is that the individual physical robot is low cost. The robot is programmed with a simple instruction and this gets replicated thousands or millions of times to support the single action the complete swarm should achieve.

Example:

> A drone can be programmed to fly between to GPS point to carry parcels between two points.
>
> When a drone experiences a backlog, the system simple programs more drones with the same task instructions.
>
> When drones experience no backlogs, they simply return to a home base for reassignment.
>
> This creates a natural flow of parcels without any complex scheduling.

The Role of Robotics in Smart Warehousing

The modern warehouse requires the clustering of many robots of different types and capabilities together into an integrated industrialized process supported by IML.

The robots learn from previous collaborations with other robots that specific activities can only occur if the other robot is in a specific pre-agreed state.

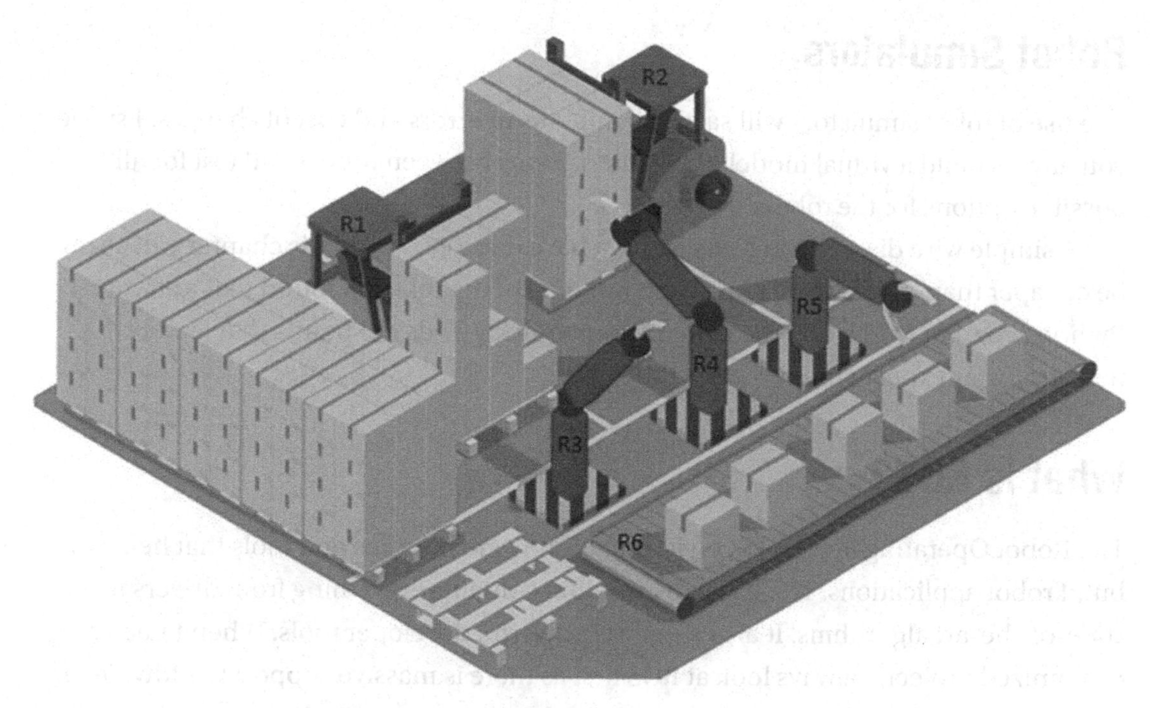

Figure 12-31. *Smart Warehousing Diagram*

Robot R1 (Figure 12-31) must be in a lower position with a not full packing rack for robot R3 to add more packages to the packing rack. Robot R3 can also only move packages between R6 and R1 if the R6 has a package ready for pickup. The same way R2 needs to not stack fill packing racks where R1 has been loaded.

The synchronized and cooperative interactions of all robots in the warehouse ensure that optimum performance, effectiveness, and efficiency are achieved.

Well-synchronized robots can:

- Help prepare Items for Shipment
- Can lift and carry loads around
- Help warehouse personnel meet staffing needs
- Robotic Drones give views of warehouse inventory

Robot Simulators

The use of robot simulators will save you millions in errors and cost of changes. I suggest you always build a virtual model of the interactions between robots and test for all possible options for the robots.

A simple wire diagram-style test like we have done earlier in this chapter will always be cheaper than the physical real-world building of the robots. I found that a simple Python program simulating the contact point quickly finds the points that are not performing as expected.

What Is ROS?

The Robot Operating System (ROS) is a set of software libraries and tools that help you build robot applications. The basic distribution contains everything from drivers to state-of-the-art algorithms. It also supports powerful developer tools. When I need a customized project, I always look at ROS first as there is massive support worldwide for the software, and it will ensure your next robotics project is a success.

```
Look at: http://www.ros.org/
```

I usually use the ROS standard as it enables the stress-free transfer of your designs to any robots that support the ROS standard. (`http://www.ros.org/core-components/`)

I accomplish my core robotics on ROS and it works well. ROS have assisted me personally well over the years to industrialize my machine learning.

Tip Start using the ROS method on smaller projects also as most small robots end up been big robots if your work is successful. Invest the time in the beginning.

What's Next?

In this chapter I have explained how with machine learning and programming you can simulate robotics from the real world.

I have explained soft robots where you create software agents that can perform work for you. I simulated a mail sorter.

I have also explained hard robots where you physically can create an agent or robot that enforces your machine learning results onto the real world. I simulated several robots.

You should now be able to use supervised or unsupervised learning to discover a desired prediction or outcome. You can then use soft or hard robots to enforce this outcome onto the real world.

Using the massive diversity of Internet-of-Things sensors, you can then get feedback and adjust your machine learning model.

In this chapter you covered the basic workings of a robot; and you were introduced to a few commercial products that use robot toolkits to achieve IML via a direct action model using soft or hard robots.

The next chapter covers the Fourth Industrial Revolution that shows you where your new knowledge enables you to change and enhance the world around you.

CHAPTER 13

Fourth Industrial Revolution (4IR)

Let's go over the first, second, and third revolutions and then dig into the fourth one.

The ***First Industrial Revolution (1IR)*** used ***water*** and ***steam power*** to ***mechanize production***. Typically agricultural, rural civilizations became manufacturing and urban. The iron and textile businesses exploded into existence thanks to the development of the steam engine. This single discovery pushed the people of the land into factories. The steam engine started in 1698 and by 1733 only 104 steam engines were in use in the United Kingdom. In 1753 it arrived in the United States of America, and we still have steam engines that are working today. It came and it keeps on changing the world around it.

The ***Second Industrial Revolution (2IR)*** used ***electric power*** to create ***mass production***. It lasted from 1870 until 1914. It simply took steel, oil, and electricity, and used electric power to create mass production on a scale that is still felt around the world. It yielded the telephone, light bulb, phonograph, and internal combustion engine. The latter resulted in even more ***rapid mobility*** for the ***unrestricted distribution*** of people and goods.

The ***Third Industrial Revolution (3IR)*** used ***electronics and information technology*** to ***automate production***. This ***Digital Revolution*** advanced technology from analog electronic and mechanical devices to digital technology. Since 1980, it has resulted in personal computers, the Internet, and the information and communications technology (ICT) industry. This resulted in ***instantaneous communication*** between people and machines.

The ***Fourth Industrial Revolution (4IR)*** is building on the Third, the digital revolution that is characterized by a ***fusion of technologies*** that is ***concealing*** the lines between the ***physical, digital, and biological spheres***.

A. F. Vermeulen, *Industrial Machine Learning*, https://doi.org/10.1007/978-1-4842-5316-8_13

You are standing on the brink of a ***technological*** and ***social revolution*** that will fundamentally alter the way humans live, work, and relate to one another. In its scale, scope, and complexity, the transformation will be contrasting everything humankind has experienced before in its history. The current issue with 4IR is the ***velocity***, ***scope***, and ***systems impact***. The speed of current breakthroughs has no historical model.

When I compare 4IR with 1IP, 2IP, or even 3IR, the 4IR is ***evolving*** at an ***exponential rate*** and not at the previous, almost linear pace. In addition, it is ***disrupting*** essentially all industries in every country. The breadth and depth of these modifications force the transformation of complete systems of production, management, and authority.

The 4R is now ***directly unsettling current real-world processes***.

The 4IR is predicted as the next step in the human race's evolution path into our future. Like the revolutions that preceded it, the 4IR has the potential to raise global income levels and improve the quality of life for populations around the world. So far, those who have gained the most from it have been consumers able to afford and access the digital world, but new activities by global organizations to tap into unconnected people will shift this lack of access quickly. The 4IR is challenging the use of Industrialized Machine Learning (IML) and will impact the way you experience and impact the world around you.

Enabler Technology

The Fourth Industrial Revolution (4IR) is marked by emerging technology breakthroughs in several fields, including robotics, artificial intelligence (AI), nanotechnology, quantum computing, biotechnology, Internet of Things (IoT), 3D printing, and autonomous vehicles. These technologies have great potential to connect billions of people to the Web in real time, radically developing the efficiency of businesses and organizations and help regenerate the natural environment through better asset management.

Manufacturers need to understand additive manufacturing such as 3D printing, robotics and autonomous production, AI, virtual reality, analytics, and advances in materials science.

The 4IR is driven by emerging technology breakthroughs in a multiplicity of arenas, covering robotics, AI, biotechnology, IoT, 3D printing, quantum computing, nanotechnology, and autonomous means of transportation.

Fully Robotic, Closed-Loop Manufacturing Cells

Closed-loop manufacturing is the technique that IML uses to monitor the numerous methods and necessities for the manufacturing procedure and accuracy requirements.

I use the following steps:

1. Planning the order and iterations of the procedures for the required method of manufacturing.
2. Premanufacturing smaller modules as a portion of the larger development. Repeatable quality is a measure of success. The whole system must be built to standards and within tolerances. Get Key Performance indicators (KPI) for each step and module.
3. Measuring the real value of every KPI in the process is a requirement. Sensors are vital at this level in the organization.
4. Calculating the outstanding gap between required KPI and measured KPI. Stop if the outstanding gap is smaller than the required accuracy. The perfection style KPIs are highly costly.
5. Keep manufacturing the product at the required quality by using adaptive learning to correct quality in near real time.
6. Repeat from step 3 as many times as required, to deliver the product you need.

Tip When you deal with manufacturing, always pin the scope down first, as achieving a closed-loop manufacturing without a fixed scope results in a solution that does not have all the required inputs and outputs. Subsequently, lack of scope results in a non-closed-loop cell.

Three-dimensional printing is a major advancement in this environment as this one type of manufacturing tool can adapt to manufacturing various components without requiring major retooling. I predict we will see more products being manufactured on request than in mass; hyper-personalization is the next step in manufacturing.

Modular Construction of Machine Learning

The application of IML, through adaptive learning and robotics, is enabled by 3D printing. You can enable yourself to analyze the required future predictions as set by the machine learning rules and at that moment produce the physical delivery.

Modular construction describes important elements of the process to standardize and create common subcomponents ready for use by your machine learning factory. The advantage of premanufacturing of common components is that you are able to predict what you need with your IML and then industrialize the required common parts. I have discovered over the years that if you prepare the majority of the IML steps with better-quality common components as a common library of tools, it is easier to deliver your solution by assembling appropriate components into individual solutions to perform the specific required machine learning tasks.

Tip Bringing a first draft of a solution to deployment as a custom solution, I have found that 80% of any high-quality IML is simply commonly predesigned solutions with minor personalization.

The global industrial manufacturing sector is predicted to spend approximately $890 billion on IoT technology by 2020. These IoT sensors will generate an increasing volume of data to empower you to develop data lakes with the IML. You will have to increase the required investment into the generation of effective and efficient IML solutions to meet this demand.

Disruptors of the Current World

You have IML with adaptive learning, AI, robotics, the IoT, 3-D printing, nanotechnology, biotechnology, materials science, energy storage, and quantum computing to modify the solutions around you and your customers. I have introduced you to a knowledge base from IML learning that will enable you to help your customers to use these disruptors to improve the world around you.

So, I will now discuss some of the most active solutions in the IML ecosystem. You simply have to start small and evolve your skills as you progress. You can only learn by doing and trying to hone your existing skills.

Here are a few current solutions I have seen that are making big changes in the domain in which I am working.

Machine-Assisted Robotic Surgery

IML, with a projected value of forty billion dollars to health care, has created robots that can analyze data from pre-op medical records to guide a surgeon's instrument (`https://www.davincisurgery.com/`) during surgery, which can lead to a 21% reduction in a patient's hospital stay. Robot-assisted surgery is considered "minimally invasive" so patients won't need to heal from large incisions. IML via AI, robots can use data from historical operations to advise on new surgical techniques. Study with orthopedic patients establishes that AI-assisted robotic procedure led to five times fewer complications compared to surgeons operating alone.

Using an IML driven robot in eye surgery has resulted in the most advanced surgical robot; the Da Vinci allows doctors to perform multifaceted procedures with greater control than conventional approaches.

Heart surgeons are assisted in operations with the use of a miniature robot that enters a small incision on the chest to perform mapping and therapy over the surface of the heart to explore the specific unique heart and adapt learning to fit the specific operation.

Virtual Nursing Assistants

The introduction of virtual nursing assistants that now actively interact with patients can direct patients to the most effective care setting; virtual nursing assistants could save the health-care industry twenty billion dollars annually.

IML with active adaptive learning now acts as virtual nurses that are available 24/7; they can now answer questions, monitor patients via IoT, and provide rapid answers. The best applications of virtual nursing assistants today allow for more regular communication between patients and care providers between office visits to prevent hospital readmission or unnecessary hospital visits. Virtual nurse assistants can deliver wellness checks through voice and AI.

I predict the application of IML in robotics as virtual nursing assistants will increase the deployment of machine learning.

Aid Clinical Judgment or Diagnosis

IML with assisted algorithms have analyzed what a person says, the tone of voice, and background noise and has detected cardiac arrests (`https://www.resuscitationjournal.com`) with a 93% success rate compared to 73% for humans.

IML uses deep learning algorithms to indicate that a diagnosis can now outperform humans when identifying breast cancer metastasis. The algorithms will examine medical records, habits, and genetic information collected from health research from historic diagnosis of patients.

I suggest together with the improved access to historic health data, IML algorithms will enhance the health services for the future.

Workflow and Administrative Tasks

IML assists the automation of administrative tasks. The IML you deploy in the future will enable the improvement of the everyday processes needed for your customers. With the understanding you have achieved up until now in this book, you can empower your customers' solutions with the use of supervised and unsupervised learning to enable a high-quality solution.

IML solutions that plan flight routes or train schedules are common. I have also completed several shift planning solutions or dynamic skill assessments matching for maintenance projects. If you need a controlled and adaptive scheduling system, IML is the answer.

The effectiveness and efficiency of the IML learning solution you can design and produce is good machine learning.

Image Analysis

Image analysis is very time consuming for human providers, but a machine learning algorithm that can analyze 3D scans up to a thousand times faster than what its capability today. This near real-time assessment can provide critical input for surgeons who are operating. Advanced health sensors with machine learning will advance the next generation of radiology tools for health-care providers and telemedicine robots. Patients can use their camera phones to send in pictures of rashes, cuts, or bruises to determine what care is necessary immediately for effective treatment.

IML eases the complex world of health care, IML tools can support human providers to be responsible for faster service, diagnose issues, and analyze data to identify trends or genetic information that would bias to a disease. Saving minutes can mean saving lives, and machine learning can be transformative not only for health care but for every single patient it attaches to IML.

Farming

The increase in population on the earth is heading to reach levels that will cause a worldwide famine. Global human population growth is projected to reach 8.5 billion in 2030, and to increase further to 9.7 billion in 2050 and 11.2 billion by 2100 (`https://www.un.org/en/sections/issues-depth/population/`).The projected resulting famine is a widespread scarcity of food, caused by several factors including war, inflation, crop failure, population imbalance, or government policies.

Land Mapping

IML uses geo-fencing of farms to capture data to develop required inputs via specific sensors in farming. Mapping land use is now possible with wide-ranging sensors that can test the quality of the land and improve the quality of farming.

The geo-fencing controls where a specific crop will and will not grow. It enables farmers to enhance the ground with extra essentials to support specific crops by informing them about what is missing from the land they are using for their crops. It shares knowledge between farmers about issues and solutions, and the farming becomes cooperation between the IML and the farmers.

Calendar of Activities

IML enables farm robots to learn at one farm how to improve the yield for better farming and then transfer the learnings to all robots. Tailor-made farming plans unique for each crop can now be formulated to adaptive learning.

Farmer Management System

An integrated farming platform enables farmers by directly connecting to buyers and sellers. This close-loop process needs proper sensors to input requirements and post-harvest feedback.

The more inputs you have for your IML, the better outcome for general food production.

Container Farm

Growing herbs and vegetables vertically in shipping containers is a rising trend in farming.

By retrofitting 40-foot-long shipping containers with the assistance of IML processes, we are now able to utilize one of these containers to yield yearly crops similar to five acres of traditional farmland; but it is faster, and it uses as little as 1% of the water the farmland needs. The use of aquaponics is a combination of aquaculture, which is growing fish and other aquatic animals; and hydroponics, which is growing plants without soil, have been combined with IML and is achieving generally good yields in improving the future capacity to grow crops required to survive. I have seen containers produce food in basements without a single cup of traditional soil. This will change the way humans survive in the future.

Finance

Finance has been developing a strong foundation on core paradigms and algorithms of machine learning (ML) as it has the funds and capacity to perform a return on investment for the empowered machine learning algorithms.

Banks are hiring first-rate data scientists who also understand how markets react to ever-changing global activities. Machines are best equipped to make trading decisions in the short and medium terms, but an army of people will be necessary to acquire, clean, and assess the data to train the machine learning.

You won't need to be a machine learning expert; you will need to be an exceptional expert and an outstanding programmer to understand how to use IML to assist your customers.

Insurance

You will find that 63% of insurers have confidence in intelligent technologies that will completely transform their industry. The insurers that will benefit to the maximum from this uptake of machine learning will be those that are prepared to rethink their approach to their people, their processes, and their data. IML will disrupt their traditional business model with an intelligent framework that augments their people's work, rethinks how they operate with intelligent automation, and unlocks growth through data.

The Trusted Robot-Advisor

Imagine having a virtual financial advisor that truly understands your personal complex needs, goals, and risk appetite. It would act in a proactive way to secure the future you deserve. It would work tirelessly, 24/7, 365 days a year; to understand your evolving lifestyle from your personal data feeds to which you've given it access. It will spot opportunities for savings on your insurance and better returns on your investments.

Almost every company I am in contact with is developing a type of chatbot to interact with their customers. Many customers do not even know when it is an IML solution or a person answering their questions or processing their interactions.

In-Stream Analytics (ISA)

Machine Learning today tends to be "open-loop" - for example, collecting masses of data offline, processing it in batches, and generating insights for eventual action, typically days later.

An emerging category of IML business use cases are called "In-Stream Analytics" (ISA). You use the IML to process data received from the real world as soon as it arrives and insights are generated quickly. Responses are milliseconds as the process are happening.

Currently action may be taken offline, and the effects of the actions are not immediately incorporated back into the learning process. The next step for IML is to go near real time, responding as and when required with the correct action to the actions just observed.

Adaptive Machine Learning

The system you are seeking is adaptive machine learning (AML) where "time" is important, real-time applications use sensors, soft robotics and even hard robotics empower you immediately by responding with the "correct" action to the ecosystem and then measures the immediate success of the "action." The ability to adapt to the changes in the ecosystem or, even better, cause the changes is paramount. A successful AML solution has 95% control over the close-loop system it is managing.

Fraud Detection

With fraud detection, rules and scoring based on historic customer transaction information, profiles, and even technical information to detect and stop a fraudulent payment transaction as it happens in near real time.

Financial Markets Trading

An automated high-frequency trading system detects patterns in the market and then counteracts them by a change in the soft robots that perform the trading as prescribed by the risk profile of the portfolio of your clients.

IoT and Capital Equipment Intensive Industries

You deploy optimization of heavy manufacturing equipment maintenance, power grids, and traffic control systems to ensure these utilities stay up and maintain its performed before any major outages or damage is incurred.

Marketing Effectiveness

This means detecting mobile phone usage patterns to trigger individualized offers for the products you want when you need them. Bulk marketing is no longer acceptable by customers.

Retail Optimization

Your in-store shopping pattern and cross-sell is directly connected to personalize in-store price checking.

The personal shopping experience is now becoming a major selling point for consumers.

Tip These systems are viable with the knowledge you already have at this point in the book. You just need to design and build them.

Real-Time Closed-Loop System

One factor that necessitates real-time interaction is the closed-loop nature of these use cases. In every case, as an external event happens, an analytics module determines a recommended action and creates a response that will impact the external event in a timely manner.

What "real time" means is case dependent. ... The rate at which data collection, analysis, and action happen could be milliseconds, hours, days, . . . The industry name for this system is Event Stream Processing or ISA.

An analytics module determines a recommended action and creates a response that will impact the external event; you should also measure the impact of the action so that the following happens:

- Your IML should know if your prescribed action was good.
- Your IML notes any shortcomings to improve the ISA so that the next ISA event will have a better outcome.
- Your IML can attribute the right portion of the result to ISA's action.

If learning is the process of generalization from experience, you can be more explicit and say learning is generalization from past experience and results of new actions - this is the true definition of learning!

If you accomplish a cause-and-effect balance of 100%, you have successfully created a Real-Time Close-Loop System.

IML has also already moved through several changes. I can identify four general generations of IML.

Four Generations of Industrialized Machine Learning

I want to inform you that there are four distinct steps you need to migrate your machine learning through to achieve 100% adaptive learning.

1st Generation: Rules

The first generation of machine intelligence meant that people manually created rules. For example, in text analytics you might create a rule that the word "*Mickey*" followed by "*Mouse*" meant that "*Mickey*" referred to a cartoon, and they would create a distinct rule that "*Mickey*" preceded by "*Rourke*" meant that "*Mickey*" referred to a real person.

The rule-based approach is time consuming and not very accurate. Even after you have exhausted all the words and phrases they can think of, there are always other contexts and new innovations that aren't captured. For one of your clients, their experts' analysts were only able to capture less than 11% of the documents they wanted to analyze using rules: this clearly is too limited.

The rules can, however, handle simple decisions at a sensor level and areas were decisions are binary.

> Door is open ... Raise the alarm.
>
> Plant is dry ... Water them.

You can use these first-generation IML solutions to handle precise and specific solutions. You will create a narrow machine learning solution that can work within a specific ecosystem. You will see an increase in these level systems as more and more IoT devices are created to meet the proliferation in automation that you will experience in the coming years.

Prediction I personally predict that as much as 90% of your future data sources will be IoT devices that do not exist yet. This area is going to explode with machine learning applications.

2nd Generation: Simple Machine Learning

The dominant form of machine intelligence today is simple machine learning. Simple machine learning uses statistical methods to make decisions about data processing. For example, a sentence might have the word "*Mickey*" labeled as a person, and the machine learning algorithm will learn by itself that the following word "*Mouse*" is evidence that "*Mickey*" is a cartoon in this context.

Simple machine learning can be fast, on the condition that that you already have labeled examples for 'supervised learning.' It also tends to be more accurate, because statistics are usually better than human intuition in deciding which features (like words and phrases) matter. The key disadvantage for supervised machine learning is that you need the labeled examples: if you have too few labels or the labels aren't representative of the entire data set, then the accuracy is low or limited to a specific domain.

Prediction Second-generation IML is the most common machine learning you will have to deal with over the next two to five years.

3rd Generation: Deep Learning

I have seen a new upsurge in the use of machine learning that learns more sophisticated relationships between features, known as deep learning. For example, if you had the sentence "The corner shop sold an apple to *Mickey Mouse* or *Mickey Rourke*," there is conflicting evidence between "*Mickey Mouse*" and "*Mickey Rourke*" about whether "*Mickey*" is a person or a cartoon.

I have discovered deep learning can regularly learn how to use permutations of features when making a decision. For simple machine learning, a human has to tell the algorithm which permutation of features to consider. Deep learning often cuts down on the amount of human time needed and typically gets up to 5% more accurate results than simple machine learning for text analytics when applied to data from the same sources as it learned from.

Warning You change the domain you apply the deep learning model to from the corner shop to movies, and your newfound deep learning loses value immediately.

Tip Successful IML is about the context of your training data. Any bias or variance you include then becomes part of its learning rules and will impact the success on unknown data it needs to process in the future.

4th Generation: Adaptive Learning

Adaptive learning involves human analysts in the process at every step. You become the human-in-the-loop that handles exceptions. This is in contrast to rule-based, simple machine learning, and deep learning approaches, where humans only create rules and label data at the start of the process. For example, if you had the sentence, "You will help *Mickey Mouse* escape from prison," and your system hadn't seen any examples of "*Mickey Mouse*" or does not understand the concept of "*cartoons,*" you will need human input to build the knowledge base.

Adaptive learning systems require the least human effort because you only require human input when it matters most and continually expand their knowledge when new information is encountered. As you set the context, these findings will become the most accurate. You successfully combine the three other types of machine intelligence, adding new types of 'unsupervised machine learning' and methods for optimizing the input from multiple, possibly disagreeing, humans.

Tip Adaptive learning is the goal you want to achieve, but I predict we will require a lot more research and progress in the field of machine learning before this is achieved.

Rapid Information Factory

The Rapid Information Factory (RIF) is a machine learning and data processing solution I have developed over the last ten years and am using in my daily work as a data scientist and university researcher (Figure 13-1).

I will share the basic processes with you in the rest of this chapter.

The solution is an evolving system that I adapt when I start any new IML. It was built out of my experience over the last thirty years so that most of the solution has a core set of work that has to be done to a specific quality, effectiveness, and efficiency to achieve success in the IML environment.

Here is the basic solution background plus some simple examples. I will also take you through a full-scale solution in Chapter 15.

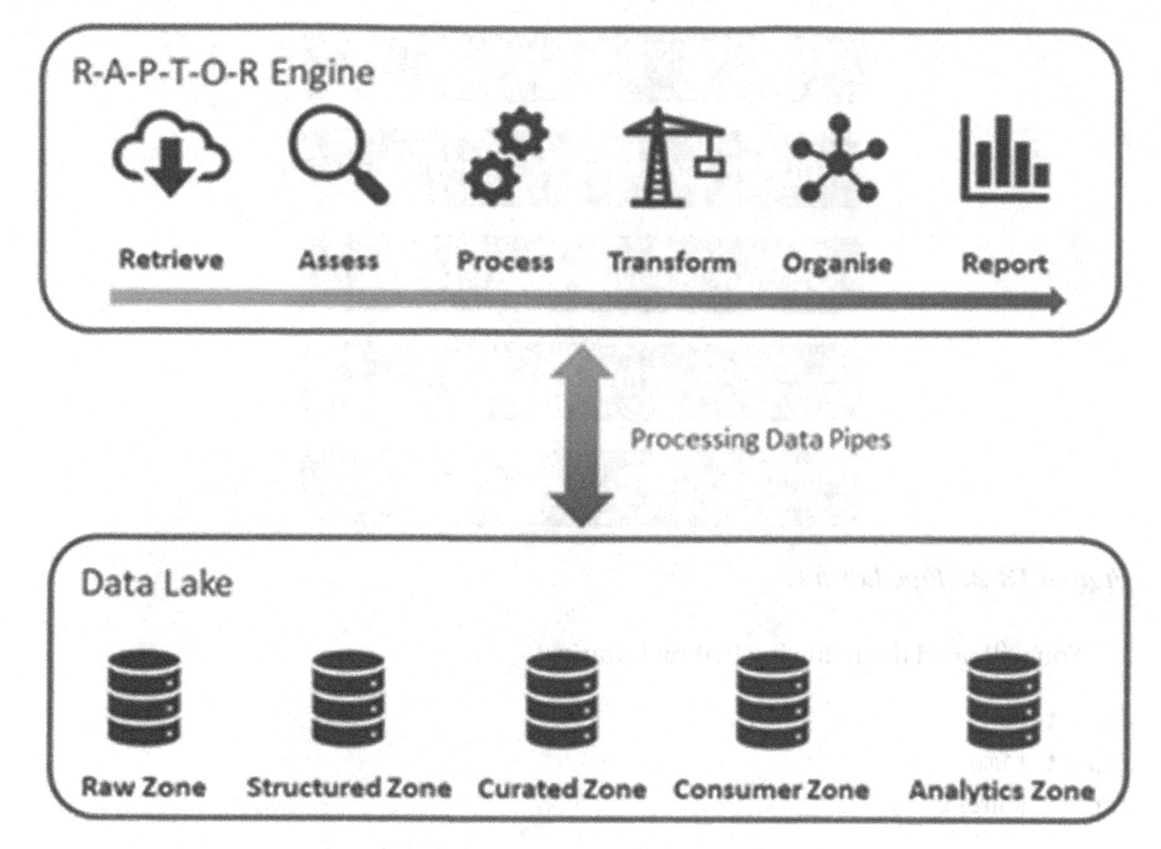

Figure 13-1. *Rapid information factory*

Five System Layers

The basic solution is designed across five core layers of machine learning processes.

You need to first load the following Jupyter Notebook from example directory Chapter 013: Chapter-013-01-Setup.ipynb

I will guide you through the creation of the basic RIF and then the Five Zone Data lake.

Note I will give you a detailed explanation of each component after we have created the required structures.

I will explain everything in Figure 13-2 to show you how to package your machine learning into a 4IR solution.

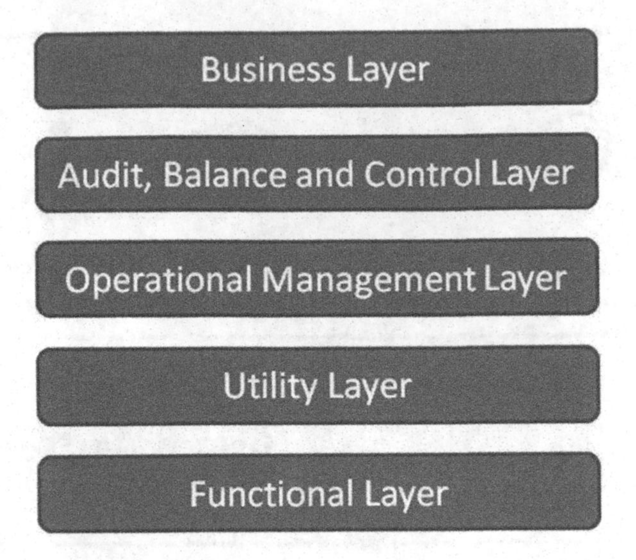

Figure 13-2. *Five Layers*

You will need the following Python libraries:

```
import os
import time
import shutil
```

You should create the RIF.

This path indicates the code where you want to create the physical RIF.

```
pathRIFname='../../Results/Chapter 013/000-RIF/'
```

Warning The next part will remove the existing RIF if you run it.

```
if os.path.exists(pathRIFname):
    shutil.rmtree(pathRIFname)
time.sleep(0.5)
```

Warning The next command is commented out because I have created a sample setup. If you create the directory, the copy command will not transfer the sample Python scripts.

```
#os.makedirs(pathRIFname)
```

Tip If you want to pre-copy the Basic Rapid Information Factory, execute the next command.

```
pathRIFInName='../../Code/Chapter 013/000-RIF/'
pathRIFOutName='../../Results/Chapter 013/000-RIF/'
if not os.path.exists(pathRIFOutName):
    shutil.copytree(pathRIFInName, pathRIFOutName)
```

Next you create the Functional Layer as

```
dirLayerpath='100-Functional-Layer'
dirname = os.path.join(pathRIFname, dirLayerpath)
if not os.path.exists(dirname):
    os.makedirs(dirname)
```

You will now expand the functional layer as per Figure 13-3.

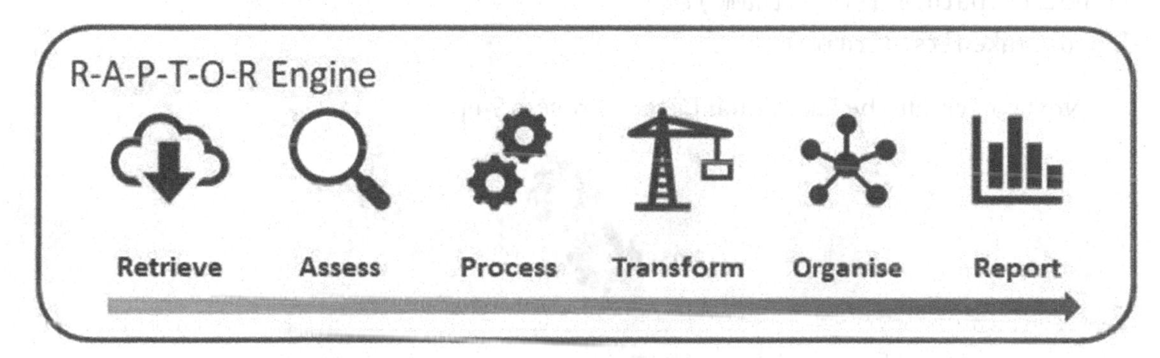

Figure 13-3. *RAPTOR Engine*

Next you create the Functional Layer – Retrieve Step.

The retrieve step enables the RIF to port data into the RIF from outside the RIF ecosystem.

```
dirpath='100-Retrieve'
dirname = os.path.join(pathRIFname, dirLayerpath, dirpath)
if not os.path.exists(dirname):
    os.makedirs(dirname)
```

Next you create the Functional Layer – Assess Step.

Assess

The assess step enables the quality assurance within the RIF.

```
dirpath='200-Assess'
dirname = os.path.join(pathRIFname, dirLayerpath, dirpath)
if not os.path.exists(dirname):
    os.makedirs(dirname)
```

Next you create the Functional Layer – Process Step.

Process

The process step performs the data amalgamation and consolidation of information for all the data in the RIF.

```
dirpath='300-Process'
dirname = os.path.join(pathRIFname, dirLayerpath, dirpath)
if not os.path.exists(dirname):
    os.makedirs(dirname)
```

Next you create the Functional Layer – Transform Step.

The transform step is where the bulk of the machine learning is performed. It uses the amalgamated data source to feed the features to the data science and machine learning.

```
dirpath='400-Transform'
dirname = os.path.join(pathRIFname, dirLayerpath, dirpath)
if not os.path.exists(dirname):
    os.makedirs(dirname)
```

Next you create the Create Functional Layer – Organize Step.

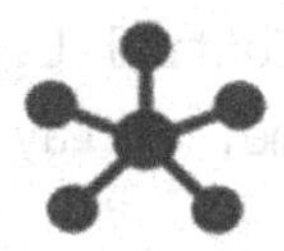

The organize step divides the machine learning results to deliver the business insights for the business users.

```
dirpath='500-Organize'
dirname = os.path.join(pathRIFname, dirLayerpath, dirpath)
if not os.path.exists(dirname):
    os.makedirs(dirname)
```

Next you create the Functional Layer – Report Step.

The report step holds all the visualization and business reports to deliver the business insights.

```
dirpath='600-Raport'
dirname = os.path.join(pathRIFname, dirLayerpath, dirpath)
if not os.path.exists(dirname):
    os.makedirs(dirname)
```

Next you create the Operational Management Layer.

```
dirLayerpath='200-Operational-Management-Layer'
dirname = os.path.join(pathRIFname, dirLayerpath)
if not os.path.exists(dirname):
    os.makedirs(dirname)
```

This layer will be discussed in detail later in the book.

Next you create the Audit, Balance, and Control Layer.

```
dirLayerpath='300-Audit-Balance-Control-Layer'
dirname = os.path.join(pathRIFname, dirLayerpath)
if not os.path.exists(dirname):
    os.makedirs(dirname)
```

This layer will be discussed in detail later in the book.

Next you create the Utility Layer.

```
dirLayerpath='400-Utility-Layer'
dirname = os.path.join(pathRIFname, dirLayerpath)
if not os.path.exists(dirname):
    os.makedirs(dirname)
```

This layer will be discussed in detail later in the book.

Next you create the Utility Layer – Maintenance Utilities.

```
dirpath='100-Maintenance-Utilities'
dirname = os.path.join(pathRIFname, dirLayerpath, dirpath)
if not os.path.exists(dirname):
    os.makedirs(dirname)
```

Next you create the Utility Layer - Data Utilities.

```
dirpath='200-Data-Utilities'
dirname = os.path.join(pathRIFname, dirLayerpath, dirpath)
if not os.path.exists(dirname):
    os.makedirs(dirname)
```

Next you create the Utility Layer - Processing Utilities.

```
dirpath='300-Processing Utilities'
dirname = os.path.join(pathRIFname, dirLayerpath, dirpath)
if not os.path.exists(dirname):
    os.makedirs(dirname)
```

Next you create the Business Layer.

```
dirLayerpath='500-Business-Layer'
dirname = os.path.join(pathRIFname, dirLayerpath)
if not os.path.exists(dirname):
    os.makedirs(dirname)
```

This layer will be discussed in detail later in the book.

Next you create the Business Layer - Functional Requirements.

```
dirpath='100-Functional-Requirements'
dirname = os.path.join(pathRIFname, dirLayerpath, dirpath)
if not os.path.exists(dirname):
    os.makedirs(dirname)
```

Next you create the Business Layer - Non-functional Requirements.

```
dirpath='200-Non-functional-Requirements'
dirname = os.path.join(pathRIFname, dirLayerpath, dirpath)
if not os.path.exists(dirname):
    os.makedirs(dirname)

dirpath='300-Data-Profiles'
dirname = os.path.join(pathRIFname, dirLayerpath, dirpath)
if not os.path.exists(dirname):
    os.makedirs(dirname)
```

```
dirpath='400-Sun-Models'
dirname = os.path.join(pathRIFname, dirLayerpath, dirpath)
if not os.path.exists(dirname):
    os.makedirs(dirname)
```

Now you create the Create Six Zone Data Lake as per Figure 13-4.

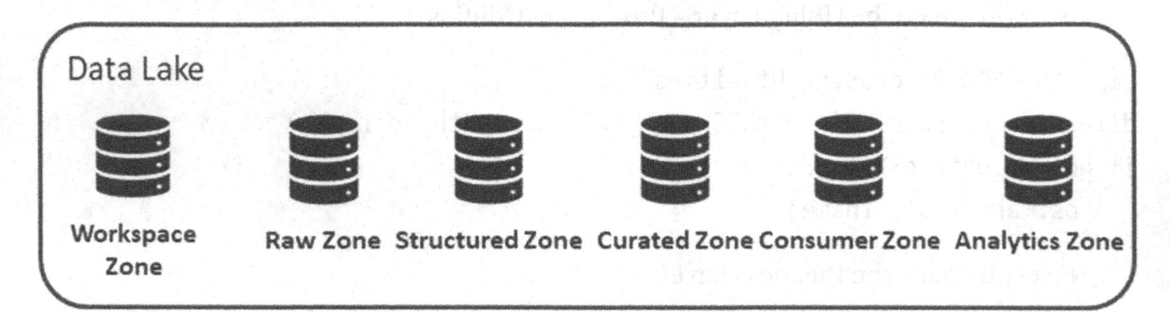

Figure 13-4. *Data Lake Structure*

Remove the existing data lake.

```
pathDLname='../../Results/Chapter 013/990-DL/'
if os.path.exists(pathDLname):
    shutil.rmtree(pathDLname)
time.sleep(0.5)
```

Tip This commented command will stop examples from transferring.

```
#os.makedirs(pathDLname)
```

Transfer the example Data Lake.

```
pathDLInName='../../code/Chapter 013/990-DL/'
pathDLOutName='../../Results/Chapter 013/990-DL/'
if not os.path.exists(pathDLOutName):
    shutil.copytree(pathDLInName, pathDLOutName)
```

Next you create the Workspace Zone as shown in Figure 13-4.

```
dirpath='000-Workspace-Zone'
dirname = os.path.join(pathDLname, dirpath)
if not os.path.exists(dirname):
    os.makedirs(dirname)
```

Next you create the Raw Zone as shown in Figure 13-4.

```
dirpath='100-Raw-Zone'
dirname = os.path.join(pathDLname, dirpath)
if not os.path.exists(dirname):
    os.makedirs(dirname)
```

Next you create the Structured Zone as per Figure 13-4.

```
dirpath='200-Structured-Zone'
dirname = os.path.join(pathDLname, dirpath)
if not os.path.exists(dirname):
    os.makedirs(dirname)
```

Next you create the Curated Zone as per Figure 13-4.

```
dirpath='300-Curated-Zone'
dirname = os.path.join(pathDLname, dirpath)
if not os.path.exists(dirname):
    os.makedirs(dirname)
```

Next you create the Consumer Zone as per Figure 13-4.

```
dirpath='400-Consumer-Zone'
dirname = os.path.join(pathDLname, dirpath)
if not os.path.exists(dirname):
    os.makedirs(dirname)
```

Next you create the Analytics Zone as per Figure 13-4.

```
dirpath='500-Analytics-Zone'
dirname = os.path.join(pathDLname, dirpath)
if not os.path.exists(dirname):
    os.makedirs(dirname)
```

Next you clean up the Python Checkpoints.

Tip When using Jupyter Notebooks, the process stores a copy of the previous saves of the notebook. I take these out of my release before I save it to the source control. This stops duplicate copies of the development.

```
pathRIFname='../../Results/Chapter 013/000-RIF/'
for root, dirs, files in os.walk(pathRIFname, topdown=True):
    for name in dirs:
        fullDir=os.path.join(root, name)
        print(fullDir)
        if name=='.ipynb_checkpoints':
            print('Remove:', fullDir)
            if os.path.exists(fullDir):
                shutil.rmtree(fullDir)
            time.sleep(0.1)
```

Well done; you now have a RIF and Five Zone Data lake (Figure 13-5).

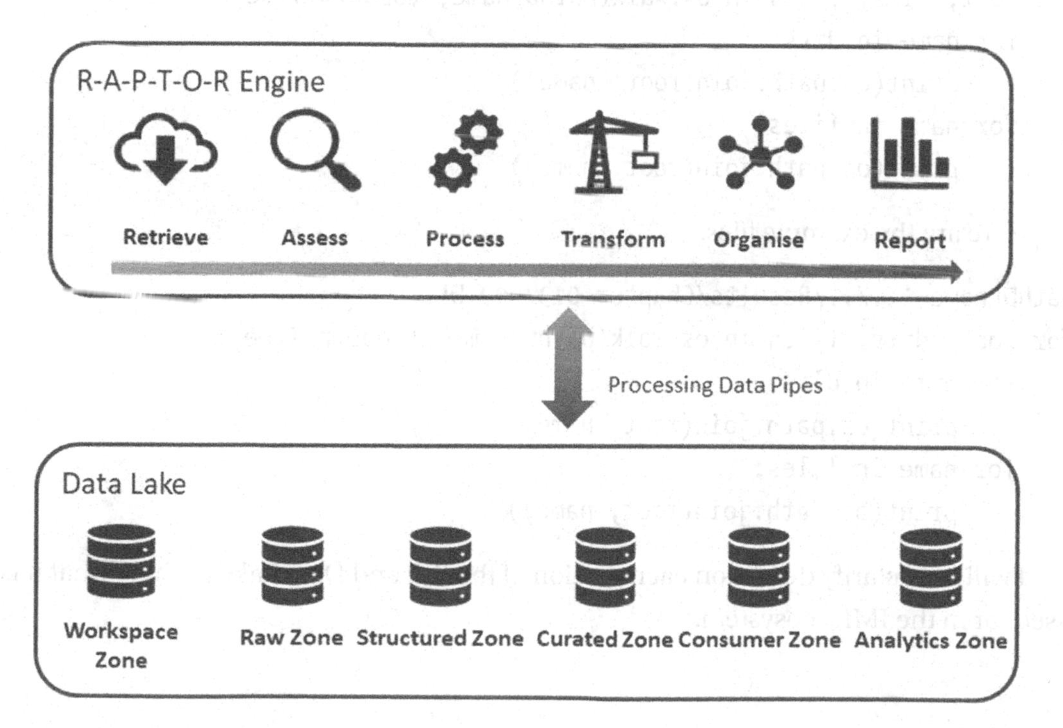

Figure 13-5. *RIF plus DL*

I will give you a quick view on the structures you created.

Let's look at the RIF:

Here are the directories you need.

```
pathRIFname='../../Results/Chapter 013/000-RIF/'
for root, dirs, files in os.walk(pathRIFname, topdown=True):
    for name in dirs:
        print(os.path.join(root, name))
```

Here are the example files.

```
pathRIFname='../../Results/Chapter 013/000-RIF/'
for root, dirs, files in os.walk(pathRIFname, topdown=True):
    for name in files:
        print(os.path.join(root, name))
```

Let's look at the Data Lake.

Here are the directories you need.

```
pathDLname='../../Results/Chapter 013/990-DL/'
for root, dirs, files in os.walk(pathDLname, topdown=True):
    for name in dirs:
        print(os.path.join(root, name))
    for name in files:
        print(os.path.join(root, name))
```

Here are the example files.

```
pathDLname='../../Results/Chapter 013/990-DL/'
for root, dirs, files in os.walk(pathDLname, topdown=True):
    for name in dirs:
        print(os.path.join(root, name))
    for name in files:
        print(os.path.join(root, name))
```

I will now clarify details on each portion of the RIF and Data Lake to show what it is used for in the IML ecosystem.

Functional Layer

The functional layer (Figure 13-2) is the core processing capability of the IML factory. The core functional data processing methodology is the R-A-P-T-O-R framework (see Figure 13-6).

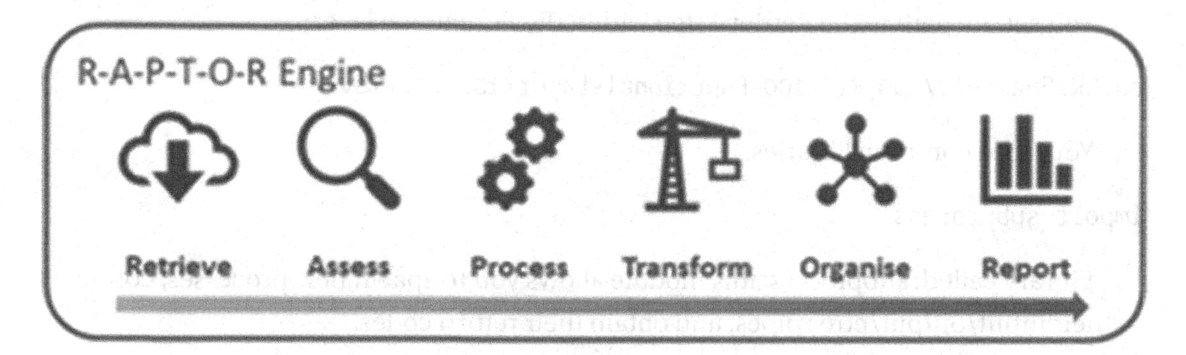

Figure 13-6. *Six Steps of RAPTOR Engine*

Retrieve Super Step

The retrieve super step (Figure 13-6) supports the interaction between external data sources (000-ExternalData) and the factory (000-RIF) that then loads the data into the data lake (990-DL) by loading the data as is into the Raw Zone (000-RIF/100-RawZone).

The purpose of this processing step is to start the data engineering of the data required by the IML.

This could be from other databases, data warehouses, sensors, or data entered by the human-in-the-loop processes.

You need to load the following Jupyter Notebook from example directory Chapter 013: Chapter-013-02-01-Retrieve.ipynb

Initially you start a clock to record how long your code will run.

```
import datetime
nowStart = datetime.datetime.now()
```

```
# ![RIF Functional Layer - Retrieve Step](../../images/RIF-FL-RET.JPG)
```

You set the path to the Retrieve step within the Functional layer:

```
pathRIFname='./000-RIF/100-Functional-Layer/100-Retrieve/'
```

You need common libraries.

```
import subprocess
```

Library called: subprocess, this module allows you to spawn new processes, connect to their input/output/error pipes, and obtain their return codes.

```
(https://docs.python.org/3/library/subprocess.html)
```

```
This assist you to execute or call other processes as part of the solution.
```

```
import os
```

Library called: os, module provides a portable way of using operating system-dependent functionality.

```
(https://docs.python.org/2/library/os.html)
```

```
import time
```

This module provides various time-related functions.

```
(https://docs.python.org/2/library/time.html)
```

```
import sys
```

This module provides access to some variables used or maintained by the interpreter and to functions that interact strongly with the interpreter.

```
(https://docs.python.org/2/library/sys.html)
```

```
import numpy as np
```

NumPy is the fundamental package for scientific computing.

(http://www.numpy.org)

To call python.exe, you need to find Python on your system.

The logical place to seek it is on the system path.

The following will find the executable "python.exe":

```
path=np.array(sys.path)
print(path.shape)
for i in range(path.shape[0]):
    pathTestPython=os.path.join(path[i],'python.exe')
    if os.path.exists(pathTestPython):
        pathPython=pathTestPython
pathPython=pathPython.replace('\\','/')
print(pathPython)

holddir=os.getcwd()
print(holddir)
```

You need a function to handle the execution of the Python programs in the solution:

```
def run_command(command, wait=True):

    try:
        if (wait):

            p = subprocess.Popen(
                [command]
                , stdout = subprocess.PIPE
                , shell = True)
            p.wait()
        else:
            p = subprocess.Popen(
                [command],
                shell = True,
                stdin = None, stdout = None, stderr = None, close_fds = True)

        (result, error) = p.communicate()
```

```
    except subprocess.CalledProcessError as e:
        sys.stderr.write(
            "common::run_command() : [ERROR]: output = %s, error code = %s\n"
            % (e.output, e.returncode))

    return result
```

You need to turn the absolute path into a real path:

```
pathRealRIFname=os.path.realpath(pathRIFname)
pathRealRIFname=pathRealRIFname.replace('\\','/')
print(pathRealRIFname)
```

Check if the setup program was run properly, and if it was, that the system finds the Python programs and will execute them via the function you created:

```
if not os.path.exists(pathRealRIFname):
    print(' Please run the Chapter-013-01-Setup program first to setup your
    examples')

else:
    os.chdir(pathRealRIFname)
    for root, dirs, files in os.walk(pathRealRIFname, topdown=True):
        for name in files:
            fullfilename=os.path.join(root, name)
            fullfilename=fullfilename.replace('\\','/')
            filename, file_extension = os.path.splitext(name)
            if file_extension.lower()=='.py':
                #print(filename)
                commandstr=name
                print(commandstr,'\n')
                res = run_command(commandstr)
                time.sleep(0.5)
                print('###################\n')
                print(res)
                print('###################\n')
```

The process ensures the working directory is reset to the same directory the program started within.

```
os.chdir(holddir)
print(os.getcwd())
```

You can now measure the time it took to complete the run of the system:

```
nowStop = datetime.datetime.now()
runTime=nowStop-nowStart
print('Start:', nowStart.strftime('%Y-%m-%d %H:%M:%S'))
print('Stop: ', nowStop.strftime('%Y-%m-%d %H:%M:%S'))
print('Time: ', runTime)
```

Congratulations with the completion of this runner program.

I want to now show you the basic Python program you are running with the above step runner.

You need to look in the example directory at the following subdirectory: IML\Code\Chapter 013\000-RIF\100-Functional-Layer\100-Retrieve

```
Open Chapter-013-02-Retrieve-01-Time.ipynb
```

Load time library:

```
import datetime
nowStart = datetime.datetime.now()
```

Set the path to Raw Zone:

```
dirname = '../../../../Chapter 013/990-DL/100-Raw-Zone/'
filename='08-02-Retrieve-Time-Date.csv'
fileZipname=filename + '.gz'
print(fileZipname)
```

Load the common libraries:

```
import pandas as pd
import os
```

Find the full Raw Zone path on the ecosystem:

```
pathRealName=os.path.realpath(dirname)
print(pathRealName)
fullRealZipName = os.path.join(pathRealName, fileZipname)
```

If the Raw Zone is missing, create the path.

```
if not os.path.exists(pathRealName):
    print('Make:', pathRealName)
    os.makedirs(pathRealName)
```

You will now start to create a new dates data set in the Raw Zone by uploading from outside of the Raw Zone.

The Pandas date_range function will create dates between 2017/01/01 and 2018/12/31.

```
dates = pd.date_range(start='2017-1-1', end='2018-12-31', freq='D')
```

Convert the range into a data frame with format '2017-01-01' and a key with format 'D20170101':

```
datesNameDF = pd.DataFrame(dates.strftime('%Y-%m-%d'))
datesNameDF.index.name='DateID'
datesNameDF.columns=['Date']

datesKey=dates.strftime('D%Y%m%d')
datesKeyDF=pd.DataFrame(datesKey)
datesKeyDF.index.name='DateID'
datesKeyDF.columns=['DateKey']

datesDF=datesKeyDF
datesDF=datesDF.merge(datesNameDF, how='inner', left_on='DateID', right_
on='DateID')
```

Store the generated data frame as a comma-separated value file compressed via GZip.

```
datesDF.to_csv(fullRealZipName, index=False, encoding='utf-8',
compression='gzip')

print(fullRealZipName)
```

Close out the run statistics.

```
nowStop = datetime.datetime.now()
runTime=nowStop-nowStart
print('Start:', nowStart.strftime('%Y-%m-%d %H:%M:%S'))
print('Stop: ', nowStop.strftime('%Y-%m-%d %H:%M:%S'))
print('Time: ', runTime)
```

```
You now have a data set with dates.
```

You now need to generate a data set for every minute in a normal day:

Build a data set for time at a minute level for a day:

```
import datetime
nowStart = datetime.datetime.now()

dirname = '../../../../Chapter 013/990-DL/100-Raw-Zone/'
filename='08-02-Retrieve-Time-Time.csv'
fileZipname=filename + '.gz'
print(fileZipname)

import pandas as pd
import os

pathRealName=os.path.realpath(dirname)
print(pathRealName)
fullRealZipName = os.path.join(pathRealName, fileZipname)

if not os.path.exists(pathRealName):
    print('Make:', pathRealName)
    os.makedirs(pathRealName)

times = pd.date_range(start='2018/01/01 00:00', end='2018/01/01 23:59',
freq='1min')
timesNameDF = pd.DataFrame(times.strftime('%H:%M'))
timesNameDF.index.name='TimeID'
timesNameDF.columns=['Time']

timesKeyDF = pd.DataFrame(times.strftime('T%H%M'))
timesKeyDF.index.name='TimeID'
timesKeyDF.columns=['TimeKey']
```

```
timesDF=timesKeyDF
timesDF=timesDF.merge(timesNameDF, how='inner', left_on='TimeID',
right_on='TimeID')

timesDF.to_csv(fullRealZipName, index=False, encoding='utf-8',
compression='gzip')

print(fullRealZipName)
nowStop = datetime.datetime.now()
runTime=nowStop-nowStart
print('Start:', nowStart.strftime('%Y-%m-%d %H:%M:%S'))
print('Stop: ', nowStop.strftime('%Y-%m-%d %H:%M:%S'))
print('Time: ', runTime)
```

You now have a time data set for one day at minute levels.

The next data set needs both dates and times.

```
import datetime
nowStart = datetime.datetime.now()

dirname = '../../../../Chapter 013/990-DL/100-Raw-Zone/'
filename='08-02-Retrieve-Time.csv'
fileZipname=filename + '.gz'
print(fileZipname)

import pandas as pd
import os

pathRealName=os.path.realpath(dirname)
print(pathRealName)
fullRealZipName = os.path.join(pathRealName, fileZipname)

if not os.path.exists(pathRealName):
    print('Make:', pathRealName)
    os.makedirs(pathRealName)
```

Here is the time portion:

```
dates = pd.date_range(start='2017-1-1', end='2018-12-31', freq='D')

datesKey=dates.strftime('D%Y%m%d')
datesKeyDF=pd.DataFrame(datesKey)
datesKeyDF.index.name='DateID'

datesKeyDF.columns=['DateKey']
datesKeyDF['Key1']='1'

times = pd.date_range(start='2018/01/01 00:00', end='2018/01/01 23:59',
freq='1min')

timesKeyDF = pd.DataFrame(times.strftime('T%H%M'))
timesKeyDF.index.name='TimeID'
timesKeyDF.columns=['TimeKey']
timesKeyDF['Key1']='1'

datetimesDF=datesKeyDF

datetimesDF=datetimesDF.merge(timesKeyDF, how='inner', left_on='Key1',
right_on='Key1')
datetimesDF.drop('Key1', axis=1, inplace=True)
datetimesDF.to_csv(fullRealZipName, index=False, encoding='utf-8',
compression='gzip')

print(fullRealZipName)
nowStop = datetime.datetime.now()
runTime=nowStop-nowStart
print('Start:', nowStart.strftime('%Y-%m-%d %H:%M:%S'))
print('Stop: ', nowStop.strftime('%Y-%m-%d %H:%M:%S'))
print('Time: ', runTime)
```

That gives you the Date Time information.

The next is the person data set.

You are required to generate a data set that has a person data set consisting of 'First Name,' 'Last Name.'

```
import datetime
nowStart = datetime.datetime.now()

dirname = '../../../../Chapter 013/990-DL/100-Raw-Zone/'
filename='08-02-Retrieve-Person.csv'
fileZipname=filename + '.gz'
print(fileZipname)

import pandas as pd
import os

pathRealName=os.path.realpath(dirname)
print(pathRealName)
fullRealZipName = os.path.join(pathRealName, fileZipname)

if not os.path.exists(pathRealName):
    print('Make:', pathRealName)
    os.makedirs(pathRealName)
```

Get the data from the external data source:

```
dirDataName = '../../../../../Data/Chapter 013/000-ExternalData/02-Person/'
pathRealDataName=os.path.realpath(dirDataName)
print(pathRealDataName)

fileDataName='FirstNamesBoy2018.csv'
fileFullDataName= os.path.join(pathRealDataName, fileDataName)
print(fileFullDataName)

firstNameDF=pd.read_csv(fileFullDataName, header=0)
firstNameDF.set_index("ID", inplace = True)
firstNameDF.columns=['FirstName']
print(firstNameDF.shape)
print(firstNameDF.head())

fileDataName='LastNames150.csv'
fileFullDataName= os.path.join(pathRealDataName, fileDataName)
print(fileFullDataName)
```

```
lastNameRawDF=pd.read_csv(fileFullDataName, header=0)
lastNameRawDF.index.name='IDKey'
lastNameRawDF.shape
lastNameRawDF.head()

lastName1DF=pd.DataFrame(lastNameRawDF.loc[:, ['ID1','Name1','Cnt1']])
lastName1DF.columns=['ID','LastName','Cnt']
lastName1DF.set_index("ID", inplace = True)
print(lastName1DF.shape)
print(lastName1DF.head())

lastName2DF=pd.DataFrame(lastNameRawDF.loc[:, ['ID2','Name2','Cnt2']])
lastName2DF.columns=['ID','LastName','Cnt']
lastName2DF.set_index("ID", inplace = True)
print(lastName2DF.shape)
print(lastName2DF.head())

lastName3DF=pd.DataFrame(lastNameRawDF.loc[:, ['ID3','Name3','Cnt3']])
lastName3DF.columns=['ID','LastName','Cnt']
lastName3DF.set_index("ID", inplace = True)
print(lastName3DF.shape)
print(lastName3DF.head())

DFs = [lastName1DF, lastName2DF, lastName3DF]
lastNameDF=pd.DataFrame(pd.concat(DFs))
lastNameDF.drop(['Cnt'], axis=1, inplace=True)

print(lastNameDF.shape)
print(lastNameDF.head())

firstNameDF['keyid']=0
lastNameDF['keyid']=0

NameDF=pd.merge(firstNameDF, lastNameDF, how='inner')
NameDF.drop(['keyid'], axis=1, inplace=True)
NameDF.reset_index(inplace=True)
NameDF.index.name='KeyID'
NameDF.columns=['ID','FirstName','LastName']
```

```
print(NameDF.shape)
print(NameDF.head())

NameDF.to_csv(fullRealZipName, index=False, encoding='utf-8',
compression='gzip')

print(fullRealZipName)
nowStop = datetime.datetime.now()
runTime=nowStop-nowStart
print('Start:', nowStart.strftime('%Y-%m-%d %H:%M:%S'))
print('Stop: ', nowStop.strftime('%Y-%m-%d %H:%M:%S'))
print('Time: ', runTime)
```

Well done; you now have an internal generated time data set and a porting from outside the data lake data set. Those two types of data are the backbone of most of the data engineering requirements within the RIF ecosystem.

Next you will be able to create the 'object' data set. It is an external data source reading process. You will import and export the data as you have done before.

```
import datetime
nowStart = datetime.datetime.now()

dirname = '../../../../Chapter 013/990-DL/100-Raw-Zone/'
filename='08-02-Retrieve-Object.csv'
fileZipname=filename + '.gz'
print(fileZipname)

import pandas as pd
import os

dirDataName = '../../../../../Data/Chapter 013/000-ExternalData/03-Object/'
pathRealDataName=os.path.realpath(dirDataName)
print(pathRealDataName)

fileDataName='Object.csv.gz'
fileFullDataName= os.path.join(pathRealDataName, fileDataName)
print(fileFullDataName)
```

```
objectFullDF= pd.read_csv(fileFullDataName, encoding='utf-8',
compression='gzip')
print(objectFullDF.shape)

pathRealName=os.path.realpath(dirname)
print(pathRealName)
fullRealZipName = os.path.join(pathRealName, fileZipname)

if not os.path.exists(pathRealName):
    print('Make:', pathRealName)
    os.makedirs(pathRealName)
```

Save Object Data set:

```
objectFullDF.to_csv(fullRealZipName, index=False, encoding='utf-8',
compression='gzip')

print(fullRealZipName)
nowStop = datetime.datetime.now()
runTime=nowStop-nowStart
print('Start:', nowStart.strftime('%Y-%m-%d %H:%M:%S'))
print('Stop: ', nowStop.strftime('%Y-%m-%d %H:%M:%S'))
print('Time: ', runTime)
```

You now have an object data set.

Next you prepare a Location Data set.

```
import datetime
nowStart = datetime.datetime.now()

dirname = '../../../../Chapter 013/990-DL/100-Raw-Zone/'
filename='08-02-Retrieve-Location.csv'
fileZipname=filename + '.gz'
print(fileZipname)

import pandas as pd
import zipfile as zp
import os
```

```
pathRealName=os.path.realpath(dirname)
print(pathRealName)
fullRealZipName = os.path.join(pathRealName, fileZipname)

if not os.path.exists(pathRealName):
    print('Make:', pathRealName)
    os.makedirs(pathRealName)

dirDataName = '../../../../../Data/Chapter 013/000-ExternalData/04-
Location/'
pathRealDataName=os.path.realpath(dirDataName)
print(pathRealDataName)

fileZipName='ukpostcodes.zip'
fileDataName='ukpostcodes.csv'
fileFullZipName= os.path.join(pathRealDataName, fileZipName)
print(fileFullZipName)
```

Tip This CSV file is within a zip file, so you need to open it up first with a zip tool and then read the CSV file.

```
with zp.ZipFile(fileFullZipName) as z:
    with z.open(fileDataName) as f:
        locationDF = pd.read_csv(f, header=0, delimiter=',')

locationDF.index.name='IDKey'
print(locationDF.shape)
print(locationDF.head())

locationDF.to_csv(fullRealZipName, index=False, encoding='utf-8',
compression='gzip')

print(fullRealZipName)
nowStop = datetime.datetime.now()
runTime=nowStop-nowStart
print('Start:', nowStart.strftime('%Y-%m-%d %H:%M:%S'))
print('Stop: ', nowStop.strftime('%Y-%m-%d %H:%M:%S'))
print('Time: ', runTime)
```

You now have a location data set ready for use.

The next data set you need is event data.

```
import datetime
nowStart = datetime.datetime.now()

dirname = '../../../../Chapter 013/990-DL/100-Raw-Zone/'
filename='08-02-Retrieve-Event.csv'
fileZipname=filename + '.gz'
print(fileZipname)

import pandas as pd
import os

dirDataName = '../../../../../Data/Chapter 013/000-ExternalData/05-Event/'
pathRealDataName=os.path.realpath(dirDataName)
print(pathRealDataName)

fileDataName='Event.csv.gz'
fileFullDataName= os.path.join(pathRealDataName, fileDataName)
print(fileFullDataName)

objectFullDF= pd.read_csv(fileFullDataName, encoding='utf-8',
compression='gzip')
print(objectFullDF.shape)

pathRealName=os.path.realpath(dirname)
print(pathRealName)
fullRealZipName = os.path.join(pathRealName, fileZipname)

if not os.path.exists(pathRealName):
    print('Make:', pathRealName)
    os.makedirs(pathRealName)

objectFullDF.to_csv(fullRealZipName, index=False, encoding='utf-8',
compression='gzip')

print(fullRealZipName)
nowStop = datetime.datetime.now()
runTime=nowStop-nowStart
```

```
print('Start:', nowStart.strftime('%Y-%m-%d %H:%M:%S'))
print('Stop: ', nowStop.strftime('%Y-%m-%d %H:%M:%S'))
print('Time: ', runTime)
```

Well done; you now have a complete Time-Person-Object-Location-Event structure.

You can describe any business activity with the five structures.

Example: John Smith pays cash money into his credit card to the value of £30-00 at 12h45 on 4 January 2018 at West Street bank of South Bank.

This gets recorded as:

- Time => 2018-01-04 12:45
- Person => John Smith
- Object => cash, credit card and South Bank
- Location => longitude and latitude of West Street Branch
- Event => money deposit

This basic structure can describe other real-world actions easily, and it gives a standard base for your IML.

You have successfully loaded the five types of data sets (T-P-O-L-E) into the data lake ecosystem.

Next you move on to the assessment of the data you loaded.

Assess Super Step

The assess super step (Figure 13-6) supports the metadata extraction and data quality clean-up in the factory.

Assess

You need to load the following Jupyter Notebook from example directory Chapter 013:

```
Chapter-013-02-Assess.ipynb
```

You set up the factory for an assess runner process, in the same manner as in the Retrieve Step (Figure 13-6).

```
import datetime
nowStart = datetime.datetime.now()
```

This is the main change. It is pointing to the Assess processes in the factory.

```
pathRIFname='./000-RIF/100-Functional-Layer/200-Assess/'

import subprocess
import os
import time
import sys
import numpy as np

path=np.array(sys.path)
print(path.shape)
for i in range(path.shape[0]):
    pathTestPython=os.path.join(path[i],'python.exe')
    if os.path.exists(pathTestPython):
        pathPython=pathTestPython
pathPython=pathPython.replace('\\','/')
print(pathPython)

holddir=os.getcwd()
print(holddir)

def run_command(command, wait=True):

    try:
        if (wait):

            p = subprocess.Popen(
                [command]
                , stdout = subprocess.PIPE
                , shell = True)
            p.wait()
```

```
        else:
            p = subprocess.Popen(
                [command],
                shell = True,
                stdin = None, stdout = None, stderr = None, close_fds = True)

        (result, error) = p.communicate()

    except subprocess.CalledProcessError as e:
        sys.stderr.write(
            "common::run_command() : [ERROR]: output = %s, error code = %s\n"
            % (e.output, e.returncode))

    return result

pathRealRIFname=os.path.realpath(pathRIFname)
pathRealRIFname=pathRealRIFname.replace('\\','/')
print(pathRealRIFname)

if not os.path.exists(pathRealRIFname):
    print(' Please run the Chapter-013-01-Setup program first to setup your
    examples')
else:
    os.chdir(pathRealRIFname)
    for root, dirs, files in os.walk(pathRealRIFname, topdown=True):
        for name in files:
            fullfilename=os.path.join(root, name)
            fullfilename=fullfilename.replace('\\','/')
            filename, file_extension = os.path.splitext(name)
            if file_extension.lower()=='.py':
                #print(filename)
                commandstr=name
                print(commandstr,'\n')
                res = run_command(commandstr)
                time.sleep(0.5)
                print('####################\n')
                print(res)
                print('####################\n')
```

```
os.chdir(holddir)
print(os.getcwd())

nowStop = datetime.datetime.now()
runTime=nowStop-nowStart
print('Start:', nowStart.strftime('%Y-%m-%d %H:%M:%S'))
print('Stop: ', nowStop.strftime('%Y-%m-%d %H:%M:%S'))
print('Time: ', runTime)
```

You now have an Assess runner for the RAPTOR process.

You use the example directory at the following subdirectory: IML\Code\Chapter 013\000-RIF\100-Functional-Layer\200-Assess

Open Chapter-013-02-Assess-01-Time.ipynb

```
import datetime
nowStart = datetime.datetime.now()

dirname = '../../../../Chapter 013/990-DL/200-Structured-Zone/'
filename='Chapter-013-02-Assess-Time.csv'
fileZipname=filename + '.gz'
print(fileZipname)

import pandas as pd
import os

dirDataName = '../../../../Chapter 013/990-DL/100-Raw-Zone/'
pathRealDataName=os.path.realpath(dirDataName)
print(pathRealDataName)

fileDataName='Chapter-013-02-Retrieve-Time.csv.gz'
fileFullDataName= os.path.join(pathRealDataName, fileDataName)
print(fileFullDataName)

timeFullDF= pd.read_csv(fileFullDataName, encoding='utf-8',
compression='gzip')
print(timeFullDF.shape)
```

You will now perform Data Quality Checks.

Warning I have limited the data set you are processing to 500. If you want to perform the complete data, set simple replace: B=timeFullDF.head(500) with B=timeFullDF in the next block of code.

```
A=timeFullDF
print('A',A.shape)
B=timeFullDF.head(500)
print('B',B.shape)

D=pd.concat([A,B], axis=0)
print('D',D.shape)

E=D.drop_duplicates(keep='last')
print('E',E.shape)

TimeFinalDF=E

pathRealName=os.path.realpath(dirname)
print(pathRealName)
fullRealZipName = os.path.join(pathRealName, fileZipname)

if not os.path.exists(pathRealName):
    print('Make:', pathRealName)
    os.makedirs(pathRealName)
```

You can now save the clean Date data set.

```
TimeFinalDF.to_csv(fullRealZipName, index=False, encoding='utf-8',
compression='gzip')

print(fullRealZipName)
nowStop = datetime.datetime.now()
runTime=nowStop-nowStart
print('Start:', nowStart.strftime('%Y-%m-%d %H:%M:%S'))
print('Stop: ', nowStop.strftime('%Y-%m-%d %H:%M:%S'))
print('Time: ', runTime)
```

The next Time process will assess the Time Data set:

```
import datetime
nowStart = datetime.datetime.now()

dirname = '../../../../Chapter 013/990-DL/200-Structured-Zone/'
filename='08-02-Assess-Time-Time.csv'
fileZipname=filename + '.gz'
print(fileZipname)
import pandas as pd
import os

dirDataName = '../../../../Chapter 013/990-DL/100-Raw-Zone/'
pathRealDataName=os.path.realpath(dirDataName)
print(pathRealDataName)

fileDataName='08-02-Retrieve-Time-Time.csv.gz'
fileFullDataName= os.path.join(pathRealDataName, fileDataName)
print(fileFullDataName)

timeFullDF= pd.read_csv(fileFullDataName, encoding='utf-8',
compression='gzip')
print(timeFullDF.shape)
```

Perform Data Quality Checks on the Time.

```
A=timeFullDF
print('A',A.shape)
B=timeFullDF.head(500)
print('B',B.shape)

D=pd.concat([A,B], axis=0)
print('D',D.shape)

E=D.drop_duplicates(keep='last')
print('E',E.shape)

TimeFinalDF=E

pathRealName=os.path.realpath(dirname)
print(pathRealName)
fullRealZipName = os.path.join(pathRealName, fileZipname)
```

```
if not os.path.exists(pathRealName):
    print('Make:', pathRealName)
    os.makedirs(pathRealName)

TimeFinalDF.to_csv(fullRealZipName, index=False, encoding='utf-8',
compression='gzip')

print(fullRealZipName)
nowStop = datetime.datetime.now()
runTime=nowStop-nowStart
print('Start:', nowStart.strftime('%Y-%m-%d %H:%M:%S'))
print('Stop: ', nowStop.strftime('%Y-%m-%d %H:%M:%S'))
print('Time: ', runTime)

import datetime
nowStart = datetime.datetime.now()

dirname = '../../../../Chapter 013/990-DL/200-Structured-Zone/'
filename='08-02-Assess-Time-Date.csv'
fileZipname=filename + '.gz'
print(fileZipname)

import pandas as pd
import os

dirDataName = '../../../../Chapter 013/990-DL/100-Raw-Zone/'
pathRealDataName=os.path.realpath(dirDataName)
print(pathRealDataName)

fileDataName='08-02-Retrieve-Time-Date.csv.gz'
fileFullDataName= os.path.join(pathRealDataName, fileDataName)
print(fileFullDataName)

timeFullDF= pd.read_csv(fileFullDataName, encoding='utf-8',
compression='gzip')
print(timeFullDF.shape)
```

Perform Data Quality Checks.

```
A=timeFullDF
print('A',A.shape)
B=timeFullDF.head(500)
print('B',B.shape)

D=pd.concat([A,B], axis=0)
print('D',D.shape)

E=D.drop_duplicates(keep='last')
print('E',E.shape)

TimeFinalDF=E

pathRealName=os.path.realpath(dirname)
print(pathRealName)
fullRealZipName = os.path.join(pathRealName, fileZipname)

if not os.path.exists(pathRealName):
    print('Make:', pathRealName)
    os.makedirs(pathRealName)

TimeFinalDF.to_csv(fullRealZipName, index=False, encoding='utf-8',
compression='gzip')

print(fullRealZipName)
nowStop = datetime.datetime.now()
runTime=nowStop-nowStart
print('Start:', nowStart.strftime('%Y-%m-%d %H:%M:%S'))
print('Stop: ', nowStop.strftime('%Y-%m-%d %H:%M:%S'))
print('Time: ', runTime)

import datetime
nowStart = datetime.datetime.now()

dirname = '../../../../Chapter 013/990-DL/200-Structured-Zone/'
filename='08-02-Assess-Person.csv'
fileZipname=filename + '.gz'
print(fileZipname)
```

The next assess process is for Person.

```
import pandas as pd
import os

dirDataName = '../../../../Chapter 013/990-DL/100-Raw-Zone/'
pathRealDataName=os.path.realpath(dirDataName)
print(pathRealDataName)

fileDataName='08-02-Retrieve-Person.csv.gz'
fileFullDataName= os.path.join(pathRealDataName, fileDataName)
print(fileFullDataName)

personFullDF= pd.read_csv(fileFullDataName, encoding='utf-8',
compression='gzip')
print(personFullDF.shape)
```

Perform Data Quality Checks:

```
A=personFullDF

E=A.drop_duplicates(keep='last')
print('E',E.shape)

PersonFinalDF=E

pathRealName=os.path.realpath(dirname)
print(pathRealName)
fullRealZipName = os.path.join(pathRealName, fileZipname)

if not os.path.exists(pathRealName):
    print('Make:', pathRealName)
    os.makedirs(pathRealName)

print(fullRealZipName)

PersonFinalDF.to_csv(fullRealZipName, index=False, encoding='utf-8',
compression='gzip')

print(fullRealZipName)
nowStop = datetime.datetime.now()
runTime=nowStop-nowStart
```

```
print('Start:', nowStart.strftime('%Y-%m-%d %H:%M:%S'))
print('Stop: ', nowStop.strftime('%Y-%m-%d %H:%M:%S'))
print('Time: ', runTime)
```

Load the Object data set:

```
import datetime
nowStart = datetime.datetime.now()

dirname = '../../../../Chapter 013/990-DL/200-Structured-Zone/'
filename='08-02-Assess-Object.csv'
fileZipname=filename + '.gz'
print(fileZipname)

import pandas as pd
import os

dirDataName = '../../../../Chapter 013/990-DL/100-Raw-Zone/'
pathRealDataName=os.path.realpath(dirDataName)
print(pathRealDataName)

fileDataName='08-02-Retrieve-Object.csv.gz'
fileFullDataName= os.path.join(pathRealDataName, fileDataName)
print(fileFullDataName)

objectFullDF= pd.read_csv(fileFullDataName, encoding='utf-8',
compression='gzip')
print(objectFullDF.shape)

Now perform Data Quality Checks:

A=objectFullDF
print('A',A.shape)

E=A.drop_duplicates(keep='last')
print('A',E.shape)

objectFinalDF=E

pathRealName=os.path.realpath(dirname)
print(pathRealName)
```

```
fullRealZipName = os.path.join(pathRealName, fileZipname)
if not os.path.exists(pathRealName):
    print('Make:', pathRealName)
    os.makedirs(pathRealName)
```

Now save Object data sets.

```
print(fullRealZipName)

objectFinalDF.to_csv(fullRealZipName, index=False, encoding='utf-8',
compression='gzip')

print(fullRealZipName)
nowStop = datetime.datetime.now()
runTime=nowStop-nowStart
print('Start:', nowStart.strftime('%Y-%m-%d %H:%M:%S'))
print('Stop: ', nowStop.strftime('%Y-%m-%d %H:%M:%S'))
print('Time: ', runTime)
```

Perform the Location data set's quality assessments:

```
import datetime
nowStart = datetime.datetime.now()

dirname = '../../../../Chapter 013/990-DL/200-Structured-Zone/'
filename='08-02-Assess-Location.csv'
fileZipname=filename + '.gz'
print(fileZipname)

import pandas as pd
import os

dirDataName = '../../../../Chapter 013/990-DL/100-Raw-Zone/'
pathRealDataName=os.path.realpath(dirDataName)
print(pathRealDataName)

fileDataName='08-02-Retrieve-Location.csv.gz'
fileFullDataName= os.path.join(pathRealDataName, fileDataName)
print(fileFullDataName)
```

```
locationFullDF= pd.read_csv(fileFullDataName, encoding='utf-8',
compression='gzip')
print(locationFullDF.shape)
```

Perform Data Quality Checks on the location data sets:

```
A=locationFullDF
print('A',A.shape)

E=A.drop_duplicates(keep='last')
print('E',E.shape)

locationFinalDF=E

pathRealName=os.path.realpath(dirname)
print(pathRealName)
fullRealZipName = os.path.join(pathRealName, fileZipname)

if not os.path.exists(pathRealName):
    print('Make:', pathRealName)
    os.makedirs(pathRealName)
```

Save the Location data set.

```
locationFinalDF.to_csv(fullRealZipName, index=False, encoding='utf-8',
compression='gzip')

print(fullRealZipName)
nowStop = datetime.datetime.now()
runTime=nowStop-nowStart
print('Start:', nowStart.strftime('%Y-%m-%d %H:%M:%S'))
print('Stop: ', nowStop.strftime('%Y-%m-%d %H:%M:%S'))
print('Time: ', runTime)
```

The next assessment is on the Event data set.

```
import datetime
nowStart = datetime.datetime.now()

dirname = '../../../../Chapter 013/990-DL/200-Structured-Zone/'
filename='08-02-Assess-Event.csv'
```

```
fileZipname=filename + '.gz'
print(fileZipname)

import pandas as pd
import os

dirDataName = '../../../../Chapter 013/990-DL/100-Raw-Zone/'
pathRealDataName=os.path.realpath(dirDataName)
print(pathRealDataName)

fileDataName='08-02-Retrieve-Time.csv.gz'
fileFullDataName= os.path.join(pathRealDataName, fileDataName)
print(fileFullDataName)

eventFullDF= pd.read_csv(fileFullDataName, encoding='utf-8',
compression='gzip')
print(eventFullDF.shape)
```

Perform Data Quality Checks on Event data set:

```
A=eventFullDF
print('A',A.shape)

E=A.drop_duplicates(keep='last')
print('E',E.shape)

EventFinalDF=E

pathRealName=os.path.realpath(dirname)
print(pathRealName)
fullRealZipName = os.path.join(pathRealName, fileZipname)

if not os.path.exists(pathRealName):
    print('Make:', pathRealName)
    os.makedirs(pathRealName)

EventFinalDF.to_csv(fullRealZipName, index=False, encoding='utf-8',
compression='gzip')

print(fullRealZipName)
nowStop = datetime.datetime.now()
runTime=nowStop-nowStart
```

```
print('Start:', nowStart.strftime('%Y-%m-%d %H:%M:%S'))
print('Stop: ', nowStop.strftime('%Y-%m-%d %H:%M:%S'))
print('Time: ', runTime)
```

That completes the assess step for the data processing.

Process Super Step

The process super step (Figure 13-6) converts data into a data vault.

You need to load the following Jupyter Notebook from example directory Chapter 013: Chapter-013-02-Process.ipynb

The next step enables the consolidation of the data sets into a single data set across the five data sets.

```
import datetime
nowStart = datetime.datetime.now()

pathRIFname='./000-RIF/100-Functional-Layer/300-Process/'

import subprocess
import os
import time
import sys
import numpy as np

path=np.array(sys.path)
print(path.shape)
for i in range(path.shape[0]):
    pathTestPython=os.path.join(path[i],'python.exe')
    if os.path.exists(pathTestPython):
        pathPython=pathTestPython
```

```
pathPython=pathPython.replace('\\','/')
print(pathPython)

holddir=os.getcwd()
print(holddir)

def run_command(command, wait=True):

    try:
        if (wait):

            p = subprocess.Popen(
                [command]
                , stdout = subprocess.PIPE
                , shell = True)
            p.wait()
        else:
            p = subprocess.Popen(
                [command],
                shell = True,
                stdin = None, stdout = None, stderr = None, close_fds = True)

        (result, error) = p.communicate()

    except subprocess.CalledProcessError as e:
        sys.stderr.write(
            "common::run_command() : [ERROR]: output = %s, error code = %s\n"
            % (e.output, e.returncode))

    return result

pathRealRIFname=os.path.realpath(pathRIFname)
pathRealRIFname=pathRealRIFname.replace('\\','/')
print(pathRealRIFname)

if not os.path.exists(pathRealRIFname):
    print(' Please run the Chapter-013-01-Setup program first to setup your
    examples')
else:
    os.chdir(pathRealRIFname)
```

```
    for root, dirs, files in os.walk(pathRealRIFname, topdown=True):
        for name in files:
            fullfilename=os.path.join(root, name)
            fullfilename=fullfilename.replace('\\','/')
            filename, file_extension = os.path.splitext(name)
            if file_extension.lower()=='.py':
                #print(filename)
                commandstr=name
                print(commandstr,'\n')
                res = run_command(commandstr)
                time.sleep(0.5)
                print('####################\n')
                print(res)
                print('####################\n')
os.chdir(holddir)
print(os.getcwd())

nowStop = datetime.datetime.now()
runTime=nowStop-nowStart
print('Start:', nowStart.strftime('%Y-%m-%d %H:%M:%S'))
print('Stop: ', nowStop.strftime('%Y-%m-%d %H:%M:%S'))
print('Time: ', runTime)
```

This process runner enables the process step of the RAPTOR flow (Figure 13-6).

You need to look in the example directory at the following subdirectory: IML\Code\Chapter 013\000-RIF\100-Functional-Layer\300-Process

Open Chapter-013-02-Process-01-Time.ipynb

```
import datetime
nowStart = datetime.datetime.now()

dirname = '../../../../Chapter 013/990-DL/300-Curated-Zone/'
filename='08-02-Process-Time.csv'
fileZipname=filename + '.gz'
print(fileZipname)

import pandas as pd
import os
```

```
dirDataName = '../../../../Chapter 013/990-DL/200-Structured-Zone/'
pathRealDataName=os.path.realpath(dirDataName)
print(pathRealDataName)

fileDataName='08-02-Assess-Time.csv.gz'
fileFullDataName= os.path.join(pathRealDataName, fileDataName)
print(fileFullDataName)

time1DF= pd.read_csv(fileFullDataName, encoding='utf-8',
compression='gzip')
print(time1DF.shape)

fileDataName='08-02-Assess-Time-Date.csv.gz'
fileFullDataName= os.path.join(pathRealDataName, fileDataName)
print(fileFullDataName)
time2DF= pd.read_csv(fileFullDataName, encoding='utf-8',
compression='gzip')
print(time2DF.shape)

fileDataName='08-02-Assess-Time-Time.csv.gz'
fileFullDataName= os.path.join(pathRealDataName, fileDataName)
print(fileFullDataName)

time3DF= pd.read_csv(fileFullDataName, encoding='utf-8',
compression='gzip')
print(time3DF.shape)

time4DF=pd.merge(time1DF,time2DF)
timeFullDF=pd.merge(time4DF,time3DF)
print(timeFullDF.shape)

pathRealName=os.path.realpath(dirname)
print(pathRealName)
fullRealZipName = os.path.join(pathRealName, fileZipname)

if not os.path.exists(pathRealName):
    print('Make:', pathRealName)
    os.makedirs(pathRealName)
```

```
timeFullDF.to_csv(fullRealZipName, index=False, encoding='utf-8',
compression='gzip')

print(fullRealZipName)
nowStop = datetime.datetime.now()
runTime=nowStop-nowStart
print('Start:', nowStart.strftime('%Y-%m-%d %H:%M:%S'))
print('Stop: ', nowStop.strftime('%Y-%m-%d %H:%M:%S'))
print('Time: ', runTime)
```

Process Person data sets.

```
import datetime
nowStart = datetime.datetime.now()

dirname = '../../../../Chapter 013/990-DL/300-Curated-Zone/'
filename='08-02-Process-Person.csv'
fileZipname=filename + '.gz'
print(fileZipname)

import pandas as pd
import os

dirDataName = '../../../../Chapter 013/990-DL/200-Structured-Zone/'
pathRealDataName=os.path.realpath(dirDataName)
print(pathRealDataName)

fileDataName='08-02-Assess-Person.csv.gz'
fileFullDataName= os.path.join(pathRealDataName, fileDataName)
print(fileFullDataName)

personFullDF= pd.read_csv(fileFullDataName, encoding='utf-8',
compression='gzip')
print(personFullDF.shape)

pathRealName=os.path.realpath(dirname)
print(pathRealName)
fullRealZipName = os.path.join(pathRealName, fileZipname)
```

```
if not os.path.exists(pathRealName):
    print('Make:', pathRealName)
    os.makedirs(pathRealName)

personFullDF.to_csv(fullRealZipName, index=False, encoding='utf-8',
compression='gzip')

print(fullRealZipName)
nowStop = datetime.datetime.now()
runTime=nowStop-nowStart
print('Start:', nowStart.strftime('%Y-%m-%d %H:%M:%S'))
print('Stop: ', nowStop.strftime('%Y-%m-%d %H:%M:%S'))
print('Time: ', runTime)
```

Process Object data sets.

```
import datetime
nowStart = datetime.datetime.now()

dirname = '../../../../Chapter 013/990-DL/300-Curated-Zone/'
filename='08-02-Process-Object.csv'
fileZipname=filename + '.gz'
print(fileZipname)

import pandas as pd
import os

dirDataName = '../../../../Chapter 013/990-DL/200-Structured-Zone/'
pathRealDataName=os.path.realpath(dirDataName)
print(pathRealDataName)

fileDataName='08-02-Assess-Object.csv.gz'
fileFullDataName= os.path.join(pathRealDataName, fileDataName)
print(fileFullDataName)

objectFullDF= pd.read_csv(fileFullDataName, encoding='utf-8',
compression='gzip')
print(objectFullDF.shape)
```

```
pathRealName=os.path.realpath(dirname)
print(pathRealName)
fullRealZipName = os.path.join(pathRealName, fileZipname)

if not os.path.exists(pathRealName):
    print('Make:', pathRealName)
    os.makedirs(pathRealName)

objectFullDF.to_csv(fullRealZipName, index=False, encoding='utf-8',
compression='gzip')

print(fullRealZipName)
nowStop = datetime.datetime.now()
runTime=nowStop-nowStart
print('Start:', nowStart.strftime('%Y-%m-%d %H:%M:%S'))
print('Stop: ', nowStop.strftime('%Y-%m-%d %H:%M:%S'))
print('Time: ', runTime)
```

Process the Location data sets.

```
import datetime
nowStart = datetime.datetime.now()

dirname = '../../../../Chapter 013/990-DL/300-Curated-Zone/'
filename='08-02-Process-Location.csv'
fileZipname=filename + '.gz'
print(fileZipname)

import pandas as pd
import os

dirDataName = '../../../../Chapter 013/990-DL/200-Structured-Zone/'
pathRealDataName=os.path.realpath(dirDataName)
print(pathRealDataName)

fileDataName='08-02-Assess-Location.csv.gz'
fileFullDataName= os.path.join(pathRealDataName, fileDataName)
print(fileFullDataName)
```

```
locationFullDF= pd.read_csv(fileFullDataName, encoding='utf-8',
compression='gzip')
print(locationFullDF.shape)

pathRealName=os.path.realpath(dirname)
print(pathRealName)
fullRealZipName = os.path.join(pathRealName, fileZipname)

if not os.path.exists(pathRealName):
    print('Make:', pathRealName)
    os.makedirs(pathRealName)

locationFullDF.to_csv(fullRealZipName, index=False, encoding='utf-8',
compression='gzip')

print(fullRealZipName)
nowStop = datetime.datetime.now()
runTime=nowStop-nowStart
print('Start:', nowStart.strftime('%Y-%m-%d %H:%M:%S'))
print('Stop: ', nowStop.strftime('%Y-%m-%d %H:%M:%S'))
print('Time: ', runTime)
```

Process the Event data sets.

```
import datetime
nowStart = datetime.datetime.now()

dirname = '../../../../Chapter 013/990-DL/300-Curated-Zone/'
filename='08-02-Hub-Event.csv'
fileZipname=filename + '.gz'
print(fileZipname)

import pandas as pd
import os

dirDataName = '../../../../Chapter 013/990-DL/200-Structured-Zone/'
pathRealDataName=os.path.realpath(dirDataName)
print(pathRealDataName)
```

```
fileDataName='08-02-Assess-Event.csv.gz'
fileFullDataName= os.path.join(pathRealDataName, fileDataName)
print(fileFullDataName)

eventFullDF= pd.read_csv(fileFullDataName, encoding='utf-8',
compression='gzip')
print(eventFullDF.shape)

pathRealName=os.path.realpath(dirname)
print(pathRealName)
fullRealZipName = os.path.join(pathRealName, fileZipname)

if not os.path.exists(pathRealName):
    print('Make:', pathRealName)
    os.makedirs(pathRealName)

eventFullDF.to_csv(fullRealZipName, index=False, encoding='utf-8',
compression='gzip')
print(fullRealZipName)
nowStop = datetime.datetime.now()
runTime=nowStop-nowStart
print('Start:', nowStart.strftime('%Y-%m-%d %H:%M:%S'))
print('Stop: ', nowStop.strftime('%Y-%m-%d %H:%M:%S'))
print('Time: ', runTime)
```

That completes the process data sets processing in the RAPTOR flow (Figure 13-6).

Transform Super Step

The transform super step (Figure 13-6) converts the data vault via sun modeling (see glossary for details) into dimensional modeling to form a data warehouse. The sun models are discussed later in this chapter in more detail.

Transform

You need to load the following Jupyter Notebook from example directory Chapter 013: Chapter-013-02-03-Transform.ipynb

```
import datetime
nowStart = datetime.datetime.now()

pathRIFname='./000-RIF/100-Functional-Layer/400-Transform/'

import subprocess
import os
import time
import sys
import numpy as np

path=np.array(sys.path)
print(path.shape)
for i in range(path.shape[0]):
    pathTestPython=os.path.join(path[i],'python.exe')
    if os.path.exists(pathTestPython):
        pathPython=pathTestPython
pathPython=pathPython.replace('\\','/')
print(pathPython)

holddir=os.getcwd()
print(holddir)

def run_command(command, wait=True):

    try:
        if (wait):
            p = subprocess.Popen(
                [command]
                , stdout = subprocess.PIPE
                , shell = True)
            p.wait()
        else:
            p = subprocess.Popen(
                [command],
```

```
                shell = True,
                stdin = None, stdout = None, stderr = None, close_fds = True)

        (result, error) = p.communicate()

    except subprocess.CalledProcessError as e:
        sys.stderr.write(
            "common::run_command() : [ERROR]: output = %s, error code = %s\n"
            % (e.output, e.returncode))

    return result

pathRealRIFname=os.path.realpath(pathRIFname)
pathRealRIFname=pathRealRIFname.replace('\\','/')
print(pathRealRIFname)

if not os.path.exists(pathRealRIFname):
    print(' Please run the Chapter-013-01-Setup program first to setup your
    examples')
else:
    os.chdir(pathRealRIFname)
    for root, dirs, files in os.walk(pathRealRIFname, topdown=True):
        for name in files:
            fullfilename=os.path.join(root, name)
            fullfilename=fullfilename.replace('\\','/')
            filename, file_extension = os.path.splitext(name)
            if file_extension.lower()=='.py':
                #print(filename)
                commandstr=name
                print(commandstr,'\n')
                res = run_command(commandstr)
                time.sleep(0.5)
                print('###################\n')
                print(res)
                print('###################\n')

os.chdir(holddir)
print(os.getcwd())
```

```
nowStop = datetime.datetime.now()
runTime=nowStop-nowStart
print('Start:', nowStart.strftime('%Y-%m-%d %H:%M:%S'))
print('Stop: ', nowStop.strftime('%Y-%m-%d %H:%M:%S'))
print('Time: ', runTime)
```

You need to look in the example directory at the following subdirectory: IML\Code\Chapter 013\000-RIF\100-Functional-Layer\400-Transform

```
Open Chapter-013-02-Transform-02-Person-01.ipynb
```

If you run the process, it will produce a new file called: Chapter-013-02-Transform-Person.csv.gz

Organize Super Step

The organize super step subdivides the data warehouse into data marts.

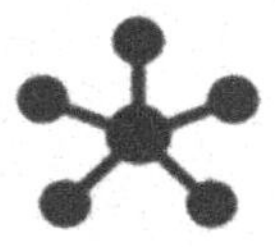

Organise

You need to load the following Jupyter Notebook from example directory Chapter 013:

```
Chapter-013-02-05-Organize.ipynb
```

If you run the process by running the complete run, that will execute all the Python programs.

You need to look in the example directory at the following subdirectory: IML\Code\Chapter 013\000-RIF\100-Functional-Layer\500-Organize

```
Open Chapter-013-02-Organize-02-Person-01.ipynb
```

If you run the process, it will produce a new file called: Chapter-013-02-Organize-Person.csv.gz

Report Super Step

The report super step is the Virtualization capacity of the factory.

> You need to load the following Jupyter Notebook from example directory Chapter 013: Chapter-013-02-Report.ipynb
>
> If you run the process by running the complete run, that will execute all the Python programs.
>
> You need to look in the example directory at the following subdirectory: IML\Code\Chapter 013\000-RIF\100-Functional-Layer\600-Report
>
> Open Chapter-013-02-Raport-02-Person-01.ipynb
>
> If you run the process, it will produce a new file called: 08-02-Raport-Person.csv.gz

Look at:

```
d=personFullDF
d['Cnt'] = 1
print(d.shape)
```

```
import numpy as np
p=pd.pivot_table(d, values='Cnt', index=['LastName'],
columns=['FirstName'], aggfunc=np.sum)
```

```
print(p.shape)
```

```
p.to_csv(fullRealZipName, index=True, encoding='utf-8', compression='gzip')
```

This report process generates a pivot table between FirstName and LastName.

Operational Management Layer

Machine Learning Definitions and Management

The section of the framework is where all the machine learning algorithms are stored, ready to be deployed into the RIF to process the data lake. This section keeps all the preconfigured techniques for the IML and is built by intense research and development by data scientists.

Parameters

The parameters section is where data scientists store all the settings for delivering effective and efficient solutions for specific types of data.

Scheduling

All the schedules for the IML are stored in one central schedule to ensure data scientists have a single view of all the IML processes that are scheduled in the RIF.

Monitoring

This section holds all the programs that provide monitoring capabilities to the RIF to ensure the IML is running effectively and efficiently with the available resources.

Communication

This section holds the central communication processes that enable the approved and centrally supported communication channels.

The type of communication can be direct via handshaking with the processes or via messaging via email, texting, or streaming using Kafka.

Alerting

This section handles all alerting within the IML process. The system handles the communication of any errors or warning as per the pre-agreed levels of tolerance.

Codes Management

This section stores all the common codes for the system. This ensures that there are consistent codes to compare the data against during the assess processing.

Work Cells

This section handles all the configurations for the physical work stations (physical or virtual) to ensure the IML creates the appropriate work cell for the specific workload of the IML algorithms and techniques.

Audit, Balance, and Control Layer

Execution Statistics

This section stores the data for the IML processes and enables the creation of machine learning algorithms to predict when the IML will fail or what work load it will require. I refer to this as being "data science on data science."

Tip Use python -m cProfile -o <test-stats>.profile <test>.py to determine the execution statistics of a program.

I will give you a quick introduction to logging in Python to show you what is possible:

Tip See (docs.python.org/3/library/logging.html) for more details.

You need to load the following Jupyter Notebook from example directory Chapter 013: 000-RIF/300-Audit-Balance-Control-Layer/100-Audit/Chapter-013-01-Audit-01.ipynb

> I will give you just the basic working of the logging. I will use some of the concepts in Chapter 9 when I discuss the real-life solution.

You need:

```
import logging
```

This library supports a complete logging capability for your solutions.

Here is a quick capability demonstration.

```
import logging
import os
import time
import datetime

logging.basicConfig(level=logging.DEBUG)

logDateTime = datetime.datetime.now()
print(logDateTime)
logFilePath = logDateTime.strftime('%Y/%m/')
logFileName = logDateTime.strftime('%Y-%m-%d-%H-%M') + '.log'
print(logFileName)

pathLog=os.getcwd() + '/0100-Logs/'
pathFixedLog=pathLog.replace('\\','/')
print(pathFixedLog)

dirname = os.path.join(pathFixedLog, logFilePath)
if not os.path.exists(dirname):
    os.makedirs(dirname)

logfile= os.path.join(dirname, logFileName)
print(logfile)
logger=logging.basicConfig(filename=logfile, filemode='w', format='%(name)
s - %(levelname)s - %(message)s')

rootLogger = logging.getLogger(")
rootLogger.debug('This is a debug message')

logging.debug('This is a debug message')
logging.info('This is an info message')
logging.warning('This is a warning message')
logging.error('This is an error message')
logging.critical('This is a critical message')
```

This will generate basic logging for your machine learning.

You need to load the following Jupyter Notebook from example directory Chapter 013:

000-RIF/300-Audit-Balance-Control-Layer/100-Audit/08-02-Audit-02.ipynb

This code demonstrates that the logging even works across treading processes that are useful in parallel processing machine learning solutions.

```
import logging
import threading
import time

def worker(arg):
    while not arg['stop']:
        logging.debug('Hi from Worker Process')
        time.sleep(0.5)

def main():
    logging.basicConfig(level=logging.DEBUG, format='%(relativeCreated)6d
    %(threadName)s %(message)s')
    info = {'stop': False}
    thread = threading.Thread(target=worker, args=(info,))
    thread.start()
    l=0
    while True:
        try:
            logging.debug('Hello from Main Process')
            time.sleep(0.75)
            l+=1
            if l > 10:
                info['stop'] = True
                break
        except KeyboardInterrupt:
            info['stop'] = True
            break
    thread.join()
```

```
if __name__ == '__main__':
    print('Start')
    time.sleep(0.2)
    main()
    time.sleep(1)
    print('Done')
```

This was only a basic introduction into logging, but I suggest you discuss with your machine learning team what level of logging you need and implement it as and when you need it.

Warning Using logging.basicConfig(level=logging.DEBUG) generates large log files!

Tip When you run your first machine learning runs of the machine learning, use logging.basicConfig(level=logging.DEBUG); but once you have a running IML solution, lower the level to the minimum that your requirements allow.

Balancing

This section actively balances the IML system resources to ensure the most effective and efficient completion of the IML processing within the allocated resources. This is an increasing important section in the IML when your cloud provider has you pay for what you use, and it is easy to bring resources online that are not effective and efficient in its executions on the cloud.

Controls

This section is the active controller of the IML processes as the modern IML systems are highly oversubscribed, and the controllers enables you to process all the IML processes with limited resources by taking 100% control of the complete process.

Rejects and Error Handling

This section handles any rejections and errors generated by the IML processes. The process in this section should be able to recover from any rejections and errors in the system.

Utility Layer

These are common components supporting other layers.

Maintenance Utilities

The maintenance utilities are designed to perform specific maintenance tasks. These utilities are normally driven by AI process that will use machine learning to activate the utilities.

Data Utilities

The data utilities are common utilities that are designed to support the data lake's specific data processing rules and acts as a common interface for all IML processes.

I will show you a data profiling utility I created for this book to explain to you how you can create common utilities and use them many times.

You need to load the following Jupyter Notebook from example directory Chapter 013: Chapter 013/000-RIF/400-Utility-Layer/200-Data-Utilities/ Chapter-013-01-Utility-Data-Profiler-01.ipynb

```
import numpy as np
import pandas as pd
import pandas_profiling as pdp
import string
```

```
rowcnt=10
columncnt=4
columnset=list(string.ascii_uppercase)
columnheader=columnset[0:columncnt]
print(columnheader)
```

```
df=pd.DataFrame(
    np.random.rand(rowcnt, columncnt),
    columns=columnheader
)
```

```
print(df.shape)
```

(10, 4)

```
print(df.describe())
```

Your results:

```
              A          B          C          D
count 10.000000  10.000000  10.000000  10.000000
mean   0.577240   0.393707   0.717630   0.595825
std    0.253877   0.234504   0.258501   0.129100
min    0.217619   0.141275   0.213244   0.401486
25%    0.329668   0.246237   0.604736   0.513982
50%    0.633584   0.321634   0.746785   0.581329
75%    0.743054   0.451565   0.929597   0.666288
max    0.943685   0.843606   0.994838   0.859670
```

```
print(df.info())
```

```
<class 'pandas.core.frame.DataFrame'>
RangeIndex: 10 entries, 0 to 9
Data columns (total 4 columns):
A     10 non-null float64
B     10 non-null float64
C     10 non-null float64
D     10 non-null float64
dtypes: float64(4)
memory usage: 400.0 bytes
```

```
profile = pdp.ProfileReport(df)
```

```
%%javascript
IPython.OutputArea.auto_scroll_threshold = 9999;
Profile
```

This profiling supports many good characteristics of the data as seen in Figures 13-7, 13-8, 13-9, 13-10, 13-11 and 13-12.

Overview

Dataset info

Number of variables	4
Number of observations	10
Total Missing (%)	0.0%
Total size in memory	400.0 B
Average record size in memory	40.0 B

Variables types

Numeric	4
Categorical	0
Boolean	0
Date	0
Text (Unique)	0
Rejected	0
Unsupported	0

Figure 13-7. *Profiling - Overview*

Variables

A
Numeric

Distinct count	10
Unique (%)	100.0%
Missing (%)	0.0%
Missing (n)	0
Infinite (%)	0.0%
Infinite (n)	0

Mean	0.57724
Minimum	0.21762
Maximum	0.94368
Zeros (%)	0.0%

Statistics Histogram Common Values Extreme Values

Quantile statistics

Minimum	0.21762
5-th percentile	0.24896
Q1	0.32967
Median	0.63358
Q3	0.74305
95-th percentile	0.89684
Maximum	0.94368
Range	0.72607
Interquartile range	0.41339

Descriptive statistics

Standard deviation	0.25388
Coef of variation	0.43981
Kurtosis	-1.4414
Mean	0.57724
MAD	0.21324
Skewness	-0.16601
Sum	5.7724
Variance	0.064453
Memory size	160.0 B

Figure 13-8. *Profiling - Variables - Statistics*

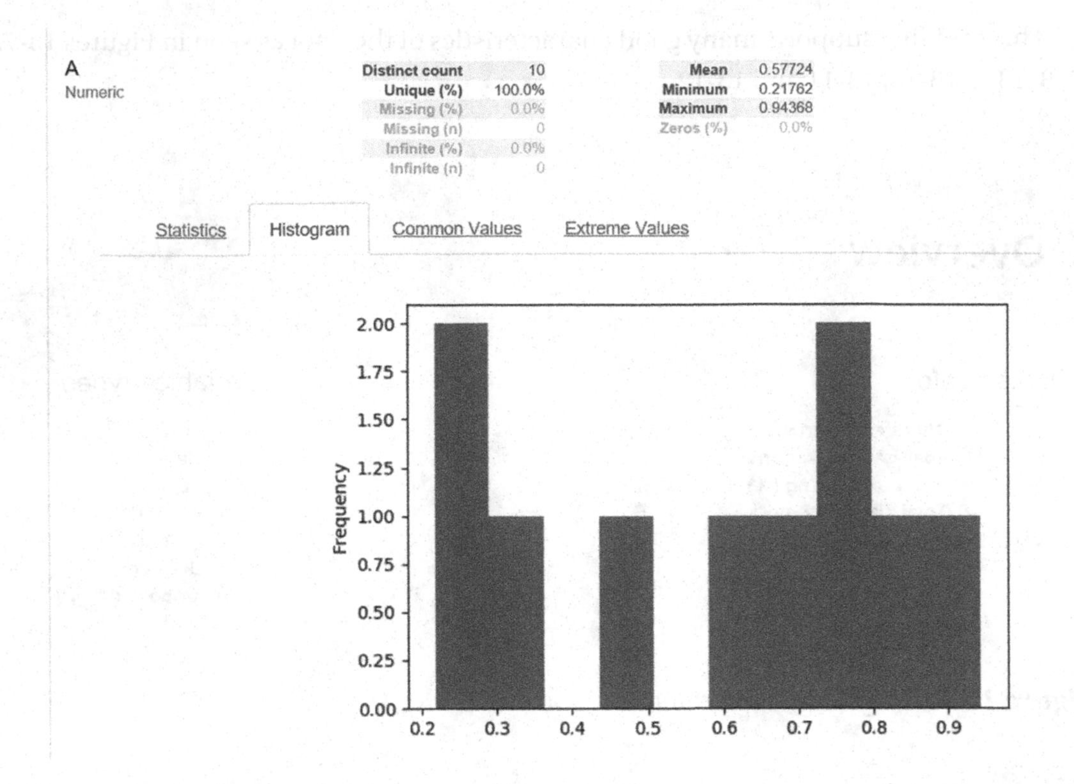

***Figure 13-9.** Profiling – Variables – Histogram*

A
Numeric

Distinct count	10
Unique (%)	100.0%
Missing (%)	0.0%
Missing (n)	0
Infinite (%)	0.0%
Infinite (n)	0

Mean	0.57724
Minimum	0.21762
Maximum	0.94368
Zeros (%)	0.0%

Toggle details

Statistics Histogram Common Values Extreme Values

Value	Count	Frequency (%)
0.2903986611372724	1	10.0%
0.7326896991410324	1	10.0%
0.6849255032952694	1	10.0%
0.21761872442914398	1	10.0%
0.74650892726829	1	10.0%
0.4474760488043278	1	10.0%
0.287262384490765	1	10.0%
0.8395944617987146	1	10.0%
0.5822425719773869	1	10.0%
0.9436845898806948	1	10.0%

***Figure 13-10.** Profiling – Variables – Common Values*

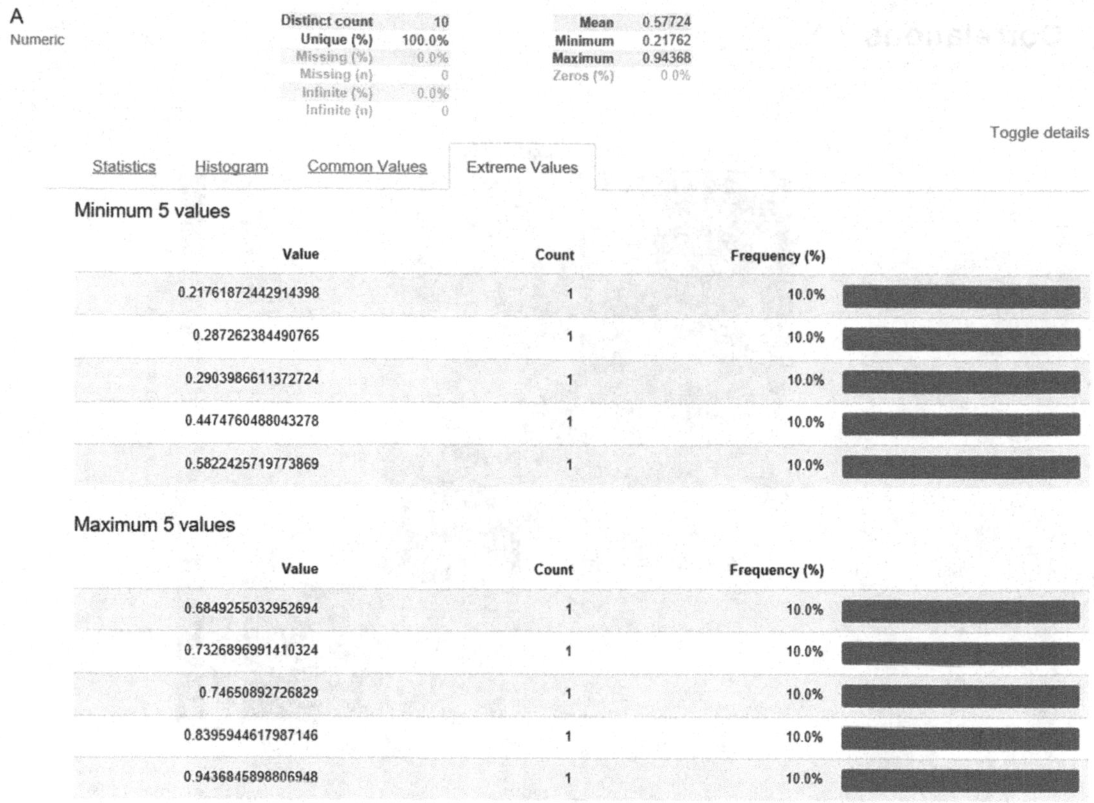

A
Numeric

Distinct count	10
Unique (%)	100.0%
Missing (%)	0.0%
Missing (n)	0
Infinite (%)	0.0%
Infinite (n)	0

Mean	0.57724
Minimum	0.21762
Maximum	0.94368
Zeros (%)	0.0%

Toggle details

Statistics Histogram Common Values Extreme Values

Minimum 5 values

Value	Count	Frequency (%)
0.21761872442914398	1	10.0%
0.287262384490765	1	10.0%
0.2903986611372724	1	10.0%
0.4474760488043278	1	10.0%
0.5822425719773869	1	10.0%

Maximum 5 values

Value	Count	Frequency (%)
0.6849255032952694	1	10.0%
0.7326896991410324	1	10.0%
0.74650892726829	1	10.0%
0.8395944617987146	1	10.0%
0.9436845898806948	1	10.0%

Figure 13-11. *Profiling – Variables – Extreme Values*

Correlations

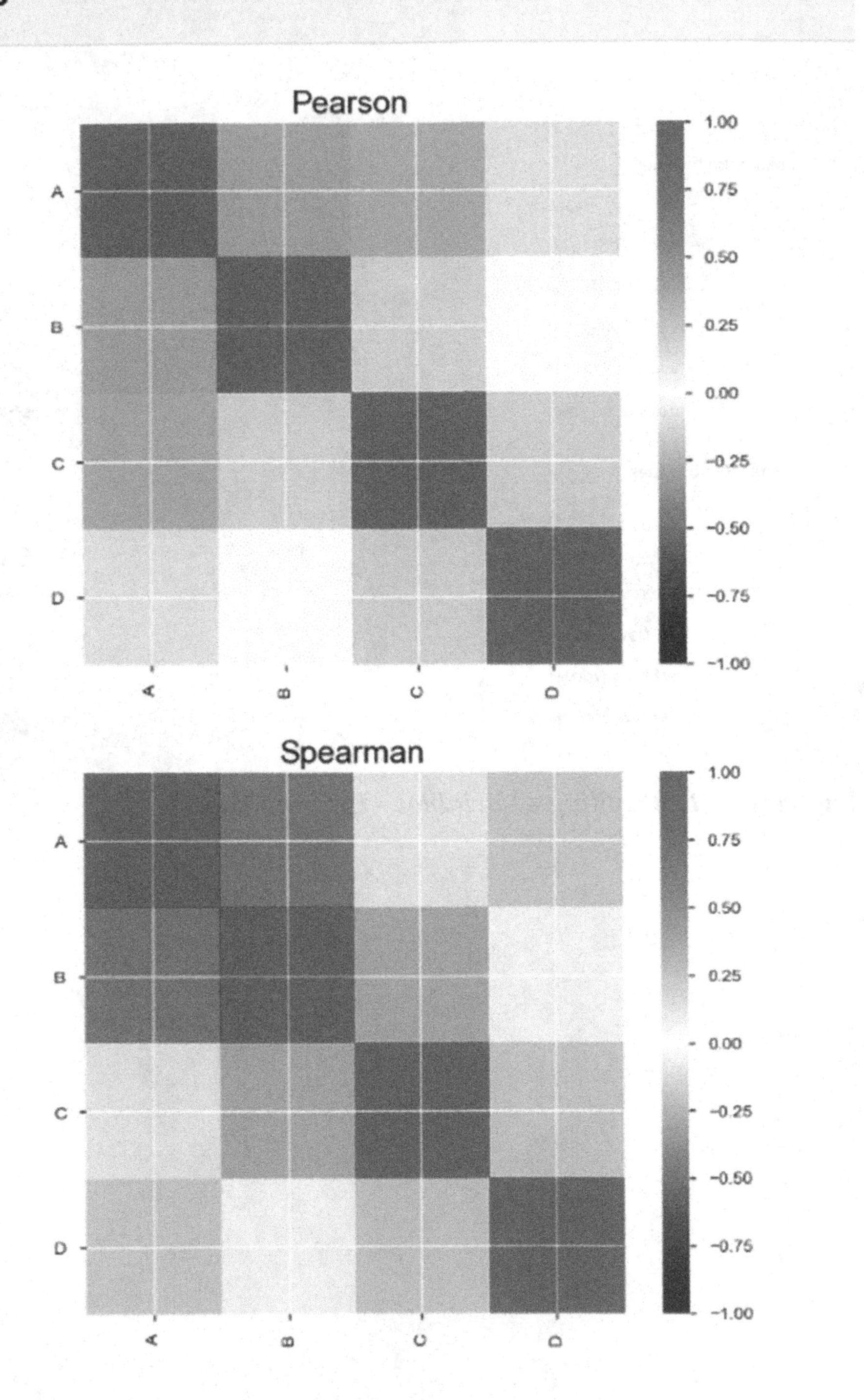

Figure 13-12. *Profiling – Correlations*

The Pearson correlation coefficient is a measure of the linear correlation between two variables X and Y.

The Spearman correlation between two variables is equal to the Pearson correlation between the rank values of those two variables; while Pearson's correlation assesses linear relationships, Spearman's correlation assesses monotonic relationships (whether linear or not). See Figure 13-13.

Sample

	A	B	C	D
0	0.839594	0.276943	0.940067	0.401486
1	0.943685	0.160689	0.898190	0.483380
2	0.732690	0.440489	0.663103	0.859670
3	0.290399	0.317968	0.213244	0.602106
4	0.684926	0.325301	0.830468	0.667925

Figure 13-13. *Profiling - Sample*

```
rejected_variables = profile.get_rejected_variables(threshold=0.9)

print(rejected_variables)

profile.to_file(outputfile="./100-Data-Profiles/08-Utility-Data-
Profiler-01.html")
```

Please check the 08-Utility-Data-Profiler-01.html using a web browser:

```
import webbrowser
import os

wwwRealfile=os.path.realpath(wwwfile)
print(wwwRealfile)

webbrowser.open_new_tab(wwwRealfile)
```

You should now have a web browser open with your results.

This profiling will show you many characteristics of your data that are useful during future processing.

Now that you understand the basic process, let's try it with some real-world data sources.

You need to load the following Jupyter Notebook from example directory Chapter 013: Chapter 013/000-RIF/400-Utility-Layer/200-Data-Utilities/08-02-Utility-Data-Profiler-01-Time.ipynb

Run the complete profile and your result should be:

```
              Date     Time
count      1051200  1051200
unique         730     1440
top     2017-09-06    08:07
freq          1440      730

<class 'pandas.core.frame.DataFrame'>
Int64Index: 1051200 entries, 0 to 1051199
Data columns (total 2 columns):
Date    1051200 non-null object
Time    1051200 non-null object
dtypes: object(2)
memory usage: 24.1+ MB
```

Warning Due to a limitation in the Profiling Library, this data set does not profile.

You need to load the following Jupyter Notebook from example directory Chapter 013: Chapter 013/000-RIF/400-Utility-Layer/200-Data-Utilities/08-03-Utility-Data-Profiler-02-Person.ipynb

Run the complete profile and your result should be as shown in Figure 13-14.

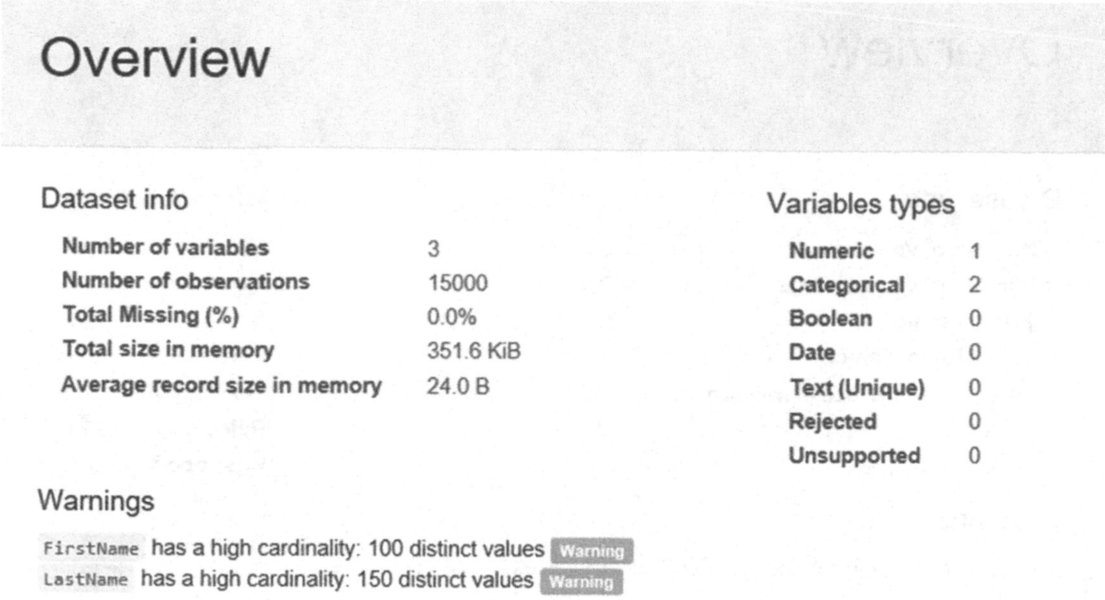

Overview

Dataset info

Number of variables	3
Number of observations	15000
Total Missing (%)	0.0%
Total size in memory	351.6 KiB
Average record size in memory	24.0 B

Variables types

Numeric	1
Categorical	2
Boolean	0
Date	0
Text (Unique)	0
Rejected	0
Unsupported	0

Warnings

`FirstName` has a high cardinality: 100 distinct values Warning
`LastName` has a high cardinality: 150 distinct values Warning

Figure 13-14. *Profile Person*

You need to load the following Jupyter Notebook from example directory Chapter 013: Chapter 013/000-RIF/400-Utility-Layer/200-Data-Utilities/08-04-Utility-Data-Profiler-03-Object.ipynb

Run the complete profile and your result should be as shown in Figure 13-15.

Overview

Dataset info

Number of variables	4
Number of observations	10000000
Total Missing (%)	0.0%
Total size in memory	305.2 MiB
Average record size in memory	32.0 B

Variables types

Numeric	2
Categorical	2
Boolean	0
Date	0
Text (Unique)	0
Rejected	0
Unsupported	0

Warnings

`SortCode` has a high cardinality: 10000 distinct values Warning

Figure 13-15. *Profile – Person*

You need to load the following Jupyter Notebook from example directory Chapter 013: Chapter 013/000-RIF/400-Utility-Layer/200-Data-Utilities/ 08-05-Utility-Data-Profiler-04-Location.ipynb

Run the complete profile and your result should be as shown in Figure 13-16.

Overview

Dataset info

Number of variables	4
Number of observations	1762397
Total Missing (%)	0.0%
Total size in memory	53.8 MiB
Average record size in memory	32.0 B

Variables types

Numeric	3
Categorical	0
Boolean	0
Date	0
Text (Unique)	1
Rejected	0
Unsupported	0

Figure 13-16. *Profiler-Location Results*

You need to load the following Jupyter Notebook from example directory Chapter 013: Chapter 013/000-RIF/400-Utility-Layer/200-Data-Utilities/08-06-Utility-Data-Profiler-05-Event.ipynb

Run the complete profile and your result should be as shown in Figure 13-17.

Overview

Dataset info

Number of variables	9
Number of observations	750000
Total Missing (%)	0.0%
Total size in memory	51.5 MiB
Average record size in memory	72.0 B

Variables types

Numeric	4
Categorical	5
Boolean	0
Date	0
Text (Unique)	0
Rejected	0
Unsupported	0

Warnings

`Date` has a high cardinality: 30000 distinct values Warning
`FirstName` has a high cardinality: 100 distinct values Warning
`LastName` has a high cardinality: 150 distinct values Warning
`SortCode` has a high cardinality: 7738 distinct values Warning

Figure 13-17. *Profiling - Event*

Processing Utilities

The processing utilities are utilities that are common interfaces with the processing capabilities in the RIF for the IML algorithms and techniques. The same IML algorithms can be ported from one processing engine to another to ensure future capabilities.

Business Layer

The layer contains the business requirements (Functional and Non-functional).

I also store all the sun models that I prepare within this layer to ensure my development of them has a complete set of business and technical documentation.

The following is stored in this layer.

Business Requirements

Business requirements are specifications that when delivered, provide value; it defines the characteristics of the proposed machine learning ecosystem from the viewpoint of the system's end user.

Warning Remember you are building your algorithms to support a bigger business goal.

Sun Models

Sun Models are a series of technical diagrams that enable the communication of data science requirements between the data scientist and his customer.

Outrigger Dimension

The outrigger dimension (Figure 13-18) enables the expansion of common data characteristics by attaching a smaller dimension directly to the main common dimension via a foreign key.

Example: All the dates like date of birth, date you left school, date of graduation, etc., now can be expanded with common characteristics that are empowering to the machine learning to find common correlations.

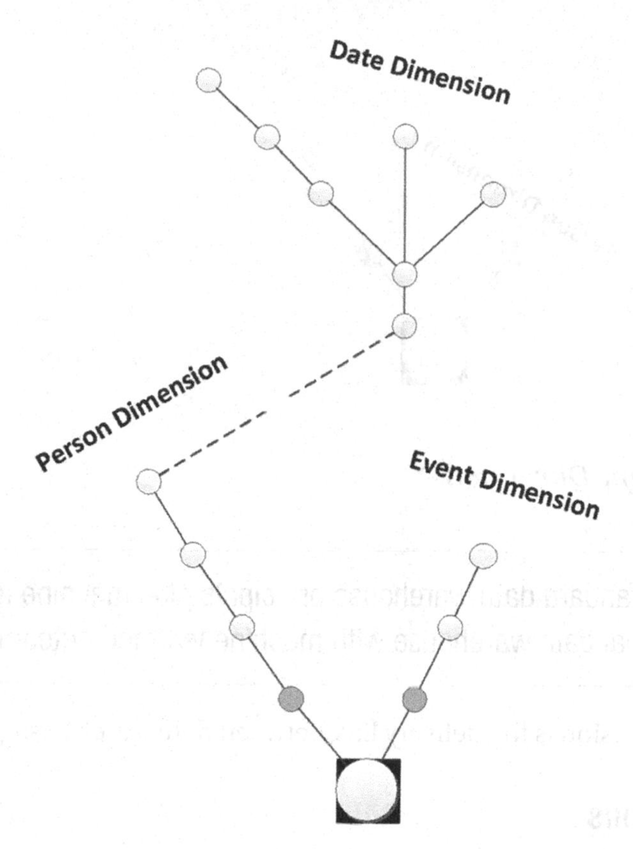

Figure 13-18. Outrigger Dimension

Mini-Dimension

The mini-dimension enables the introduction of a new set of characteristics of the data dimension by linking it to the dimension via a primary key of the dimension.

A mini-dimension is used to enable data science via machine learning for classifications, clusters, or predictions. See Figure 13-19.

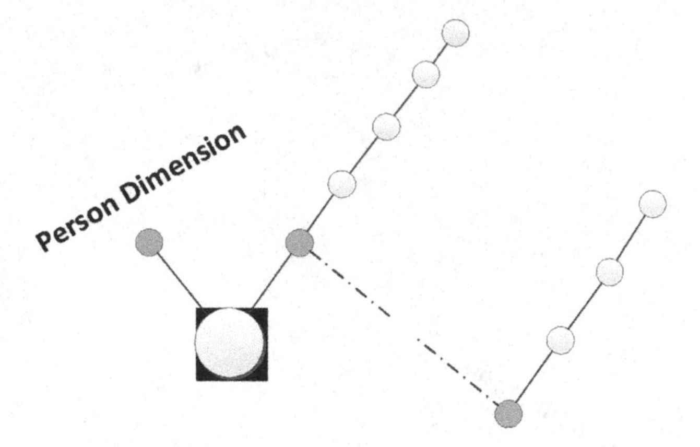

Figure 13-19. Mini-Dimension

Tip I use the standard data warehouse principals plus machine learning to enhance traditional data warehouse with machine learning outcomes.

The mini-dimension is the delivery link between data warehousing and data science.

Bridge Dimensions

The bridge dimension enables the creation of a many-to-many relationship between two dimensions to handle characteristics where one dimension is linked more than once via the same set of keys. See Figure 13-20.

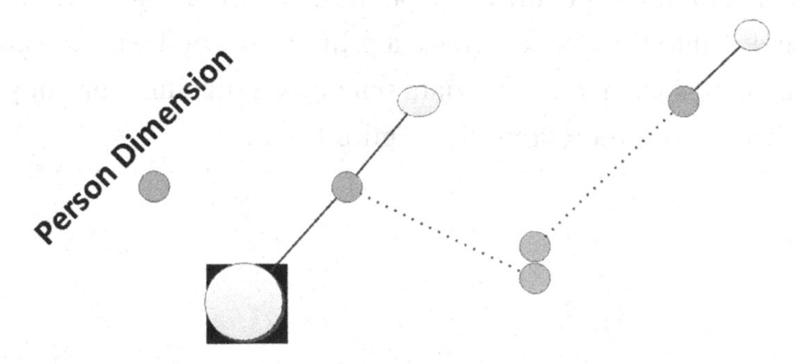

Figure 13-20. Bridge Dimension

This bridge is common where there is a complex relationship between the data dimensions. I found this to be common where there is an overlap in the classification of clustering or due to uncertainty if the specific entity belongs to more than one dimension at any given point in time.

Warning Use these sparingly as they can expand into massive dimensional spaces that are heavy in its demand for processing.

That was a quick basic introduction to sun models.

If you want more information, please look at the following books:

- *Practical Hive: A Guide to Hadoop's Data Warehouse System* by Scott Shaw, Andreas François Vermeulen, Ankur Gupta, and David Kjerrumgaard (Apress, 2016).
- *Practical Data Science: A Guide to Building the Technology Stack for Turning Data Lakes into Business Assets* by Andreas François Vermeulen (Apress, 2018).

I now cover these concepts in more detail.

Data Profiling

Data profiling is the process of exploratory investigation of the data accessible from a current information source and collecting statistics or revealing summaries about that data created via a series of data science processes.

Tip This data profiling is the core process that will feed your true information about the data.

The data profiling will also point out insights into the data that can determine if a specific category of machine learning algorithm is suitable to resolve the data that is presented to it by data scientists.

Data Traceability

Traceability, in processing chain traceability, is the ability to identify, track, and trace elements of a data process as it moves along the processing chain from raw data to complete data science insights.

Two important reports are generated by this process.

Data Provenance

This explains precisely where the data originated, and it clarifies to the business person using your results what was the validity of your original source.

Example: Extracted 2019-01-10 06:00 from CRM London - means no data from New York or data from London after 06:00 is included.

Warning Lack of Data Provenance could render great machine learning results useless due to a lack of trust.

Data Lineage

Data lineage refers to the origin and transformations that data goes through over time. Basically, data lineage tells the story of a specific piece of data. This allows you to understand where the data comes from as well as when and where it separates and merges with other data.

Tip Remember to include the Provenance and Lineage of your machine learning insights as this will give your results a base of trust.

That is the five system layers completed. I will explain in Chapter 15 how you use these in a real-life solution.

I will now introduce you the data lake where the data is stored to persist your IML's data engineering and data science.

Six Data Lake Zones

There are six data lake zones (Figure 13-21).

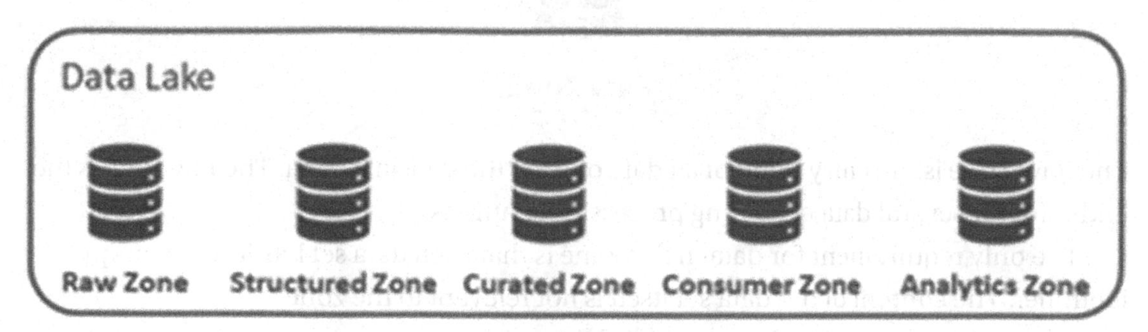

Figure 13-21. Data lake zones

Workspace Zone

The workspace zone is a temporary area in the data lake where IML processes could store results on a short-term basis while processing the data into insights.

Warning Data in this area should always be deemed as incorrect, incomplete, and inaccurate unless you created it yourself during the current running batch of the IML ecosystem.

Tip Always perform a housekeeping action before and after each batch run in the workspace as this ensures an effective and efficient data lake.

I have a personal ratio that I apply that states if the workspace exceeds 10% of the data lake, you are not using it correctly. If I find this quota bridge, I will start investigating why so much transient space is used.

Raw Zone

Raw Zone

The Raw Zone is the entry point of all data outside the data lake data. The raw zone is the endpoint for several data extracting processing solutions.

The only requirement for data in this zone is that each data set has its own unique identifier. The content of the data set itself is not relevant to the zone.

All data in this zone is schema-on-read.

Nice-to-have extra features are the following:

- A unique identifiable source system ID included with every data set.
- A complete metadata description for each data set.
- A full audit of the activities in the zone.

The data sets can arrive by any of the following routes:

- Bulk uploads - This is the most common route into the data lake.
- Streaming via a messaging system like Apache Kafka (`http://kafka.apache.org/`), Apache NIFI (`https://nifi.apache.org/`). or Apache Flink (`https://flink.apache.org/`).
- Through Internet-of-Things processes.

Note The raw zone should always work independently from the rest of the data engineering, that is, it should never hold or block the processing of data.

Warning Never remove data from the raw zone while performing the IML processing. Use a separate independent weeding policy.

Remember specific types of data may only have a limited life span as being valid. Example: If a security card allows a person into a specific room and the action is logged as the person is now in that room. It would be logically accepted that if a week later the person is still in the room, there is an error in the data.

Structured Zone

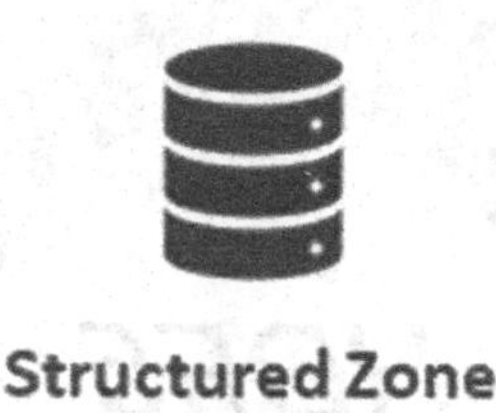

The structured zone is used to convert the raw data into enhanced data sources. This zone's data is in harmonized formats to assist the next zone's processing capability. Any data quality issues are resolved in this zone.

The structured zone is the storage space for the raw data that has been processed by identifying the metadata related to the data set being processed. This data now carries a unique ID plus detailed data types and field names on a data set by data set level.

The data format of the data sets in the structured zone are typically one or as few as possible formats.

The design concept is to reduce the data to a common format for the complete structured zone.

Tip Data formats like HDF5 (`https://www.hdfgroup.org/`) works well in this environment as it stores both data and metadata.

HDF5

HDF5 is a data model, library, and file format for storing and managing data. It supports an unlimited variety of data types and is designed for flexible and efficient I/O and for high volume and complex data. HDF5 is portable and extensible, allowing applications to evolve in their use. See Figure 13-22.

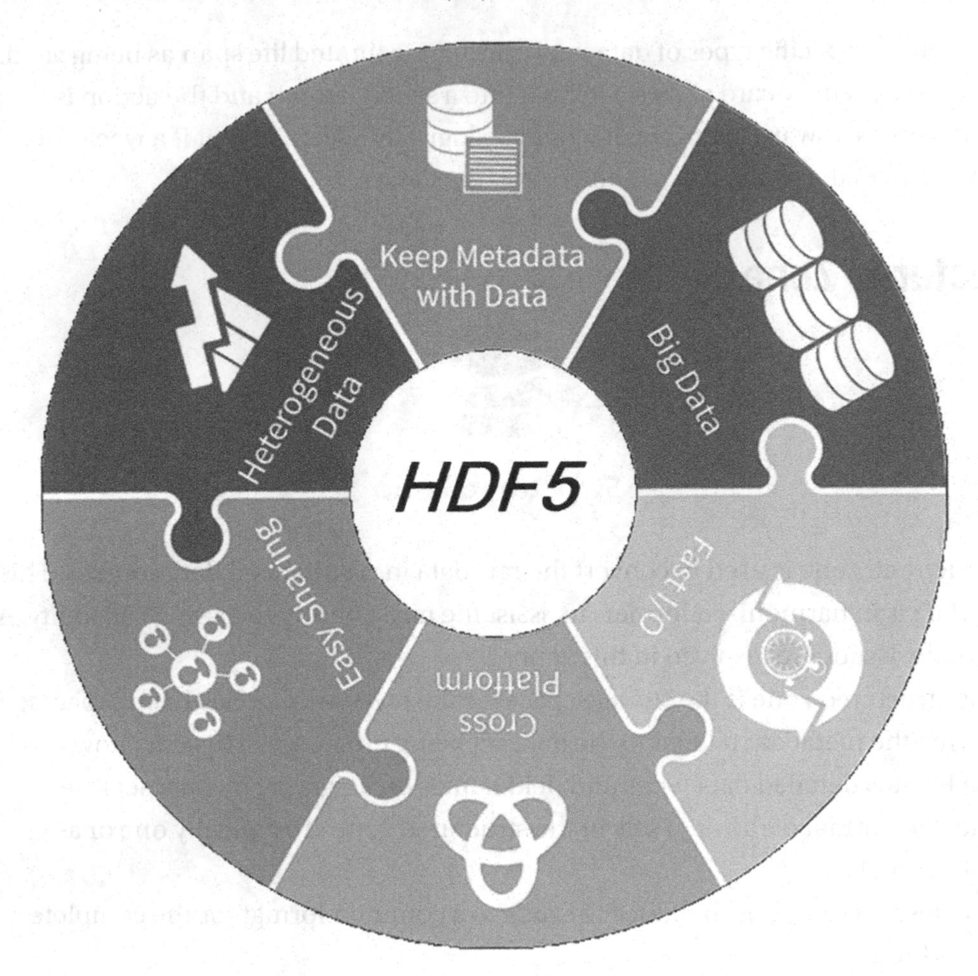

Figure 13-22. *HDF5 Ecosystem*

You need to load the following Jupyter Notebook from example directory Chapter 013: Open Chapter 013/000-RIF/100-Functional-Layer/100-Retrieve/08-03-Retrieve-04-Location-HDF5-01.ipynb

You can see the simple Python command:

```
locationDF.to_hdf(fullRealName1, format='fixed', key='location',
index=False, encoding='utf-8')
```

This enables the HDF5 formatting for the system.

Parquet

Apache Parquet is a columnar storage format available to any project in the Hadoop ecosystem.

(https://parquet.apache.org/)

The main reason for you to investigate this format is that is widely use by the Hadoop ecosystem.

Using it will ensure in the future that you can migrate from on-premises to cloud ecosystems with no format changes.

You may need an extra library to enable the Parquet engine. I normally use pyarrow as it is the most common engine in the cloud ecosystems.

```
conda install -c conda-forge pyarrow
```

I will guide you through the two options for Parquet.

You need to load the following Jupyter Notebook from example directory Chapter 013:

Open Chapter 013/000-RIF/100-Functional-Layer/100-Retrieve/08-03-Assess-04-Location-Parquet-01.ipynb

You can see the simple Python command:

```
locationDF.to_parquet(fullRealName1, engine='auto', compression='snappy')
```

and

```
locationDF.to_parquet(fullRealName2, engine='auto', compression='gzip')
```

This enables the Parquet formatting for the system.

Warning Snappy file size is 1.496 times bigger than GZip file size. So, if you can use GZip. This reduction in file size will enable you to store more data in the data lake before you need more disks.

That finishes my introduction to the structured zone until Chapter 15.

Curated Zone

The Curated Zone is the current "***single truth***" across the data lake. The data vault and data warehouse in the zone enable the consolidation and amalgamation of data sources from the structured zone. The "live" data science models' results are stored in this zone.

Consumer Zone

Consumer Zone

The consumer zone is the area where different data marts for the business insights are stored, ready for end users to visualize their business insights.

This is the primary zone for the majority of businesspeople's questions.

Analytics Zone

Analytics Zone

The analytics zone is the "sandbox" of the data lake. This zone is used to design, develop, and train new and innovative data science and machine learning solutions before DevOps hands over the new version of the IML to the Curated Zone.

Tip Set the Curated Zone to the Analytic Zone on a planned and on-request update cycle as you do not want to have an unexpected update in the middle of a piece of innovative IML work. I normally perform an update at the start of every second 10-day sprint cycle that is not in the same week as the month's end because that gives my team a month to complete the work and we are not pressured with an update during our busy month-end sprint.

That is the last of the data lake zones used by IML

Delta Lake

I will quickly discuss a new innovative processing engine that has been released to open source.

Delta Lake is an open source storage layer that brings ACID transactions to Apache Spark™ and big data workloads. (See `https://delta.io/`)

Just to remind you, ACID (Atomicity, Consistency, Isolation, and Durability) is a set of properties of Data Lake transactions intended to guarantee validity even in the processing event that causes errors or fails.

This is a major boost to the use of data lakes as core data storage for IML ecosystems.

Tip I will be discussing the next concepts using standard file directories, but I will also reference how to perform the same using Delta Lake.

You will need:

> pip install --upgrade pyspark and then pyspark --packages io.delta:delta-core_2.12:0.1.0
>
> Open Chapter 013/000-RIF/100-Functional-Layer/100-Retrieve/08-03-Assess-04-Location-Parquet-01.ipynb

Warning New innovative machine learning concept coming!

RAPTOR/QUBE

The ***RAPTOR/QUBE*** is an innovative processing framework I have been developing over the last eight years. It is not an industry standard but my own combination of knowledge I have gathered over the last ten years.

Note I will explain the concepts to enable you to develop your own automatic processing engine.

Rapid Information Framework

The ***Rapid Information Framework*** is the processing engine of the comprehensive ecosystem, and I use the similar framework for my entire data science, machine learning, and data warehousing work.

It is a common ***platform*** for processing data for my customers and my company.

You need to load the following Jupyter Notebook from example directory Chapter 013: Open Chapter 013/000-RIF/100-Functional-Layer/100-Retrieve/08-03-Setup-RAPTORQUBE.ipynb

You have seen this structure before but there are two small changes:

- RIF is numbered as: 010-RIF
- Data Lake (DL) is numbered as: 995-DL

This was done to create a new structure for the upcoming work in this chapter.

At this point I suggest you perform a run all steps in the notebook.

If you want to follow the steps, you could also run the cells top to bottom to understand the workings of the process.

Warning When this setup runs, it removes any previous data from both the RIF and the DL.

Tip I typically use more than one RIF to enable more granular processing capabilities for the customer and an easy swap to new releases.

Tip I only use multiple data lakes when there are requirements for high security or privacy separation of the data.

You can close the Chapter-013-03-Setup-RAPTORQUBE now as you need to proceed to the next phase of building your processing micro-processing engines.

You need to load the following Jupyter Notebook from example directory Chapter 013: Open Chapter 013/000-RIF/100-Functional-Layer/100-Retrieve/08-04-Build-RAPTORQUBE.ipynb

You need our core libraries again:

```
import datetime
nowStart = datetime.datetime.now()
import os
import time
import shutil
import pandas as pd
```

Set up some defaults:

```
pathDL='../../Results/Chapter 013/995-DL/'
pathRIFInName='../../Data/Chapter 013/100-Kafka-Data/'
pathRIFOutName='../../Results/Chapter 013/995-DL/100-Raw-Zone/100-Kafka-Data/'
```

Perform some housekeeping by clearing out the data lake.

```
if os.path.exists(pathRIFOutName):
    shutil.rmtree(pathRIFOutName)
    time.sleep(5)
```

Reset data lake core data sets:

```
if not os.path.exists(pathRIFOutName):
    shutil.copytree(pathRIFInName, pathRIFOutName)
```

Get the full path for Data Lake:

```
pathRealDLname=os.path.realpath(pathRIFOutName)
print(pathRealDLname)
```

Show the existing structure:

```
for root, dirs, files in os.walk(pathRealDLname, topdown=True):
    for name in dirs:
        print(os.path.join(root, name))
```

You should see results like:

```
.\Results\995-DL\100-Raw-Zone\100-Kafka-Data\01-Time
.\Results\995-DL\100-Raw-Zone\100-Kafka-Data\02-Person
.\Results\995-DL\100-Raw-Zone\100-Kafka-Data\03-Object
.\Results\995-DL\100-Raw-Zone\100-Kafka-Data\04-Location
.\Results\995-DL\100-Raw-Zone\100-Kafka-Data\05-Event
```

```
for root, dirs, files in os.walk(pathRealDLname, topdown=True):
    for name in files:
        print(os.path.join(root, name))
```

```
.\Results\995-DL\100-Raw-Zone\100-Kafka-Data\01-Time\Time-Date.csv
.\Results\995-DL\100-Raw-Zone\100-Kafka-Data\01-Time\Time-Time.csv
.\Results\995-DL\100-Raw-Zone\100-Kafka-Data\02-Person\Person.csv
.\Results\995-DL\100-Raw-Zone\100-Kafka-Data\03-Object\Object.csv.gz
.\Results\995-DL\100-Raw-Zone\100-Kafka-Data\04-Location\Location.csv.gz
.\Results\995-DL\100-Raw-Zone\100-Kafka-Data\05-Event\Event.csv.gz
```

Retrieve

The retrieve IML scripts are stored here for each combination of machine learning algorithms use for extracts the data from outside the factory into the data lake.

```
kafkaName=os.path.join(pathRealDLname, '01-Time','Time-Date.csv')
print(kafkaName)

DateDF=pd.read_csv(kafkaName, header=0)
print(DateDF.shape)

retrieveName = os.path.join(pathDL, '100-Raw-Zone', '08-02-Time-Date.csv.gz')
DateDF.to_csv(retrieveName, index=False, encoding='utf-8',
compression='gzip')
```

```
kafkaName=os.path.join(pathRealDLname, '01-Time','Time-Time.csv')
print(kafkaName)

DateDF=pd.read_csv(kafkaName, header=0)
print(DateDF.shape)

retrieveName = os.path.join(pathDL, '100-Raw-Zone', '08-02-Time-Time.csv.gz')
DateDF.to_csv(retrieveName, index=False, encoding='utf-8',
compression='gzip')
```

Assess

The assess IML scripts for each combination of machine learning algorithms are used for evaluating the data against preapproved data quality rules to improve the quality of the data.

```
retrieveName = os.path.join(pathDL, '100-Raw-Zone', '08-02-Time-Date.csv.gz')
DateDF=pd.read_csv(retrieveName, header=0, encoding='utf-8',
compression='gzip')

print(DateDF.shape)
(730, 2)
print(DateDF.info())

<class 'pandas.core.frame.DataFrame'>
RangeIndex: 730 entries, 0 to 729
Data columns (total 2 columns):
DateKey    730 non-null object
Date       730 non-null object
dtypes: object(2)
memory usage: 11.5+ KB
None

print(DateDF.describe())

          DateKey        Date
count         730         730
unique        730         730
top     D20170410  2017-05-24
freq            1           1
```

```
assessName = os.path.join(pathDL,'200-Structured-Zone', 'Time-Date.csv.gz')
DateDF.to_csv(assessName, index=False, encoding='utf-8',
compression='gzip')
```

```
retrieveName = os.path.join(pathDL, '100-Raw-Zone', '08-02-Time-Time.csv.gz')
DateDF=pd.read_csv(retrieveName, header=0, encoding='utf-8',
compression='gzip')
```

```
print(DateDF.shape)
```

```
(1440, 2)
```

```
print(DateDF.info())
```

```
<class 'pandas.core.frame.DataFrame'>
RangeIndex: 1440 entries, 0 to 1439
Data columns (total 2 columns):
TimeKey    1440 non-null object
Time       1440 non-null object
dtypes: object(2)
memory usage: 22.6+ KB
None
```

```
print(DateDF.describe())
```

```
       TimeKey   Time
count     1440   1440
unique    1440   1440
top      T1939  13:35
freq         1      1
```

```
assessName = os.path.join(pathDL,'200-Structured-Zone', 'Time-Time.csv.gz')
DateDF.to_csv(assessName, index=False, encoding='utf-8',
compression='gzip')
```

Process

The process IML scripts are used for each combination of machine learning algorithms used for loading the data into the Time-Person-Object-Location-Event (T-P-O-L-E) data vault to create uniform data structures.

```
retrieveName1 = os.path.join(pathDL, '100-Raw-Zone', '08-02-Time-Date.csv.gz')
retrieveName2 = os.path.join(pathDL, '100-Raw-Zone', '08-02-Time-Time.csv.gz')

Date1DF=pd.read_csv(retrieveName1, header=0, encoding='utf-8',
compression='gzip')
Date2DF=pd.read_csv(retrieveName2, header=0, encoding='utf-8',
compression='gzip')

print(Date1DF.info())

<class 'pandas.core.frame.DataFrame'>
RangeIndex: 730 entries, 0 to 729
Data columns (total 2 columns):
DateKey    730 non-null object
Date       730 non-null object
dtypes: object(2)
memory usage: 11.5+ KB
None

print(Date2DF.info())

<class 'pandas.core.frame.DataFrame'>
RangeIndex: 1440 entries, 0 to 1439
Data columns (total 2 columns):
TimeKey    1440 non-null object
Time       1440 non-null object
dtypes: object(2)
memory usage: 22.6+ KB
None

Date1DF['Key'] = 1
Date2DF['Key'] = 1

DateDF=Date1DF.merge(Date2DF, how='outer')
DateDF.drop('Key',axis=1, inplace=True)
DateDF.set_index(['DateKey','TimeKey'], inplace=True)

print(DateDF.info())
```

```
<class 'pandas.core.frame.DataFrame'>
Int64Index: 1051200 entries, 0 to 1051199
Data columns (total 4 columns):
DateKey     1051200 non-null object
Date        1051200 non-null object
TimeKey     1051200 non-null object
Time        1051200 non-null object
dtypes: object(4)
memory usage: 40.1+ MB
None
```

```
processName = os.path.join(pathDL,'300-Curated-Zone', 'Hub', 'Time', 'Hub-
Time.csv.gz')
DateDF.to_csv(processName, index=False, encoding='utf-8',
compression='gzip')
print(processName)
```

```
TimeMax=DateDF.shape[0]
TimeMax=1000
for i in range(TimeMax):
    d0 = str(DateDF['Date'][i] + ' ' + DateDF['Time'][i])
    DateDF0 = datetime.datetime.strptime(d0,'%Y-%m-%d %H:%M')
    DateDF1 = str(DateDF0)
    fulldatetime=DateDF0.strftime('%Y-%m-%d %H:%M:%S')
    yearweek = DateDF0.strftime('%Y-%W')
    weekdaylong = DateDF0.strftime('%A')
    weekdayshort = DateDF0.strftime('%a')
    monthlong = DateDF0.strftime('%B')
    monthshort = DateDF0.strftime('%b')
    ampm = DateDF0.strftime('%p')
    timezone = DateDF0.strftime('%Z')
    year = DateDF0.strftime('%Y')
    quarter = (int(DateDF0.strftime('%m')) % 4)
    month = DateDF0.strftime('%m')
    day = DateDF0.strftime('%d')
```

```
    if i==0:
        s1=pd.DataFrame([[DateDF1,fulldatetime]])
        s2=pd.DataFrame([[DateDF1,yearweek]])
        s3=pd.DataFrame([[DateDF1,weekdaylong,weekdayshort]])
        s4=pd.DataFrame([[DateDF1,monthlong,monthshort]])
        s5=pd.DataFrame([[DateDF1,ampm]])
        s6=pd.DataFrame([[DateDF1,timezone]])
        s7=pd.DataFrame([[DateDF1,year]])
        s8=pd.DataFrame([[DateDF1,month]])
        s9=pd.DataFrame([[DateDF1,day]])
        s10=pd.DataFrame([[DateDF1,quarter]])
    else:
        s1a=pd.DataFrame([[DateDF1,fulldatetime]])
        s2a=pd.DataFrame([[DateDF1,yearweek]])
        s3a=pd.DataFrame([[DateDF1,weekdaylong,weekdayshort]])
        s4a=pd.DataFrame([[DateDF1,monthlong,monthshort]])
        s5a=pd.DataFrame([[DateDF1,ampm]])
        s6a=pd.DataFrame([[DateDF1,timezone]])
        s7a=pd.DataFrame([[DateDF1,year]])
        s8a=pd.DataFrame([[DateDF1,month]])
        s9a=pd.DataFrame([[DateDF1,day]])
        s10a=pd.DataFrame([[DateDF1,quarter]])
        s1=s1.append(s1a)
        s2=s2.append(s2a)
        s3=s3.append(s3a)
        s4=s4.append(s4a)
        s5=s5.append(s5a)
        s6=s6.append(s6a)
        s7=s7.append(s7a)
        s8=s8.append(s8a)
        s9=s9.append(s9a)
        s10=s10.append(s10a)

s1.columns = (['DateTime','FullDateTime'])
s2.columns = (['DateTime','YearWeek'])
s3.columns = (['DateTime','WeekDayLong','WeekDayShort'])
```

```
s4.columns = (['DateTime','MonthDayLong','MonthDayShort'])
s5.columns = (['DateTime','Am-PM'])
s6.columns = (['DateTime','Time-Zone'])
s7.columns = (['DateTime','Year'])
s8.columns = (['DateTime','Month'])
s9.columns = (['DateTime','Day'])
s10.columns = (['DateTime','Quarter'])

s0=s1
s0=s0.merge(s2, on='DateTime')
s0=s0.merge(s3, on='DateTime')
s0=s0.merge(s4, on='DateTime')
s0=s0.merge(s5, on='DateTime')
s0=s0.merge(s6, on='DateTime')
s0=s0.merge(s7, on='DateTime')
s0=s0.merge(s8, on='DateTime')
print(s0.shape)

DateDF=s0
processName = os.path.join(pathDL,'300-Curated-Zone', 'Satellite', 'Hub',
'Time', 'Satellite-Hub-Time-00.csv.gz')
DateDF.to_csv(processName, index=False, encoding='utf-8',
compression='gzip')
print(processName)

DateDF=s1
processName = os.path.join(pathDL,'300-Curated-Zone', 'Satellite', 'Hub',
'Time', 'Satellite-Hub-Time-01.csv.gz')
DateDF.to_csv(processName, index=False, encoding='utf-8',
compression='gzip')
print(processName)

DateDF=s2
processName = os.path.join(pathDL,'300-Curated-Zone', 'Satellite', 'Hub',
'Time', 'Satellite-Hub-Time-02.csv.gz')
DateDF.to_csv(processName, index=False, encoding='utf-8',
compression='gzip')
```

```
print(processName)

DateDF=s3
processName = os.path.join(pathDL,'300-Curated-Zone', 'Satellite', 'Hub',
'Time', 'Satellite-Hub-Time-03.csv.gz')
DateDF.to_csv(processName, index=False, encoding='utf-8',
compression='gzip')
print(processName)

DateDF=s4
processName = os.path.join(pathDL,'300-Curated-Zone', 'Satellite', 'Hub',
'Time', 'Satellite-Hub-Time-04.csv.gz')
DateDF.to_csv(processName, index=False, encoding='utf-8',
compression='gzip')
print(processName)
DateDF=s3
processName = os.path.join(pathDL,'300-Curated-Zone', 'Satellite', 'Hub',
'Time', 'Satellite-Hub-Time-04.csv.gz')
DateDF.to_csv(processName, index=False, encoding='utf-8',
compression='gzip')
print(processName)

DateDF=s5
processName = os.path.join(pathDL,'300-Curated-Zone', 'Satellite', 'Hub',
'Time', 'Satellite-Hub-Time-05.csv.gz')
DateDF.to_csv(processName, index=False, encoding='utf-8',
compression='gzip')
print(processName)

DateDF=s6
processName = os.path.join(pathDL,'300-Curated-Zone', 'Satellite', 'Hub',
'Time', 'Satellite-Hub-Time-06.csv.gz')
DateDF.to_csv(processName, index=False, encoding='utf-8',
compression='gzip')
print(processName)
```

```
DateDF=s7
processName = os.path.join(pathDL,'300-Curated-Zone', 'Satellite', 'Hub',
'Time', 'Satellite-Hub-Time-07.csv.gz')
DateDF.to_csv(processName, index=False, encoding='utf-8',
compression='gzip')
print(processName)

DateDF=s8
processName = os.path.join(pathDL,'300-Curated-Zone', 'Satellite', 'Hub',
'Time', 'Satellite-Hub-Time-02.csv.gz')
DateDF.to_csv(processName, index=False, encoding='utf-8',
compression='gzip')
print(processName)

DateDF=s9
processName = os.path.join(pathDL,'300-Curated-Zone', 'Satellite', 'Hub',
'Time', 'Satellite-Hub-Time-02.csv.gz')
DateDF.to_csv(processName, index=False, encoding='utf-8',
compression='gzip')
print(processName)

DateDF=s10
processName = os.path.join(pathDL,'300-Curated-Zone', 'Satellite', 'Hub',
'Time', 'Satellite-Hub-Time-10.csv.gz')
DateDF.to_csv(processName, index=False, encoding='utf-8',
compression='gzip')
print(processName)
```

Let's investigate the Raw Zone you created:

```
pathReportname= '../../Results/Chapter 013/995-DL/100-Raw-Zone/'
for root, dirs, files in os.walk(pathReportname, topdown=True):
    for name in files:
        print(os.path.join(root, name))
```

```
../../Results/Chapter 013/995-DL/100-Raw-Zone/08-02-Time-Date.csv.gz
../../Results/Chapter 013/995-DL/100-Raw-Zone/08-02-Time-Time.csv.gz
```

```
../../Results/Chapter 013/995-DL/100-Raw-Zone/100-Kafka-Data\01-Time\Time-
Date.csv
../../Results/Chapter 013/995-DL/100-Raw-Zone/100-Kafka-Data\01-Time\Time-
Time.csv
../../Results/Chapter 013/995-DL/100-Raw-Zone/100-Kafka-Data\02-Person\
Person.csv
../../Results/Chapter 013/995-DL/100-Raw-Zone/100-Kafka-Data\03-Object\
Object.csv.gz
../../Results/Chapter 013/995-DL/100-Raw-Zone/100-Kafka-Data\04-Location\
Location.csv.gz
../../Results/Chapter 013/995-DL/100-Raw-Zone/100-Kafka-Data\05-Event\
Event.csv.gz
```

Let's investigate the Structured Zone you created:

```
../../Results/Chapter 013/995-DL/200-Structured-Zone/Time-Date.csv.gz
../../Results/Chapter 013/995-DL/200-Structured-Zone/Time-Time.csv.gz
```

Let's investigate the Curated Zone you created:

```
pathReportname= '../../Results/Chapter 013/995-DL/300-Curated-Zone/Hub/'
for root, dirs, files in os.walk(pathReportname, topdown=True):
    for name in files:
        print(os.path.join(root, name))
```

```
../../Results/Chapter 013/995-DL/300-Curated-Zone/Hub/Time\Hub-Time.csv.gz
```

```
pathReportname= '../../Results/Chapter 013/995-DL/300-Curated-Zone/
Satellite/Hub/Time'
for root, dirs, files in os.walk(pathReportname, topdown=True):
    for name in files:
        print(os.path.join(root, name))
```

```
../../Results/Chapter 013/995-DL/300-Curated-Zone/Satellite/Hub/Time\
Satellite-Hub-Time-00.csv.gz
../../Results/Chapter 013/995-DL/300-Curated-Zone/Satellite/Hub/Time\
Satellite-Hub-Time-01.csv.gz
```

```
../../Results/Chapter 013/995-DL/300-Curated-Zone/Satellite/Hub/Time\
Satellite-Hub-Time-02.csv.gz
../../Results/Chapter 013/995-DL/300-Curated-Zone/Satellite/Hub/Time\
Satellite-Hub-Time-03.csv.gz
../../Results/Chapter 013/995-DL/300-Curated-Zone/Satellite/Hub/Time\
Satellite-Hub-Time-04.csv.gz
../../Results/Chapter 013/995-DL/300-Curated-Zone/Satellite/Hub/Time\
Satellite-Hub-Time-05.csv.gz
../../Results/Chapter 013/995-DL/300-Curated-Zone/Satellite/Hub/Time\
Satellite-Hub-Time-06.csv.gz
../../Results/Chapter 013/995-DL/300-Curated-Zone/Satellite/Hub/Time\
Satellite-Hub-Time-07.csv.gz
../../Results/Chapter 013/995-DL/300-Curated-Zone/Satellite/Hub/Time\
Satellite-Hub-Time-08.csv.gz
../../Results/Chapter 013/995-DL/300-Curated-Zone/Satellite/Hub/Time\
Satellite-Hub-Time-09.csv.gz
../../Results/Chapter 013/995-DL/300-Curated-Zone/Satellite/Hub/Time\
Satellite-Hub-Time-10.csv.gz
```

Transform

The transform IML scripts for each combination of machine learning algorithms used for extracts the data vault into the data lake.

The portion of the processing sets up the machine learning against the data that was loaded.

```
processName = os.path.join(pathDL, '300-Curated-
Zone','Satellite','Hub','Time', 'Satellite-Hub-Time-00.csv.gz')

DateDF=pd.read_csv(processName, header=0, encoding='utf-8',
compression='gzip')

print(DateDF.shape)

(5000, 11)

print(DateDF.describe())
```

```
       Time-Zone    Year   Month
count        0.0  5000.0  5000.0
mean         NaN  2017.0     1.0
std          NaN     0.0     0.0
min          NaN  2017.0     1.0
25%          NaN  2017.0     1.0
50%          NaN  2017.0     1.0
75%          NaN  2017.0     1.0
max          NaN  2017.0     1.0
```

```
print(DateDF.info())
```

```
<class 'pandas.core.frame.DataFrame'>
RangeIndex: 5000 entries, 0 to 4999
Data columns (total 11 columns):
DateTime         5000 non-null object
FullDateTime     5000 non-null object
YearWeek         5000 non-null object
WeekDayLong      5000 non-null object
WeekDayShort     5000 non-null object
MonthDayLong     5000 non-null object
MonthDayShort    5000 non-null object
Am-PM            5000 non-null object
Time-Zone        0 non-null float64
Year             5000 non-null int64
Month            5000 non-null int64
dtypes: float64(1), int64(2), object(8)
memory usage: 429.8+ KB
None
```

```
import pandas_profiling as pdp
```

```
profile=pdp.ProfileReport(DateDF)
```

```
%%javascript
IPython.OutputArea.auto_scroll_threshold = 9999;
```

```
Profile
```

Organize

The organize IML scripts for combinations of machine learning algorithms splits the data warehouse into appropriate subject-specific data marts.

```
wwwfile= os.path.join(pathDL, '300-Curated-Zone','Satellite','Hub','Time',
'Satellite-Hub-Time-00.html')
profile.to_file(outputfile=wwwfile)
```

You have stored the outcome of the data profiling that includes correlations. This action has now organized the data ready for the customer to access.

Report

The report IML scripts for each combination of machine learning algorithms create visualizations for the factory to report the knowledge it formulated.

```
import webbrowser
import os

wwwRealfile=os.path.realpath(wwwfile)
print(wwwRealfile)
```

You will now open the report (Figure 13-23).

```
webbrowser.open_new_tab(wwwRealfile)
```

Overview

Dataset info

Number of variables	11
Number of observations	5000
Total Missing (%)	9.1%
Total size in memory	429.8 KiB
Average record size in memory	88.0 B

Variables types

Numeric	0
Categorical	4
Boolean	0
Date	0
Text (Unique)	2
Rejected	5
Unsupported	0

Warnings

`Month` has constant value 1 Rejected
`MonthDayLong` has constant value January Rejected
`MonthDayShort` has constant value Jan Rejected
`Time-Zone` has 5000 / 100.0% missing values Missing
`Time-Zone` has constant value Rejected
`Year` has constant value 2017 Rejected

***Figure 13-23.** Report Profiling Time*

The results are that fields like Month and Time-Zone are not useful for another machine learning.

Tip By performing simple machine learning or data science, you can quickly exclude problem fields from your data engineering feature creation.

Well done; you have completed a basic data discovery process using simple machine learning analysis.

Deep Learning Engine

The deep learning engine is the core of the automatic IML that monitors and controls several rapid information factories to process the data science and data engineering requirements.

Quantum

The basic engine consists of ***numerous*** optimized processors running as serverless, cloud-based data integration agents that use ***swarming principles*** that enable the smallest data action required to manipulate the data entities across factory. The processing engine uses a ***quantum principle*** to enable the processing of more than one flow of data through more than one factory in parallel at the same time.

This results in a ***Directed acyclic graph (DAG)*** that ***amplifies*** the net processing throughput by sharing data among ***rapid information factories***.

Tip I have created a processing engine that creates Python code as an output. Yes, I have a Python system that writes data science and machine learning code.

I will mention this evolutionary code generator in concept only as it is currently still under development and testing before I can share it with a general audience.

Note Once I have it resolved, I will update this section with more details.

Universal

The processing is guided by a ***systematic*** and ***precise*** set of rules that is established by you as a data scientist and machine learning engineer.

The core design is that you only enable specific types of machine learning or design a ***reinforcement learning*** algorithm that uses ***evolutionary programming*** to discover the required machine learning that works for each data set in the raw zone of your data lake.

Bounded

The processing domain is only supporting a ***set of rules*** that is ***bounded*** as it has a ***finite population***. This finite population results in a domain that is predictable in processing requirements as every step in the domain is individually optimized for ***effectiveness*** and ***efficiency***.

Engine

The principal core is a ***data processing appliance*** that converts data into knowledge and uses a human-in the loop to learn how to learn new data sets. The advantage is that the conforming of the data via the ***bounded rules*** and the ***quantum processing plans*** ensure a ***highly effective*** and ***efficient processing*** profile for the numerous data factories under the control of the engine.

I want you to take note of these automatic coding platforms as I want to predict that within the next five years, a larger portion of the machine learning will be industrialized and then created by these agents.

However, until then you will need human-in-the-loop help to resolve business insights.

I suggest you read the rest of the chapter to indicate what type of human help you need for your requirements. It will also indicate where you fit in the overall machine learning ecosystem.

What Type of Machine Learning?

Machine learning needs highly skilled specialists to make it work perfectly. I will now introduce you to some of these generalized job descriptions that you will hear about in the machine learning ecosystems.

Data Analyst

Job Description

Your job will be to translate data into actionable business insights. You'll often be the go-between for technical teams and business strategy, sales, or marketing teams. Data visualization is going to be a big part of your day to day.

Why It's Important

Highly technical people often have a hard time understanding why data analysts are so important, but they really are the connection with the business. Somebody needs to adapt a trained and tested model and large volumes of user data into a digestible format so that business strategies and insights can be designed around them. Data analysts help to make sure that data science teams don't waste their time solving problems that don't deliver business value.

Requirements

The technologies you'll be working with include Python, SQL, Tableau, and Excel. You'll also need to be a good communicator.

Questions

The questions you'll be dealing with sound like this:

- "What's driving our bankcard growth figures?"
- "How can we clarify to management that the new increase in banking fees is turning people away?"

Data Engineer

Job Description

You'll be managing data pipelines for companies that deal with large volumes of data. That means making sure that your data is being efficiently collected and retrieved from its source when needed, and then cleaned and preprocessed.

Why It's Important

If you've only constantly worked with relatively small (<5 Gb) data sets stored in .csv or .txt files, it might be hard to understand why there would exist people whose full-time job it is to build and maintain data pipelines.

Here are a couple of reasons: 1) A 100 Gb dataset won't be suitable in your computer's RAM, so you generally need other ways to feed it into your model; and 2) that much data can take a silly amount of time to process and regularly must be stored unnecessarily in swap space to complete the required processing profiles.

Managing that volume of storage takes specialized technical know-how. This is where you will operate.

Requirements

The technologies you'll be working with include Apache Spark, Hadoop and/or Hive, as well as Kafka. You'll most likely need to have a solid foundation in SQL and database design. You will be the builder of systems and ecosystems.

Questions

The questions you'll be dealing with sound like this:

- "How do I build a pipeline that can handle 100 000 requests per minute?"
- "How can I clean this data set without loading it all in RAM?"
- "Do I need Cuda GPU's to process my data?"

Data Scientist

Job Description

Your job will be to clean and explore data sets and make predictions that deliver business value. Your day-to-day job will involve training and optimizing models, and often deploying them to production.

Why It's Important

When you have a pile of data that's too big for a human to parse, and too valuable to be ignored, you need some way of pulling digestible insights from it. That's the basic job of a data scientist: to convert data sets into digestible conclusions.

Requirements

The technologies you'll be working with include Python, scikit-learn, Pandas, SQL, and possibly Flask, Spark and/or TensorFlow/PyTorch. Some data science positions are purely technical, but the majority will require you to have some business sense, so that you don't end up solving problems that no one has.

You need to understand the detailed responses of every machine learning algorithm you use, as you need to know what parameters at what settings give optimum solutions with high effectiveness and efficiencies.

Questions

The questions you'll be dealing with sound like this:

- "How many different user types do we really have?"
- "Can we build a model to predict which products will sell to which users?"
- "Can you perform sentiment analysis with my webchat data?

Tip The role of a data scientist will become so important or relevant that a board member will be responsible for machine learning and their impact on the business and their customers.

Machine Learning Researcher

Job Description

Your job will be to find new ways to solve challenging problems in data science and deep learning. You won't be working with out-of-the-box solutions, but rather will be making your own. They are the research and development part of the IML workshop.

Tip I will predict that these people will be highly in demand as ML goes mainstream within the next five years.

Requirements

The technologies you'll be working with include Python, TensorFlow/PyTorch (and/or enterprise deep learning frameworks), and SQL.

Questions

The questions you'll be dealing with are wide ranging:

> "How do I improve the accuracy of our model to some degree closer to the state of the art?"
>
> "Would a custom optimizer help decrease training time?"
>
> "How do I change the data engineering requirements to yield better results?"

The five job descriptions I've laid out here definitely don't stand alone in all cases. At an early-stage startup, for instance, a data scientist might have to be a data engineer and/or a data analyst, too. But most jobs will fall more neatly into one of these categories than the others—and the larger the company, the more these categories will tend to apply.

Overall, the thing to remember is that in order to get hired, you'll usually be better off building a more focused skill set: don't learn TensorFlow if you want to become a data analyst, and don't prioritize learning PySpark if you want to become a machine learning researcher.

Machine Learning Engineer

Job description: Your job will be to build, optimize, and deploy machine learning models to production. You'll generally be treating machine learning models as APIs or components, which you'll be plugging into a full-stack app or hardware of some kind, but you may also be called upon to design models yourself.

Requirements: The technologies you'll be working with include Python, JavaScript, scikit-learn, TensorFlow/PyTorch (and/or enterprise deep learning frameworks), and SQL or MongoDB (typically used for app DBs).

The questions you'll be dealing with sound like this:

- "How do I integrate this Keras model into our JavaScript app?"
- "How can I reduce the prediction time and prediction cost of our recommender system?"

What Have You Learned?

You should have a basic design idea for the RIF.

You need to understand that the processing has six distinct processing super steps to ensure proper machine learnings that are effective, efficient, and reproducible.

The changing of traditional machine learning into IML is simply the application of a repeatable standard process that you and all you fellow workers can easily follow with success for your customers and yourselves.

What's Next?

You now understand what the fourth revolution is.

The next chapter is "Industrialized Artificial Intelligence." I will summarize the previous chapters to supply you with a practical implementation of all the knowledge in this book.

So, I suggest that you have a cup of coffee or tea and then we will take the final phase in Chapter 15.

CHAPTER 14

Industrialized Artificial Intelligence

Industrialized Artificial Intelligence (IAI) in the real world is a complex process. You have the capacity to simulate basic building blocks of the real world in your own environment to test your machine learning applications against your model.

Where Does Machine Learning Fit?

Artificial intelligence (AI) is made up of the principal technologies that will one day be accomplished to formulate an all-purpose methodology that can drive our industrialized developments. Machine Learning fits completely within the knowledge territory of AI.

Deep Learning is the ultimate machine learning process as discussed earlier in the book.

Now I will guide you to enhance your knowledge of fundamental data science to understand machine learning. This will empower you with IAI that with our present-day abilities can assist you in accomplishing the goal of self-governing machine learning.

A. F. Vermeulen, *Industrial Machine Learning*, https://doi.org/10.1007/978-1-4842-5316-8_14

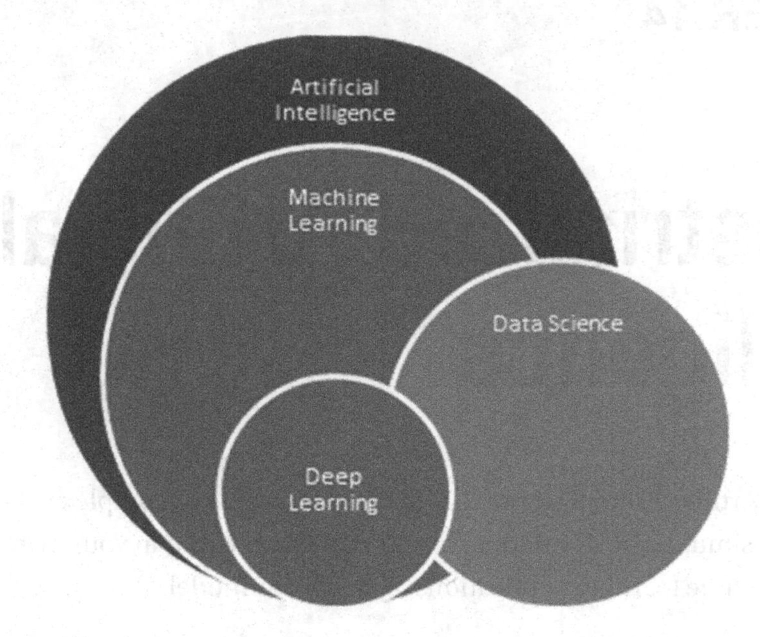

Figure 14-1. *Machine Learning Diagram*

Machine learning overlaps with AI, Deep Learning, and some data science.

The complete skills set in Figure 14-1 would supply you with most of the skills you would need to be a great data scientist who performs good machine learning.

Big Data Impact

The enormous growth in big data is predicted to grow to 175 zettabytes (ZB) worldwide by 2025.

The growth of this data will be the result of the incorporation of intelligent agents that use machine learning and other forms of AI to analyze the growing amount of data generated by the digital possessions in our lives. Our equipment around us is mapping our lives to a detail like never in the history of humankind.

This data is generated in the course of building driver assistance and autonomous vehicle technologies; IoT devices including sensors in our bodies, homes, factories and cities; creating high resolution content for 360 video and augmented reality; and 5G communications. It is enabled by building the edge networks and centralized data centers that help to analyze, communicate, and store the resulting data. The creation of this complete set of digital technology is called "*digital transformation*."

By 2025 Forbes (`www.forbes.com`) projects that 49% of the world's stored data will reside in public cloud environments and 30% of the global data volume will be real time by 2025. At this point in time, every connected person in the world (about 75% of the total population at that time) will have a digital data engagement over 4,900 times per day, about once every 18 seconds. The IoT devices generating much of this data will generate over 90 ZB of data in 2025.

The four main areas of growth are discussed next.

Health Care

The IDC (`www.idc.com`) predicts growth from 2019 of less than 1.5 ZB to more than 12 ZB by 2025 for health-care data.

Sizing Predictions

More than 80% of health-care providers are planning to use a cloud solution as their primary storage.

More than 75% of health-care providers are planning to use edge computing to handle decisions.

More than 65% of health-care providers will be using big data and machine learning to handle current data requirements until 2025.

More than 9 ZB will be secure and sensitive data by 2025.

More than 2 ZB will be near real time by 2025.

More than 1.5 ZB will be analytics data to benefit decision-making by 2025.

Data access is skewed as less than 3% of any data stored that is active at a specific point of time, despite the fact the required access turnaround time for a dormant patient's data in an emergency is less than 15 minutes for less than 2% of critical life-saving information.

Challenges

The shift to value-based care, the rise of consumerism, and the promise of personalized medicine are driving the health-care industry.

Health-care organizations must be ready for the digital age, yet the industry has traditionally, and admittedly, lagged behind others in the adoption of technology.

The right identity and security controls are critical for organizations to achieve effective data governance, prevent data misuse, and accelerate process automation.

The IT skill set gap is more pronounced in health care and is not likely to close anytime soon. Health-care companies must seek external assistance.

Failures will cost lives or cause serious long-term damage.

Solution

Heath care is going online, and your IoT devices are becoming part of the solution as you opt-in on services.

In exchange for better health care, we are giving permission to systems to monitor your body's vital measurements 24/7.

Warning More than 35% of health-care data custodians were uncertain of the knowledge of which data required what level of protection and how to protect it when the IML processes it.

Example:

Open the Jupyter Notebook in examples under Chapter 14 named: Chapter-014-01-Setup.ipynb

On execution you will create a complete Rapid Information Factory for Chapter 14's processing.

Open the Jupyter Notebook in examples under Chapter 14 named: Chapter-014-02-01-Retrieve.ipynb

You should now have a file '14-01-Retrieve-Healthcare.csv.gz' called in the directory.

..\990-DL\100-Raw-Zone

You have successfully loaded your health-care data.

Close the notebook.

Open the Jupyter Notebook in examples under Chapter 14 named: Chapter-014-02-01-Assess.ipynb

You should now have a file called 14-01-Assess-Healthcare.csv.gz in ..\990-DL\200-Structured-Zone

Your results are shown in Figure 14-2.

Overview

Dataset info

Number of variables	20
Number of observations	4793
Total Missing (%)	0.2%
Total size in memory	749.0 KiB
Average record size in memory	160.0 B

Variables types

Numeric	3
Categorical	17
Boolean	0
Date	0
Text (Unique)	0
Rejected	0
Unsupported	0

Warnings

`Address` has a high cardinality: 4767 distinct values Warning

`City` has a high cardinality: 2938 distinct values Warning

`County Name` has a high cardinality: 1562 distinct values Warning

`Hospital Name` has a high cardinality: 4608 distinct values Warning

`Meets criteria for meaningful use of EHRs` has 178 / 3.7% missing values Missing

`State` has a high cardinality: 56 distinct values Warning

***Figure 14-2.** Health-Care Data Profile*

You can now close your notebook.

Tip When working with health-care data, make sure you have all the correct security and ethical approvals as these types of data are protected by privacy protections that carry severe fines.

Financial Services

Financial services are predicted to growth from 2019 at less than 2.5 ZB to more than 10 ZB by 2025.

That is a 400% growth if the market stays at current growth predictions.

Sizing Predictions

More than 80% of financial services are planning to use a cloud solution as their primary storage.

More than 75% of financial services are planning to use edge computing to handle decisions.

More than 65% of financial services will be using big data and machine learning to handle current data requirements until 2025.

More than 9.5 ZB will be secured and sensitive data by 2025.

More than 5.5 ZB will be near real time by 2025.

More than 0.3 ZB will be analytics data to benefit decision-making by 2025.

Challenges

The biggest challenge is the management and creation of an effective data environment within the silo structure of most organizations in the financial services industry. The divided lines of business within each subindustry have no motivation to share data among themselves or to collaborate toward an enterprise solution.

Having solved the critical challenge of data consolidation, the organization must next address security, privacy, and compliance.

Financial services companies are stewards of some of the most sensitive and private data associated with customers. Digital trust is a complex relationship between a service provider and a customer, and it is underpinned by protecting privacy, securing data, and preventing fraud. Customer loyalty depends upon digital trust.

Managing the customer experience across all digital platforms is vital to improve customer satisfaction and maintain customer loyalty.

Solution

Open the Jupyter Notebook in examples under Chapter 14 named:Chapter-014-05-Utility-Data-Profiler-02.ipynb

> You should now have a file called 14-01-Assess-Healthcare.csv.gz in (..\990-DL\200-Structured-Zone)

Your results should be those shown in Figure 14-3.

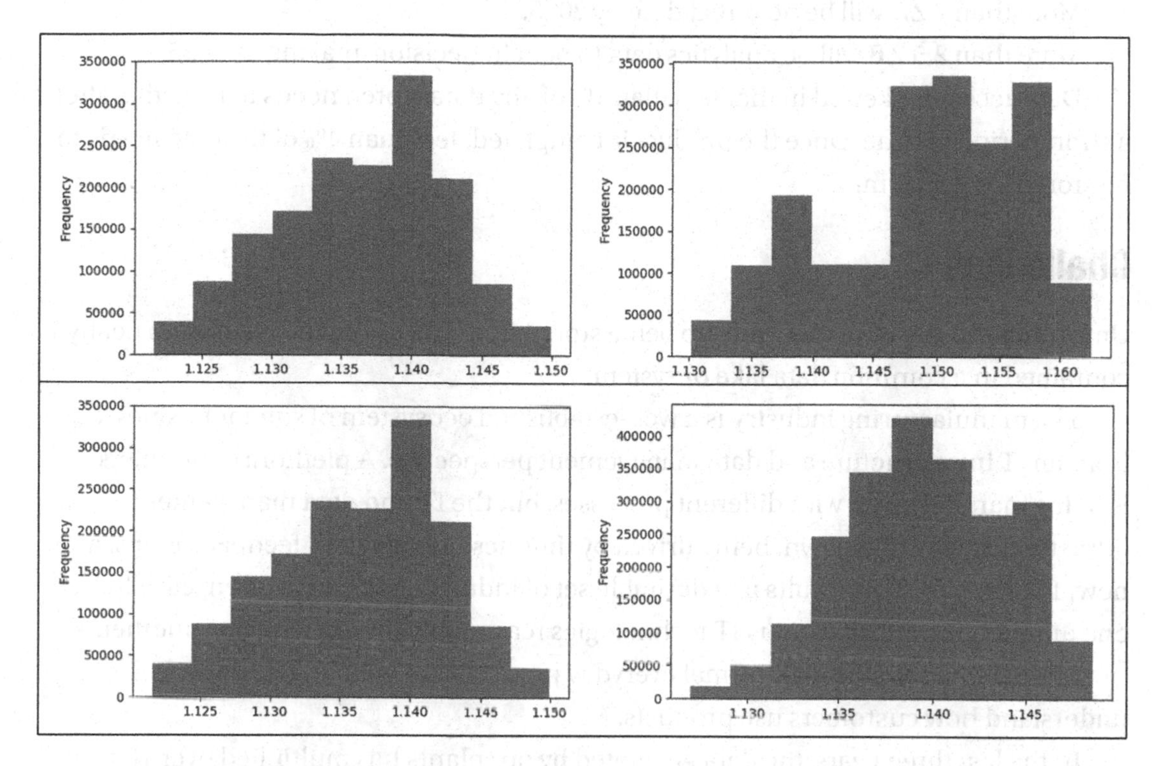

***Figure 14-3.** Four Histograms for FinTechs*

Manufacturing

Predict growth from 2019 of less than 2.5 ZB to more than 15 ZB by 2025

Sizing Predictions

More than 90% of manufacturing is planning to use cloud solution as their primary storage.

More than 95% of manufacturing is planning to use edge computing to handle decisions.

More than 95% of manufacturing will be using big data and machine learning to handle current data requirements until 2025.

More than 9 ZB will be secured and sensitive data by 2025.

More than 7 ZB will be near real time by 2025.

More than 2.5 ZB will be analytics data to benefit decision-making by 2025.

Data access is skewed in that less than 1% of any data stored needs to be active after a short period of time. Once the product is completed, less than 4% of the data needs to be stored longer term.

Challenges

Unfortunately, the data that ends up being stored from IoT's growth isn't always neatly contained in a common data lake or system.

The manufacturing industry is a well-established ecosystem of similar businesses, from an IT infrastructure and data management perspective. A plethora of products is being manufactured with different processes, but the IT and data management infrastructure is well known, being driven by time-tested analytics, feedback loops, and, now, IoT devices. This results in a desirable set of industry peers who often learn from one another and adopt various IT technologies reasonably in step with one another.

Embedding sensors into normal everyday products are enabling new ways to understand how customers use products.

In the last three years, the data generated by our plants has multiplied over 100 times. New sensors, processes, and more ubiquitous connectivity have allowed our engineers to embed sensors into anything and everything. We are still struggling with how to get the data out of the siloes, but we are on the verge of a major transformation in how we use the data.

Solution

Robots are communicating across the cloud to ensure a consistent, effective, and efficient manufacturing process is supported. The biggest challenges are the amalgamation of the correct sensors data while maintaining the flexibility of the inclusive system.

Warning More than 65% of robotics feedback data never leaves the robot ecosystem, so the data custodians have no knowledge of which data is not used by their IML processes.

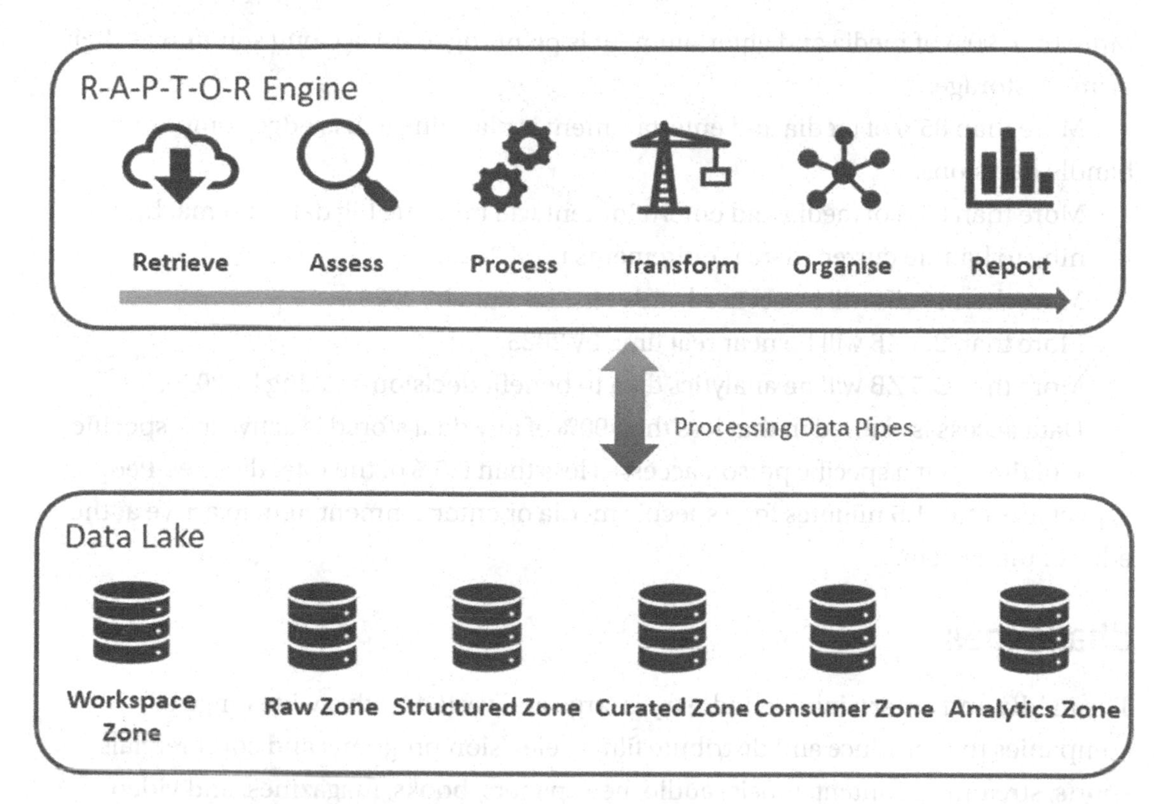

Figure 14-4. RAPTOR with Data Lake Structure

Tip I have found a RAPTOR process with a Structured Data Lake (Figure 14-4) working very well on these IoT solutions as you only need to modify the specific retrieval process that reads that specific type of sensor to adapt to changes.

I will predict that the Internet of Things will expand faster than any of the current predictions.

This single market will become the biggest data generator in our history of humans.

Media and Entertainment

There is predicted growth from 2019 of less than 3.5 ZB to more than 20 ZB by 2025.

Sizing Predictions

More than 90% of media and entertainment is planning to use a cloud solution as their primary storage.

More than 85% of media and entertainment is planning to use edge computing to handle decisions.

More than 80% of media and entertainment will be using big data and machine learning to handle current data requirements until 2025.

More than 19 ZB will be secured and sensitive data by 2025.

More than 2.5 ZB will be near real time by 2025.

More than 3.5 ZB will be analytics data to benefit decision-making by 2025.

Data access is skewed in that less than 90% of any data stored is active at a specific point of time, but a specific person accesses less than 0.5% of the total data set. People expect less than 1.5 minutes for a specific media or entertainment item to arrive at the edge of the system.

Challenges

The media and entertainment industry is large and varied, with a wide range of companies that produce and distribute films, television programs and commercials, sports, streaming content, music, audio, newspapers, books, magazines, and video games.

There is a lot of data, especially as our marketing groups have digitized. A lot of that has been done by third parties, and there's a lot of data we own, but it sits out with other people. A big part of trying to figure out that data environment ... is how we might bring back that data and store it.

Anything a CPU touches has a deflationary impact (i.e., creating something virtually is cheaper than creating the real thing) ... eventually the actors and actresses ... the building out of sets ... all of a sudden instead of millions of dollars ... it's hundreds of thousands.

Getting control of digital media assets by moving them to the cloud is an important first step for many media and entertainment companies. Digital rights management (DRM) and anti-pirating technologies can be enhanced effectively.

Solution

Warning The volumes of media and entertainment contained are a major challenge for IML.

Congratulations; you have reached the end of the Big Data impact.

Games

There is a new type of entertainment that is gathering massive amounts of data and is generating major machine learning opportunities.

Open the Jupyter Notebook in examples under Chapter 14 named: Chapter-014-06-Games-01.ipynb

You should look at PyGame as it is a Python wrapper module for the SDL multimedia library that enables you to simulate games that can be used to solve complex real-world problems.

Execute the notebook step by step to achieve insight into the techniques the code uses to achieve the end goal.

The basic requirement is to solve this chess challenge:

> You have one knight and must visit every square of the massive chessboard of 18 x 18 blocks.

Tip The effective processing of ML against games data is a highly sought-after skill at the moment in the world.

Simulations

A simulation is a model of the real world that you are studying by creating a virtual representation of the processes to enable your machine learning to test its capabilities in a world that can be reset many times to enable the application of reinforced learning using deep learning processes.

The art of simulations is formulated from the field of operational research, data science, and machine learning.

SimPy

SimPy is a process-based discrete-event simulation framework based on standard Python.

Processes in SimPy are defined by Python generator functions and may, for example, be used to model active components like customers, vehicles, or agents. SimPy also provides various types of shared resources to model limited capacity congestion points (like servers, checkout counters, and tunnels).

Simulations can be performed "as fast as possible," in real time (wall clock time) or by manually stepping through the events.

SimPy library is install by the following commands:

```
Installing SimPy
conda install -c mutirri/label/anaconda simpy
```

Basic Simulator Setup

I will now guide you on how to formulate and test your first simulation.

The basic simulations will simulate four processes.

- Very Fast Processor - runs every quarter second
- Fast Processor- runs every half second
- Slow Processor - runs every second
- Very Slow Processor - runs every two seconds

Open the Jupyter Notebook in examples under Chapter 9 named: Chapter-014-07-Basic Sim-01. ipynb

```
import simpy

def clock(env, name, tick):
    while True:
        print('%6.2f > %s' % (env.now, name))
        yield env.timeout(tick)

env = simpy.Environment()
env.process(clock(env, 'very fast', 0.25))
env.process(clock(env, 'fast', 0.5))
env.process(clock(env, 'slow', 1))
```

```
env.process(clock(env, 'very slow', 2))
env.run(until=3)
```

Your results are:

```
0.00 > very fast
  0.00 > fast
  0.00 > slow
  0.00 > very slow
  0.25 > very fast
  0.50 > fast
  0.50 > very fast
  0.75 > very fast
  1.00 > slow
  1.00 > fast
  1.00 > very fast
  1.25 > very fast
  1.50 > fast
  1.50 > very fast
  1.75 > very fast
  2.00 > very slow
  2.00 > slow
  2.00 > fast
  2.00 > very fast
  2.25 > very fast
  2.50 > fast
  2.50 > very fast
  2.75 > very fast
```

Tip Investigate what happens if you change env.run(until=100).

See Chapter-014-07-Basic-Sim-02.ipynb

Bank Simulation

I will guide you into how to apply the knowledge you acquired to industrialize the banking system by simulating a banking queueing system for 2000 customers arriving every 5 seconds.

Open the Jupyter Notebook in examples under Chapter 9 named: Chapter-014-07-Bank-Sim-01.ipynb

This notebook simulates a bank queue with people only waiting a while and then leaving. This simulation will result in more realistic data to base your data science and machine learning to show the near-real behavior of the ecosystem.

Your Result is:

```
Bank Withdraw Simulation One
 0.0000 Customer-0001-Ready: I have arrived
 0.0000 Customer-0001-Ready: Waited for 0.0000
12.6687 Customer-0001-Ready: Finished
16.0343 Customer-0002-Ready: I have arrived
16.0343 Customer-0002-Ready: Waited for 0.0000
18.0321 Customer-0003-Ready: I have arrived
20.0008 Customer-0003-Ready: left bank unserved at 1.9687
41.4728 Customer-0002-Ready: Finished
60.0490 Customer-0004-Ready: I have arrived
60.0490 Customer-0004-Ready: Waited for 0.0000
61.3580 Customer-0005-Ready: I have arrived
62.4971 Customer-0005-Ready: left bank unserved at 1.1391
64.9048 Customer-0004-Ready: Finished
```

Tip Try with NEW_CUSTOMERS = 500.

You are now serving 500 customers and not doing well!

Tip Try with counter = simpy.Resource(env, capacity=10).

You are now serving 500 customers and 10 tellers - You are doing better!

Tip Try with NEW_CUSTOMERS = 500 and counter = simpy.Resource(env, capacity=20) but customers are now leaving quicker MAX_PATIENCE = 3

```
Wow, It is working. I think your bank is going to work.
```

Can you see the value of these simulations as you can formulate a real-world bank easily by using these simple building blocks?

Machine Shop Simulation

I will guide you into how to apply the knowledge you acquired to industrialize the machine shop system and simulate a workshop.

Open the Jupyter Notebook in examples under chapter 9 named Chapter-014-08-Workshop-Sim-01.ipynb

This simulation supports the processing of three different shops:

```
Two staff start up
Multi-desk - not enough staff
Multi-desk - to many staff
```

```
Run the simulation and see how your shops are doing.
```

Results 1:

Small Shop - One-to-one Workflow.

At time 6: Worker completed work: Front Desk book in work at 6.

LATE Work Order: at time 12: Worker not completed work: Front Desk book in work at 10.

LATE Work Order: at time 17: Worker not completed work: Front Desk book in work at 16.

At time 21: Worker completed work: Front Desk book in work at 21.

Results 1:

LATE Work Order: at time 22: Supplier person B not completed work: Front desk - Suppliers book in work at 21.

```
Result  3:
Still having issues?

Emergency Change:

Train the workers better.
```

Multi-flow Workshop with Training

At time 4: Workshop Worker 000 completed work: Front desk - Workshop book in work at 4.

At time 4: Workshop Worker 001 completed work: Front desk - Workshop book in work at 4.

At time 6: Salesperson A completed work: Front desk - Sales book in work at 6.

At time 6: Salesperson B completed work: Front desk - Sales book in work at 6.

At time 6: Supplier person A completed work: Front desk - Suppliers book in work at 6.

At time 6: Supplier person B completed work: Front desk - Suppliers book in work at 6.

At time 8: Workshop Worker 000 completed work: Front desk - Workshop book in work at 8.

At time 8: Workshop Worker 001 completed work: Front desk - Workshop book in work at 8.

At time 12: Salesperson A completed work: Front desk - Sales book in work at 12.

At time 12: Salesperson B completed work: Front desk - Sales book in work at 12.

At time 12: Supplier person A completed work: Front desk - Suppliers book in work at 12.

At time 12: Supplier person B completed work: Front desk - Suppliers book in work at 12.

At time 14: Workshop Worker 000 completed work: Front desk - Workshop book in work at 14.

At time 14: Workshop Worker 001 completed work: Front desk - Workshop book in work at 14.

At time 16: Salesperson A completed work: Front desk - Sales book in work at 16.

At time 16: Salesperson B completed work: Front desk - Sales book in work at 16.

At time 16: Supplier person A completed work: Front desk - Suppliers book in work at 16.

At time 16: Supplier person B completed work: Front desk - Suppliers book in work at 16.

At time 19: Workshop Worker 000 completed work: Front desk - Workshop book in work at 19.

At time 19: Workshop Worker 001 completed work: Front desk - Workshop book in work at 19.

At time 20: Supplier person A completed work: Front desk - Suppliers book in work at 20.

At time 20: Supplier person B completed work: Front desk - Suppliers book in work at 20.

At time 22: Salesperson A completed work: Front desk - Sales book in work at 22.

At time 22: Salesperson B completed work: Front desk - Sales book in work at 22.

Now you have a successful workshop with just the right amount of trained staff.

Movies Simulation

I will guide you into how to apply the knowledge you acquired to industrialize the movies system to simulate a movie publisher or streaming services.

Open the Jupyter Notebook in examples under Chapter 14 named: Chapter-014-09-Movies-Sim-01.ipynb

This simulation demonstrates how to model three movie channels transmitted to several movie boxes. If you look at the results you can see some people do not get their requested movies on time. See if you can find the correct combination of parameters that will resolve the issues.

Fuel Station Simulation

I am happy to report that we made it to space and you won the contract to refuel all our spacecrafts.

Can you solve the issue with the client not getting fuel?

Car Wash Simulation

I will now guide you to build a car wash simulator.

Open the Jupyter Notebook in examples under Chapter 9 named: Chapter-014-12-Car-Wash-Sim-01.ipynb

You are now running the local car wash and you want to grow by 50%; what can you expect will happen to your ability to serve you customers?

This is better than building the business bigger first.

Tip If you add a simple write file to the code, you can use these simulations to generate data for your IML.

Now you have the knowledge you need to perform the IAI.

Restrictions on Industrialized Artificial Intelligence

The IAI capability is the biggest evolution in the modern world as it is changing the way we interact with our world. In human history, we needed other people to perform intelligent work for us.

This new evolution has also generated a new set of laws and policies that governs the IAI ecosystem.

The biggest of these new laws is the General Data Protection Regulation GDPR (see appendix for more information) that controls the way our IAI can interact with people's personal data for Europe. These rules are causing disruptions of the processing capability of the current IAI ecosystems.

Here is a quick summary of the basic guidelines.

The Right to Be Informed

Individuals have the right to be informed about the collection and use of their personal data. This is a key transparency requirement under the GDPR.

You must provide individuals with information including: your purposes for processing their personal data, your retention periods for that personal data, and who it will be shared with. We call this 'privacy information'.

Warning If you apply Artificial Intelligence (AI) to personal data:

Be upfront about it and explain your purposes for using AI.

If the purposes for processing are unclear at the outset, give people an indication of what you are going to do with their data. As your processing purposes become clearer, update your privacy information and actively communicate this to the people.

Inform people about any new uses of personal data before you actually start the processing.

If you use AI to make solely automated decisions about people with legal or similarly significant effects, tell them what information you use, why it is relevant, and what the likely impact is going to be.

The Right of Access

Individuals have the right to access their personal data.

Individuals have the right to obtain the following from you:

- confirmation that you are processing their personal data. You must list the precise IML techniques you are applying to the data.
- a copy of their personal data. You must supply the complete set of data that belongs to the requester.

Tip I suggest you store any data you process with the metadata on what you used to process in an archive that will enable you to quickly respond to these requests as under the law, you cannot charge fees to send this data. So, I suggest you store it in the most cost-effective manner.

The Right to Rectification

The GDPR includes a right for individuals to have inaccurate personal data rectified, or completed if it is incomplete.

An individual can make a request for rectification verbally or in writing.

You have one calendar month to respond to a request.

In certain circumstances, you can refuse a request for rectification.

Tip I use the Assess step from Six Steps of RAPTOR Engine in Chapter 13 to ensure data quality is achieved by default. This normally resolves any inaccurate personal data issues.

So make sure the quality of the data you process is always good and that you can show proof that you have ensured that this data stays good.

The Right to Erasure

The GDPR introduces a right for individuals to have personal data erased.

The right to erasure is also known as 'the right to be forgotten'.

Individuals can make a request for erasure verbally or in writing.

You have one month to respond to a request.

The right is not absolute and only applies in certain circumstances. (See section on when not to erase.)

Warning The right to erasure and children's personal data carries an extra complexity that is still enforced even in the case when the data subject is no longer a child, because a child may not have been fully aware of the risks involved in the processing at the time of consent.

This right is a major issue for companies as most of the data they hold has cost them money to collect or in some cases even was provided at an extra cost by information brokers. The erasure of data will have financial impacts.

Note Do not always erase!

There is specific data that is protected under other laws and should not be erased.

The right to erasure does not apply if processing is necessary for one of the following reasons:

- to exercise the right of freedom of expression and information;
- to comply with a legal obligation;
- for the performance of a task carried out in the public interest or in the exercise of official authority;
- for archiving purposes in the public interest, scientific research, historical research, or statistical purposes where erasure is likely to render impossible or seriously impair the achievement of proper processing;
- for the establishment, implementation, or justification of legal claims.

Note If a valid erasure request is received and no exemption applies, then you will have to take steps to ensure erasure from backup systems as well as live systems.

The Right to Restrict Processing

Individuals have the right to request the restriction or suppression of their personal data. When processing is restricted, you are permitted to store the personal data but not use it.

That means you may have the data but cannot apply your machine learning against the data.

Warning Ensure you have full unrestricted processing before you start any ML processing.

Note This right has close links to the right to rectification and the right to object.

The Right to Data Portability

The right to data portability allows individuals to obtain and reuse their personal data for their own purposes across different services. It allows you to copy or transfer personal data easily from one IT environment to another in a safe and secure way, without affecting its usability.

Doing this enables individuals to take advantage of applications and services that can use this data to find them a better deal or help them understand their spending habits.

Open Banking is a prime example of where personal data publishing is officially allowed to improve the person's services.

I personally also use data portability agreement to obtain data that I use for research though my university and other research projects we are sponsoring to enable progressing in the fields of health care and improvements for smart cities.

Tip I suggest you seek out opportunities where you can apply your ML for the improvement of people in general. You will mostly find support for these projects and also sometimes even funding.

The Right to Object

The GDPR gives individuals the right to object to the processing of their personal data in certain circumstances.

Individuals have an absolute right to stop their data being used for direct marketing.

You must tell individuals about their right to object.

This means you must have settings in your data processing to stop the processing of data for specific purposes.

I personally use the organize step of the RAPTOR Engine in Chapter 13 to limit the use of the data for only the specific purpose.

Rights in Relation to Automated Decision-Making and Profiling

Warning IML is explicitly distinguished by the GDPR.

The processes of automated individual decision-making (making a decision solely by automated means without any human involvement) and profiling (automated processing of personal data to evaluate certain things about an individual) can only be performed if:

You can only carry out this type of decision-making where the decision is:

- necessary for the entry into or performance of a contract.
- authorized by the law applicable to the controller.
- with the individual's explicit consent.

Warning Profiling is seen part of the automated decision-making process.

That means the training of your ML models is subject to the GDPR rules.

You have now completed the majority of the background information for IML.

Well done!!!

What's Next?

We have now covered industrialized AI. The next chapter is a final project. You will now be able to apply all your new knowledge against a project to practice the new skills and techniques you have mastered during this book.

So, I suggest have a cup of coffee or tea, and then we will take the final phase in Chapter 15.

CHAPTER 15

Final Industrialization Project

In this book we investigated the tools you need to create a full-scale industrialized project. The project is the creation of a mining company that will mine minerals on Mars and then transport the minerals back to Earth and distribute it to the centers that require the specific resource.

Requirements

Mars has abundant resources to mine, and they can be shipped back to Earth:

Silicon, hydrogen, oxygen, iron, magnesium, manganese, aluminum, calcium, gold, silver, nickel, titanium, platinum, and copper are ready for mining.

Your Costs

A mining spaceport is £100 000-00 and can only support 25 miners and 250 transports. It needs two rockets to get to Mars.

A rocket costs £ 20 000-00 per trip and a trip to Mars takes between 150–300 days. Each rocket is only capable of sending four pieces of mining equipment per rocket.

An ore miner is £ 150 000-00 and costs £ 5000-00 per day to run. It can mine 1000 pounds per day.

An ore transport is £ 50 000-00 and costs £ 3500-00 per day to run. It travels 100 miles per day, carries 250 pounds per trip, and loads/offloads in 10 minutes.

Hydrogen and oxygen extractors are £ 10 000-00 each and cost £ 1500-00 per day to run. An extractor produces 500 pounds of hydrogen or oxygen per day.

A. F. Vermeulen, *Industrial Machine Learning*, https://doi.org/10.1007/978-1-4842-5316-8_15

Each piece of mining equipment uses 200 pounds of hydrogen and 100 pounds of oxygen per day for power.

A 3D printing plant can use 1000 pounds of silicon to print extra miner, 750 pounds for an ore transport, 300 pounds for an extractor, and 5000 per rocket. A new 3D printing plant takes 15 days and 7500 pounds of silicon.

It takes 5 days to make a miner, 3 days for an ore transport, a day for an extractor.

Three-dimensional printing uses 400 pounds of hydrogen and 300 pounds of oxygen per day.

You are only funded to a maximum of £ 1 000 000-00.

During a return flight, each rocket can only carry 500 000 pounds of ore and uses 6000 pounds of hydrogen and 2000 pounds of oxygen per day.

You must return 20% profit to your shareholders per year.

Your Income

Current Trading prices for 1000 pounds of ore on delivery to Earth:

> Silicon(£1300), iron (£80), magnesium(£5000), manganese (£40), aluminum(£800), calcium(£220), gold(£1200), silver(£15), nickel(£1500), titanium(£30000), platinum(£870), and copper (£2500).
>
> Now that you have the industrialized problem, you need to change the requirements into an IML solution.
>
> I will guide you through the process of turning your new knowledge into a full working solution.

Basic Solution

The Rapid Information Factory (RIF) and R-A-P-T-O-R engine is an industrial hardened processing methodology that enables the structured design and build of a data engineering processing engine (Figure 15-1).

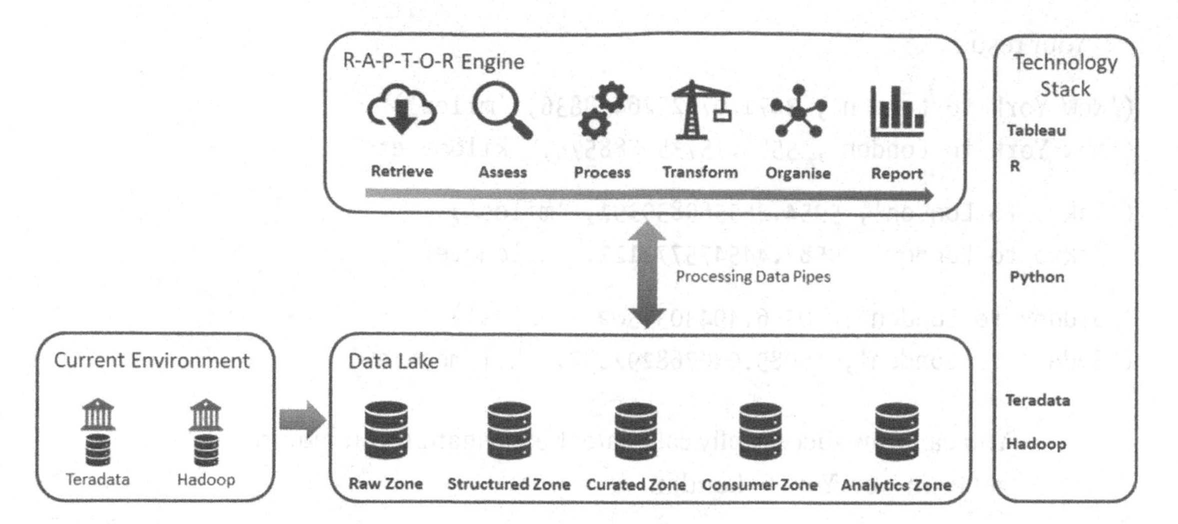

***Figure 15-1.** The RAPTOR engine and RIF*

This Mars factory will assist you to keep you on track to become the Mars Industrialist.

So, open the Jupyter Notebook in examples under Chapter 15 named: Chapter-015-01-Setup-Mars.ipynb

Run the complete notebook to set up the new ecosystem.

This will setup a Rapid Information Factory by number ***010-RIF*** and a Data Lake by number ***999-DL***.

Geospatial Knowledge

The following will introduce you to basic geospatial solutions before you start your space mining colony.

> Open the Jupyter Notebook in examples under Chapter 15 named: Chapter-015-02-GeoSpace-01.ipynb
>
> Let's start with traveling on our own planet. I suggest we travel from New York to London to demonstrate the geospatial principals.
>
> Execute the notebook.
>
> Take note of the concept of Latitude and Longitude to navigate the planet.

Your results:

```
('New York to London', 3471.0772526498836, 'miles')
('New York to London', 5586.157350088574, 'kilometers')

('Tokyo to London', 5954.255569830391, 'miles')
('Tokyo to London', 9582.445475773122, 'kilometers')

('Sydney to London', 10556.4943034509, 'miles')
('Sydney to London', 16989.030768292887, 'kilometers')
```

You can now successfully calculate the navigation you require to get from New York to London.

As well as Tokyo and Sydney ... Well done, Earthling!

Now we can move to outer space. Let's look at our solar system.

Open the Jupyter Notebook in examples under Chapter 15 named: Chapter-015-03-GeoSpace-02.ipynb

The planets are of the following size:

```
Planets Smallest to Largest
Pluto      diameter is:   2376.600 kilometers
Moon       diameter is:   3474.200 kilometers
Mercury    diameter is:   4879.400 kilometers
Mars       diameter is:   6779.000 kilometers
Venus      diameter is:  12104.000 kilometers
Earth      diameter is:  12742.000 kilometers
Neptune    diameter is:  49244.000 kilometers
Uranus     diameter is:  50724.000 kilometers
Saturn     diameter is: 116460.000 kilometers
Jupiter    diameter is: 139820.000 kilometers
```

If we test by placing two important pieces of equipment on the surface:

```
mine_crusher = (0, 0)
mine_shaft = (1, 1)
MineRoute(4)
```

We get a result:

```
Mine 004 on Mercury  is     60216.812979 meters (44.996 degrees) from mine
to crusher
Mine 004 on Venus    is    149375.805283 meters (44.996 degrees) from mine
to crusher
Mine 004 on Earth    is    157249.381272 meters (44.996 degrees) from mine
to crusher
Mine 004 on Moon     is     42875.200158 meters (44.996 degrees) from mine
to crusher
Mine 004 on Mars     is     83659.830140 meters (44.996 degrees) from mine
to crusher
Mine 004 on Jupiter  is   1725522.562348 meters (44.996 degrees) from mine
to crusher
Mine 004 on Saturn   is   1437236.143693 meters (44.996 degrees) from mine
to crusher
Mine 004 on Uranus   is    625986.314208 meters (44.996 degrees) from mine
to crusher
Mine 004 on Neptune  is    607721.592478 meters (44.996 degrees) from mine
to crusher
Mine 004 on Pluto    is     29329.687610 meters (44.996 degrees) from mine
to crusher
```

I suggest we move on to comparing Earth against Mars as we are already planning missions.

Open the Jupyter Notebook in examples under Chapter 15 named: Chapter-015-04-GeoSpace-03.ipynb

Great news; once on Mars, we can travel quicker as the planet is smaller.

Result:

```
The ratio Earth to Mars is: 1.83333
```

This means on Mars the locations on a standard longitude and latitude grid are workable, but the distance is simply 1.8 times shorter.

This is useful for the standard location generation.

Mars Mission Simulator Project

Congratulations; you have just won the contract to mine Mars. You are required to perform your best machine learning to ensure your profits are high and your mission is an overall success.

You will now build a Mars mission simulator to assist you with your data science to prepare for your machine learning during the Transform step of the Mars mission project.

Important Business Fact:

Journey time from Earth to Mars takes between 150–300 days depending on the speed of the launch.

That means you need a minimum of 600 days for an exciting round trip.

I am suggesting an extra 300 for mining operations.

Your model's mission is planned for ***1000 days***. That is well within the parameters of your real-world model.

Important Business Fact:

On Mars, a solar day lasts 24 hours, 39 minutes, and 35 seconds.

On Earth, a solar day lasts 23 hours, 56 minutes, and 4 seconds.

That means that your mission requires ***two distinct time clocks***, that is, Earth time and Mars time.

As nobody has yet lived on Mars from Earth, I recommend we use a concept of mission time, which means on the first day our rocket launches from Earth. You simply synchronize Earth and Mars by setting ***zero*** on both mission time Earth and mission time Mars.

Earth Time and Mars Time

Let's build these interesting clocks for your mission.

Open the Jupyter Notebook in examples under Chapter 15 named: Chapter-015-05-Earth-Mars-Time-01.ipynb

I suggest you use the pandas library to resolve these clocks.

You have successfully created a valid clock for the mission:

```
Results:

             EarthTime
0 2018-10-20 09:15:30
1 2018-10-20 09:16:30
2 2018-10-20 09:17:30
3 2018-10-20 09:18:30
4 2018-10-20 09:19:30
```

Here is just before you return to Earth with your riches:

```
Results:

                   EarthTime
1439995 2021-07-16 09:10:30
1439996 2021-07-16 09:11:30
1439997 2021-07-16 09:12:30
1439998 2021-07-16 09:13:30
1439999 2021-07-16 09:14:30
```

You can use this shift of the index value to a column value to add the minutes since launch to your data set.

Tip I suggest you always look at innovative ways to add extra value through data engineering.

You simply reset the index and you have the data; next you perform a rename of the column and you have a new feature as part of the process.

It works, and here is the result:

```
   MinutesSinceLaunch            EarthTime
ID
0                    0 2018-10-20 09:15:30
1                    1 2018-10-20 09:16:30
2                    2 2018-10-20 09:17:30
```

```
3                      3 2018-10-20 09:18:30
4                      4 2018-10-20 09:19:30
```

Result:

```
        MinutesSinceLaunch           EarthTime
ID
1439995             1439995 2021-07-16 09:10:30
1439996             1439996 2021-07-16 09:11:30
1439997             1439997 2021-07-16 09:12:30
1439998             1439998 2021-07-16 09:13:30
1439999             1439999 2021-07-16 09:14:30
```

You need to create a new feature to add the seconds since launch. The precision of space travel requires that everything is done in seconds.

One more feature is completed; see the results:

```
   MinutesSinceLaunch           EarthTime  SecondsSinceLaunch
ID
0                   0 2018-10-20 09:15:30                   0
1                   1 2018-10-20 09:16:30                  60
2                   2 2018-10-20 09:17:30                 120
3                   3 2018-10-20 09:18:30                 180
4                   4 2018-10-20 09:19:30                 240
```

Mars Clock

The next feature is more complex but a similar principle is used to calculate the Mars Clock:

```
def marstime(s):
    #set to 24 hours, 39 minutes, and 35 seconds
    marsdayseconds = (24*60*60)+(39*60)+35
    marsday = round(s/marsdayseconds,6)
    return marsday

dateDF['MarsTime'] = dateDF.apply(lambda row: marstime(row['SecondsSince
Launch']), axis=1)
```

Earth Clock

The next feature is a correction about our own planet's day length. Our days are, on average, shorter than 24 hours, hence our leap years. The mission must take this into account.

```
def earthtime(s):
    # set to 23 hours, 56 minutes, and 4 seconds
    earthdayseconds = (23*60*60)+(59*60)+4
    earthday = round(s/earthdayseconds,6)
    return earthday

dateDF['EarthTime'] = dateDF.apply(lambda row:earthtime(row['SecondsSince
Launch']), axis=1)
```

Success; you have the results:

```
    MinutesSinceLaunch  EarthTime  SecondsSinceLaunch  MarsTime
ID
0                    0   0.000000                   0  0.000000
1                    1   0.000695                  60  0.000676
2                    2   0.001390                 120  0.001352
3                    3   0.002085                 180  0.002028
4                    4   0.002780                 240  0.002703
```

Earth Mars Gap

You must understand that the gap in the time on the day counts will separate during your trip, so you need to track this:

I suggest you add a feature called "Earth Mars Gap" to keep track of this shift in time due to the two planets' different day lengths.

```
def findgap(et,mt):
    gaptime = et-mt
    return gaptime

dateDF['EarthMarsGap'] = dateDF.apply(lambda row:findgap(row['EarthTime'],
row['MarsTime']), axis=1)
```

You will see that the features are perfect but due to the Human-in-the-Loop, you are now required to move the columns to match as a requirement by your boss:

He wants to check the numbers.

So create Boss' file in ../../Results/Chapter 09/999-DL/400-Consumer-Zone/015-04-Retrieve-Mars-Time.csv

---

Note The file is in normal CSV and not compressed. This is because humans cannot read compressed files, and most people will be able to load a CSV. To adapt to the human, you need to uncompress the file.

Now you close out the process by completing the Audit.

You can now close the notebook.

Mars Mines

Next you will simulate the mining operation that you suggest for Mars.

You can open the Jupyter Notebook in examples under Chapter 15 named:

Chapter-015-06-Mars-Mines-01.ipynb

This code creates a simulation of a loading hub on Mars for four different mining sites.

The core library you need for this simulation is the "***SimPy***" process.

You used this previously in this book.

I will help you enhance the previous knowledge plus some creations of data sets while running the data simulation to enable an output that matches the proposed Mars mining process.

Requirements:

- Four loading bays one per mine,
- One million hoppers with ore to mine,
- Every 20 seconds a new hopper is loaded,
- A hopper can only be in a loading bay for between 15 and 20 seconds,
- The next hopper will then simply push the current hopper off the loading pad,
- The reason your drivers are autonomous robots and their simple programming causes this process.

Your settings are in Part A:

```
LOADBAYCNT = 4              # Amount of Load Bays
NEW_HOPPERS = 1000000       # Total number of Mars Hoppers
#NEW_HOPPERS = 1000          # Total number of Mars Hoppers
INTERVAL_HOPPERS = 20.0     # Generate new Hopper roughly every x seconds
MIN_WAITLIMIT = 15          # Min. Hopper WAITLIMIT
MAX_WAITLIMIT = 20          # Max. Hopper WAITLIMIT
#DISPLAYRESULTS = True
DISPLAYRESULTS = False
```

Your hopper data is stored in this file:

```
../../Results/Chapter 15/999-DL/100-Raw-Zone/015-06-Retrieve-Mars-Hopper.
csv
```

Mars Ore

Wait ... What are we mining?

You need to classify the ore:

```
  HopperType
0    silicon
1       iron
2  magnesium
3   aluminum
4    calcium
5  potassium
```

You can now mine six different ores.

Mars Hopper

Part B of the simulation creates a Mars Hopper:

```
hoppertypes = pd.DataFrame(['silicon',
                            'hydrogen',
                            'oxygen',
```

```
                                        'iron',
                                        'magnesium',
                                        'manganese',
                                        'aluminum',
                                        'calcium',
                                        'gold',
                                        'silver',
                                        'nickel',
                                        'titanium',
                                        'platinum',
                                        'copper']
                                    )
hoppertypes.columns = ['HopperType']
```

Hopper Types: 14

Start Mining

Part C sets the simulation parameters.

Part D runs the simulation.

Simulation Time

The biggest requirement is you only have 300 days to mine.

Does your simulation meet that requirement?

> ***Simulation of 20019146.275066 seconds => yields 225.504323 of real days of Mars Mining time.***

You are successful. Well done.

You should, as per your simulation, complete in 256 days.

Mine Locations

You need to create the four Mine Locations and your Mars landing pad, that is, Home for next 300 days.

I suggest you just copy the proposed location directly into your code:

ID	MineID	Longitude	Latitude
0	0	0	0
1	1	-88	-27
2	2	77	77
3	3	-19	68
4	4	20	19

Save Mars Mines into ../../Results/Chapter 15/999-DL/100-Raw-Zone/015-06-Retrieve-Mars-Mines.csv

Now you have a mine that can mine ore for you.

Machine Learning – Mars Mission

Now that you have a simulation set of data, you will now want to apply some of the knowledge from this book against this data.

I have prepared a full IML process with data engineering and data science to use your knowledge against your Mars Mission.

Open the Jupyter Notebook in examples under Chapter 15 named:

Chapter-015-07-Build-RAPTOR-Mars-01.ipynb

Set up the Rapid Information Factory (RIF) and the Data Lake (DL).

I suggest you use RIF: ***010-RIF*** and DL: ***999-DL.***

Note You will get a basic Data Lake copied to your workspace to assist with your progress through the rest of this chapter.

Warning The command will remove the previous data lake with number ***999-DL***.

Set Up Data Lake

Your results:

```
..\Results\Chapter 09\999-DL\100-Raw-Zone\015-04-Retrieve-Mars-Hopper.csv.gz
..\Results\Chapter 09\999-DL\100-Raw-Zone\015-04-Retrieve-Mars-Mines.csv.gz
..\Results\Chapter 09\999-DL\100-Raw-Zone\015-04-Retrieve-Mars-Time.csv.gz
```

These three files are the core of the information you have for your Mars mission.

Tip You can investigate the three files or just continue with this notebook. That is totally up to you.

File: 015-04-Retrieve-Mars-Hopper.csv.gz

You can use the Jupyter Notebook in examples under Chapter 15 named:

Chapter-015-07-Build-RAPTOR-Mars-01.ipynb to perform data profiles on the file.

I have included the results (Figure 5) in: 015-04-Retrieve-Mars-Hopper.html in Raw Zone of the Data Lake.

Standard "***describe***" command gives you this result:

```
              Time       Tonnage      LoadBayID      WaitTime
count  2.919739e+06  2.919739e+06  2.919739e+06  2.919739e+06
mean   1.000863e+07  5.997769e+01  2.501647e+00  1.701870e+01
std    5.779345e+06  2.339773e+01  1.118402e+00  3.660686e+01
min    0.000000e+00  2.000000e+01  1.000000e+00  7.933378e-05
25%    5.003417e+06  4.000000e+01  2.000000e+00  1.000000e-04
50%    1.000357e+07  6.000000e+01  3.000000e+00  1.000000e-04
75%    1.501119e+07  8.000000e+01  4.000000e+00  1.680835e+01
max    2.001915e+07  1.000000e+02  4.000000e+00  6.345246e+02
```

More insight from an "***info***" command:

```
<class 'pandas.core.frame.DataFrame'>
RangeIndex: 2919739 entries, 0 to 2919738
Data columns (total 7 columns):
Time            float64
Event           object
```

```
ObjectHopper     object
ObjectOre        object
Tonnage          int64
LoadBayID        int64
WaitTime         float64
dtypes: float64(2), int64(2), object(3)
memory usage: 155.9+ MB
None
```

Warning This file has a large volume of 155.9+ MB, and I have used a sample process to only process 0.01% of the data.

This sample profile still shows you great insight into the data.

You can see from the smaller sample's "***info***" that you now only use 1.8 mb of the file:

```
<class 'pandas.core.frame.DataFrame'>
Int64Index: 29197 entries, 1201809 to 1197298
Data columns (total 7 columns):
Time             29197 non-null float64
Event            29197 non-null object
ObjectHopper     29197 non-null object
ObjectOre        29197 non-null object
Tonnage          29197 non-null int64
LoadBayID        29197 non-null int64
WaitTime         29197 non-null float64
dtypes: float64(2), int64(2), object(3)
memory usage: 1.8+ MB
```

This provides a good profile (Figure 5).

Look at notebook: Chapter-015-10-Utility-Mines-Profiler-01 to data profile (Figure 15-2), your financials for your mines.

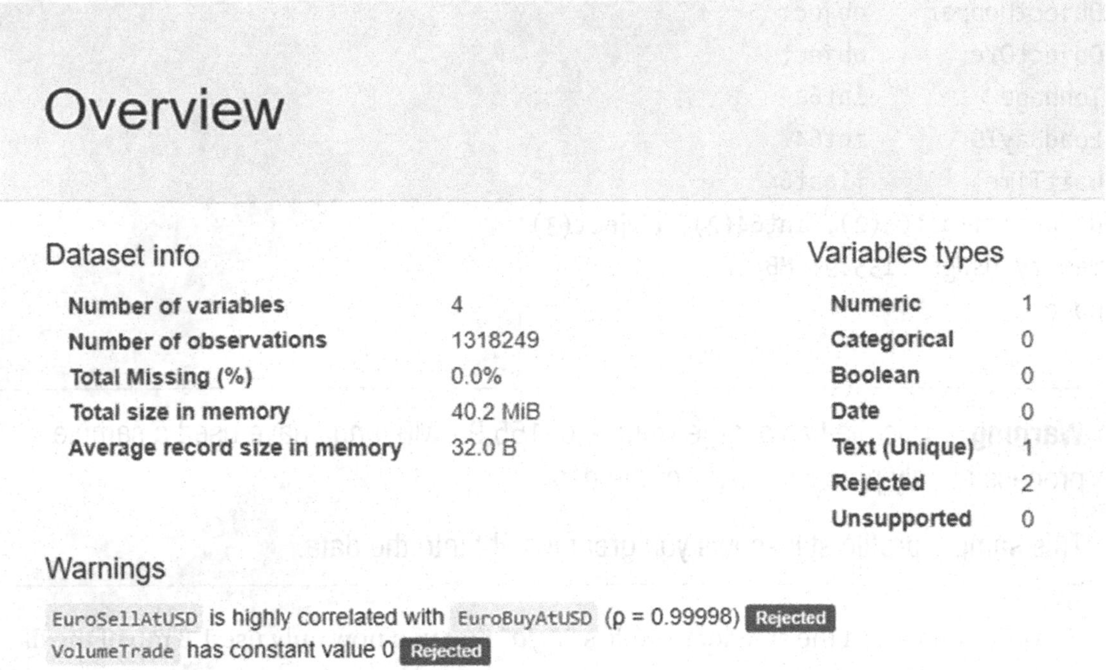

Overview

Dataset info

Number of variables	4
Number of observations	1318249
Total Missing (%)	0.0%
Total size in memory	40.2 MiB
Average record size in memory	32.0 B

Variables types

Numeric	1
Categorical	0
Boolean	0
Date	0
Text (Unique)	1
Rejected	2
Unsupported	0

Warnings

`EuroSellAtUSD` is highly correlated with `EuroBuyAtUSD` (ρ = 0.99998) Rejected
`VolumeTrade` has constant value 0 Rejected

Figure 15-2. *Financials data profile for mines*

Look at notebook: Chapter-015-11-Utility-Data-Profiler-01 to data profile (Figure 15-3), your hoppers information.

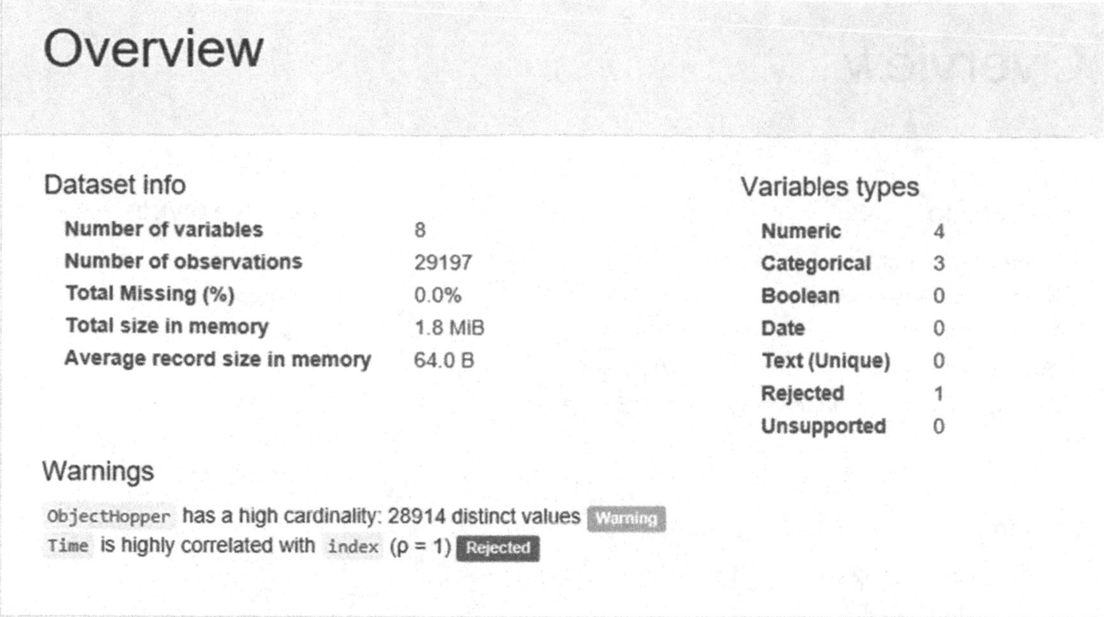

Overview

Dataset info

Number of variables	8
Number of observations	29197
Total Missing (%)	0.0%
Total size in memory	1.8 MiB
Average record size in memory	64.0 B

Variables types

Numeric	4
Categorical	3
Boolean	0
Date	0
Text (Unique)	0
Rejected	1
Unsupported	0

Warnings

`ObjectHopper` has a high cardinality: 28914 distinct values Warning
`Time` is highly correlated with `index` (ρ = 1) Rejected

Figure 15-3. *Retrieve-Mars-Hopper Profile*

For file: 15-04-Retrieve-Mars-Mines.csv.gz use Chapter-015-12-Utility-Data-Profiler-02.ipynb

I have included the results in: 015-04-Retrieve-Mars-Mines.html in Raw Zone of the Data Lake.

Note An insights is that you have only four mines in the data.

Here is the result (Figure 15-4).

Overview

Dataset info

Number of variables	3
Number of observations	5
Total Missing (%)	0.0%
Total size in memory	200.0 B
Average record size in memory	40.0 B

Variables types

Numeric	3
Categorical	0
Boolean	0
Date	0
Text (Unique)	0
Rejected	0
Unsupported	0

Warnings

`Latitude` has 1 / 20.0% zeros Zeros
`Longitude` has 1 / 20.0% zeros Zeros
`MineID` has 1 / 20.0% zeros Zeros

Figure 15-4. *015-04-Retrieve-Mars-Mines Profile*

For file: 15-04-Retrieve-Mars-Time.csv.gz use Chapter-015-13-Utility-Data-Profiler-03.ipynb

Well done; you now have complete knowledge of your data.

I have included the results in: 015-04-Retrieve-Mars-Hopper.htmlin Raw Zone of the Data Lake.

You will see due to the volume of 54.9 MB, I suggest you apply a 10% sample:

```
<class 'pandas.core.frame.DataFrame'>
RangeIndex: 1440000 entries, 0 to 1439999
Data columns (total 5 columns):
MinutesSinceLaunch    1440000 non-null int64
SecondsSinceLaunch    1440000 non-null int64
EarthTime             1440000 non-null float64
MarsTime              1440000 non-null float64
EarthMarsGap          1440000 non-null float64
```

```
dtypes: float64(3), int64(2)
memory usage: 54.9 MB
None
```

The insights are as follows (Figure 15-5):

The data set has ***highly correlated*** data that will cause issues with feature engineering as the relationship is due to the mathematical formula that relates the columns and not as cause-and-effect of the real-world activities.

Overview

Dataset info

Number of variables	6
Number of observations	144000
Total Missing (%)	0.0%
Total size in memory	6.6 MiB
Average record size in memory	48.0 B

Variables types

Numeric	1
Categorical	0
Boolean	0
Date	0
Text (Unique)	0
Rejected	5
Unsupported	0

Warnings

`EarthMarsGap` is highly correlated with `MarsTime` (ρ = 1) Rejected
`EarthTime` is highly correlated with `SecondsSinceLaunch` (ρ = 1) Rejected
`MarsTime` is highly correlated with `EarthTime` (ρ = 1) Rejected
`MinutesSinceLaunch` is highly correlated with `index` (ρ = 1) Rejected
`SecondsSinceLaunch` is highly correlated with `MinutesSinceLaunch` (ρ = 1) Rejected

Figure 15-5. *015-04-Retrieve-Mars-Time Profile*

Tip You are now back to the process of making your Mars mission work.

Retrieve Step

You need to load the data from outside the ecosystem.

But let's first check what you already have loaded:

```
for root, dirs, files in os.walk(pathRealDLname, topdown=True):
    for name in files:
        print(os.path.join(root, name))
```

Good news!

Mission control has already supplied you with the Raw Zone. So very nice of them! You have the following:

Time, Person, Objects, Location, and Event Retrieve Data.

The Mars communication system has delivered the information from Mars to your Raw Zone.

That system is outside your control.

This means no processing is needed to get the data into the RIF ecosystem.

Warning It also means data can arrive unexpectedly and in a changed format.

Assess Step

The assess step enables you to apply the data quality and fix any unexpected shortfalls in the data.

You will work your way through the standard Time-Person-Object-Location-Event (TPOLE) steps to ensure all the required data is in the ecosystem and in the correct format.

Time

This part of the data set contains all data related to the time that anything happened.

Tip This is the base for most of your time series machine learning.

In the data you have created, you have several parallel timekeeping features that you can use for your machine learning.

I will assist you in exposing these to your machine learning knowledge.

So, let's load the data and start your data engineering to clean the data ready for the process step later that will consolidate and amalgamate the data ready for the transform step where you will deploy your machine learning against it.

But first we need to get the data ready:

```
retrieveTimeName = os.path.join(pathDL, '100-Raw-Zone', '015-04-Retrieve-
Mars-Time.csv.gz')
TimeDF=pd.read_csv(retrieveTimeName, header=0, encoding='utf-8',
compression='gzip')

print(TimeDF.shape)
```

Results: (1440000, 5)

The insight is that the time data has 5 columns and you have 1.44 million records.

You can investigate them using the "info" command:

```
print(TimeDF.info())

<class 'pandas.core.frame.DataFrame'>
RangeIndex: 1440000 entries, 0 to 1439999
Data columns (total 5 columns):
MinutesSinceLaunch    1440000 non-null int64
SecondsSinceLaunch    1440000 non-null int64
EarthTime             1440000 non-null float64
MarsTime              1440000 non-null float64
EarthMarsGap          1440000 non-null float64
dtypes: float64(3), int64(2)
memory usage: 54.9 MB
None

You can also "describe" the data set:

print(TimeDF.describe())
```

```
       MinutesSinceLaunch  SecondsSinceLaunch    EarthTime     MarsTime \
count        1.440000e+06        1.440000e+06 1.440000e+06 1.440000e+06
mean         7.199995e+05        4.319997e+07 5.003239e+02 4.866231e+02
std          4.156923e+05        2.494154e+07 2.888625e+02 2.809523e+02
min          0.000000e+00        0.000000e+00 0.000000e+00 0.000000e+00
25%          3.599998e+05        2.159998e+07 2.501620e+02 2.433116e+02
50%          7.199995e+05        4.319997e+07 5.003239e+02 4.866231e+02
75%          1.079999e+06        6.479996e+07 7.504859e+02 7.299347e+02
max          1.439999e+06        8.639994e+07 1.000648e+03 9.732463e+02

       EarthMarsGap
count  1.440000e+06
mean   1.370079e+01
std    7.910162e+00
min    0.000000e+00
25%    6.850394e+00
50%    1.370079e+01
75%    2.055118e+01
max    2.740158e+01
```

This explains the basic statistical information of the data set.

You can save the data as it is good enough.

```
assessTimeName = os.path.join(pathDL,'200-Structured-Zone', '015-04-Assess-
Mars-Time.csv.gz')
TimeDF.to_csv(assessTimeName, index=False, encoding='utf-8',
compression='gzip')
print(assessTimeName)
```

You now have all the data for the Time portions of the process.

Please progress to the next phase as you can now handle time-related data.

Person

This part of the data covers information about the people that are involved in the data for the machine learning.

Tip This data is normally used in machine learning for requirements like the following:

Know your customer, Personal Preferences, and Personal Shopping Choices.

You must add the staff you are sending on the mission:

```
personDF = pd.DataFrame([
    ['Martin','Marsman','Operations Manager'],
    ['Angus','Hundfoot','VR Miner'],
    ['Jock','Mc Bite','VR Driver']
])

print(personDF.shape)

personDF.columns = ['Firstname', 'Lastname', 'Skills']
personDF.index.name = 'ID'

print(personDF.head())
```

You can see how this creates a good data set ready for your machine learning.

```
   Firstname  Lastname               Skills
ID
0     Martin   Marsman  Operations Manager
1      Angus  Hundfoot            VR Miner
2       Jock   Mc Bite           VR Driver
```

Tip I have determined over the years that I have processed data that the best way to identify a person is the combination of FirstName, MiddleName, Lastname, Date-of-Birth, and Place-of-Birth, which statistically reduces the likelihood of two people having the same entries but being two different people.

This person's data can be stored as these three people will most likely be the only people on Mars.

You can now store the person's data in the '200-Structured-Zone' for future loading by your machine learning:

```
assessPersonName = os.path.join(pathDL, '200-Structured-Zone',
'015-04-Assess-Mars-Person.csv.gz')
personDF.to_csv(assessPersonName, index=False, encoding='utf-8',
compression='gzip')
print(assessPersonName)
```

```
You have just completed the person step.
```

Objects and Events

You can now load the Mars mining simulation data as it contains a complex set of data that is spread across two of the main areas, that is, Objects and Events.

Please proceed loading the Mars Hopper data.

```
retrieveObjectName = os.path.join(pathDL, '100-Raw-Zone', '015-04-Retrieve-
Mars-Hopper.csv.gz')
objectDF=pd.read_csv(retrieveObjectName, header=0, encoding='utf-8',
compression='gzip')
```

You can now start the standard step that I showed you before:

- Shape - Basic size of the data (columns and records)
- Info -
- Describe

```
print('Data has %0d columns and %0d records' % (objectDF.shape[1],objectDF.
shape[0]))
```

Data has 7 columns and 2919739 records

```
print(objectDF.info())
```

```
<class 'pandas.core.frame.DataFrame'>
RangeIndex: 2919739 entries, 0 to 2919738
Data columns (total 7 columns):
Time                float64
Event               object
```

```
ObjectHopper    object
ObjectOre       object
Tonnage         int64
LoadBayID       int64
WaitTime        float64
dtypes: float64(2), int64(2), object(3)
memory usage: 155.9+ MB
None
```

```
Insight you can get here are the data set is 155.9+ megabytes
```

You have two objects: Mars Hopper (small robot-driven ore truck) and the Ore.

There is a relationship between the two objects as a hopper carries a specific amount of tonnage of ore.

There is a relationship between the Loading Bay and the hopper in form of a wait time during each load of the ore.

Tip You will find that most features creation is part of your data engineering. I have found that as much as 90% of the work required for good machine learning is data engineering because most data science for machine learning is standard and stable machine algorithms.

You can also look at the description of results for the data for insights:

```
print(objectDF.describe())
```

```
               Time       Tonnage     LoadBayID      WaitTime
count  2.919739e+06  2.919739e+06  2.919739e+06  2.919739e+06
mean   1.000863e+07  5.997769e+01  2.501647e+00  1.701870e+01
std    5.779345e+06  2.339773e+01  1.118402e+00  3.660686e+01
min    0.000000e+00  2.000000e+01  1.000000e+00  7.933378e-05
25%    5.003417e+06  4.000000e+01  2.000000e+00  1.000000e-04
50%    1.000357e+07  6.000000e+01  3.000000e+00  1.000000e-04
75%    1.501119e+07  8.000000e+01  4.000000e+00  1.680835e+01
max    2.001915e+07  1.000000e+02  4.000000e+00  6.345246e+02
```

```
The insight here is from the Tonnage:

from scipy.stats import kurtosis, skew
x=objectDF['Tonnage']

print( 'Excess kurtosis of normal distribution (should be 0): {}'.format
( kurtosis(x) ))

Excess kurtosis of normal distribution (should be 0): -1.2010454999869449

print( 'Skewness of normal distribution (should be 0): {}'.format( skew(x) ))

Skewness of normal distribution (should be 0): 0.002058618196855117

print( 'Skewtest: {}'.format( skewtest(x) ))

Skewtest : SkewtestResult(statistic=1.4360608846207914,
pvalue=0.15098501718218116)
```

The *skewness* is a parameter to measure the symmetry of a data set, and the *kurtosis* is to measure how heavy its tails are compared to a normal distribution.

A skewness value > 0 means that there is more weight in the left tail of the normal distribution.

Negative kurtosis: A distribution with a negative kurtosis value indicates that the distribution has **lighter tails and a flatter peak than the normal distribution**.

The skewtest tests the null hypothesis that the skewness of the population that the sample was drawn from is the same as that of a corresponding normal distribution.

You can plot the data to get a better look at the distributions:

```
%matplotlib inline

t = np.array(objectDF['Tonnage'])
w = np.array(objectDF['WaitTime'])

import matplotlib.pyplot as plt

f, (ax1, ax2) = plt.subplots(1, 2)
ax1.hist(t, bins='auto')
ax1.set_title('Tonnage')
ax2.hist(w, bins='auto')
ax2.set_title('Wait Time')
plt.tight_layout()
```

You can see the histogram for Tonnage and Wait Time (Figure 15-6).

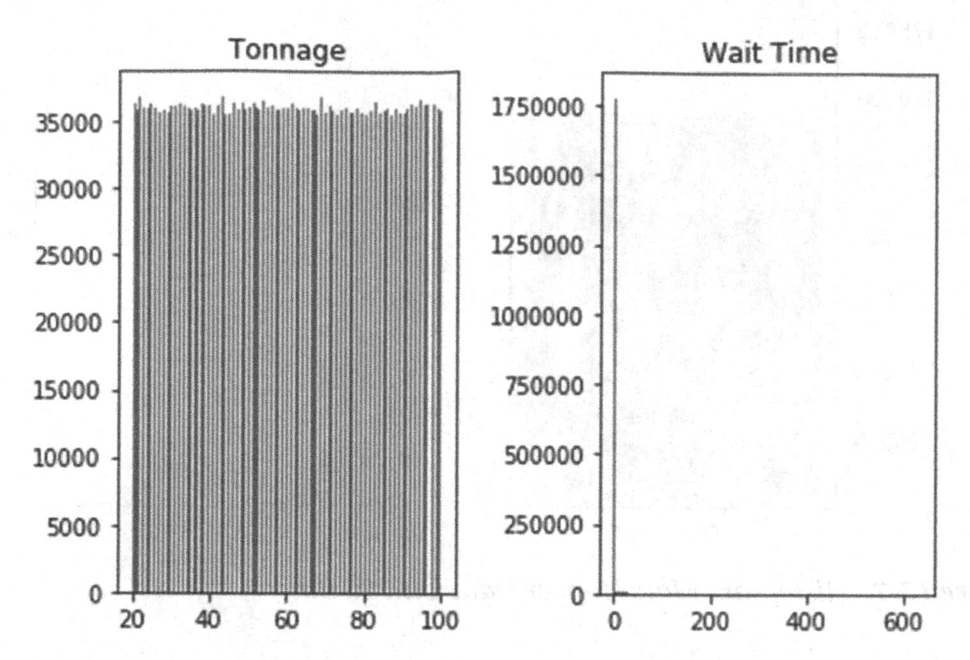

Figure 15-6. *Histogram Mars Hopper Data with Auto Bins*

```
%matplotlib inline

t = np.array(objectDF['Tonnage'])
w = np.array(objectDF['WaitTime'])

import matplotlib.pyplot as plt

f, (ax1, ax2) = plt.subplots(1, 2)
ax1.hist(t, bins=20)
ax1.set_title('Tonnage')
ax2.hist(w, bins=20)
ax2.set_title('Wait Time')
plt.tight_layout()
```

You can see in the result (Figure 15-7) that:

- Tonnage is mostly a flat distribution of load size.
- Wait Time is mostly short wait times.

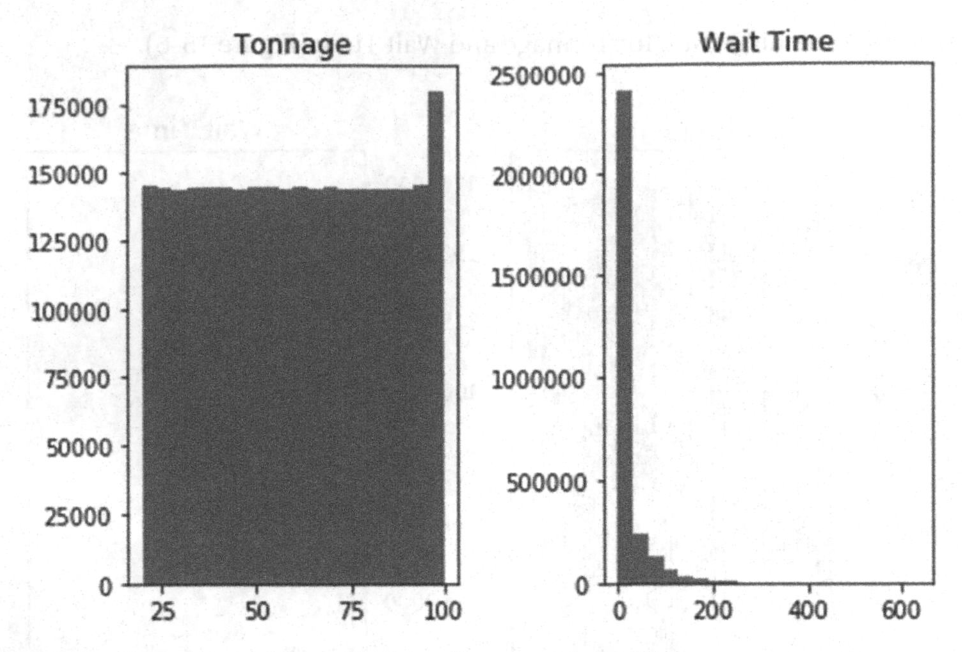

Figure 15-7. *Histogram Mars Hopper Data with 20 Bins*

You can agree there is nothing out of order here, so let's save the data.

```
assessObjectName1 = os.path.join(pathDL,'200-Structured-Zone',
'015-04-Assess-Mars-Mine-Delivery.csv.gz')
objectDF.to_csv(assessObjectName1, index=False, encoding='utf-8',
compression='gzip')
```

Object – Equipment

You have just discovered you have no data on the equipment you are planning to take to Mars. After a quick email, you got the information.

Using the default fallback of human-in-the-loop, you enter the data as follows:

```
equipmentDF = pd.DataFrame([
    ['T101','Robot','Track Puller Robot'],
    ['T102','Robot','Track Puller Robot'],
    ['T103','Robot','Track Ruller Robot'],
    ['T104','Robot','Track Puller Robot'],
    ['T200','Robot','Human-in-Loop Walker Robot'],
```

```
    ['T300','Robot','6x6 Wheeled Transport Robot'],
    ['T400','Robot','Drone Scout Robot']
])
```

You can now save the equipment.

```
assessObjectName2 = os.path.join(pathDL, '200-Structured-Zone', '015-04-
Assess-Mars-Robots.csv.gz')
equipmentDF.to_csv(assessObjectName2, index=False, encoding='utf-8',
compression='gzip')
print(assessObjectName2)
```

Location

The location data is stored in the Mars Mines data set. You can load that now.

```
retrieveName = os.path.join(pathDL, '100-Raw-Zone', '015-04-Retrieve-Mars-
Mines.csv.gz')
locationDF=pd.read_csv(retrieveName, header=0, encoding='utf-8',
compression='gzip')
```

You should follow the same investigation path:

```
print('Data has %0d columns and %0d records' % (locationDF.
shape[1],locationDF.shape[0]))
```

Data has 3 columns and 5 records

```
print(locationDF.info())

<class 'pandas.core.frame.DataFrame'>
RangeIndex: 5 entries, 0 to 4
Data columns (total 3 columns):
MineID       5 non-null int64
Longitude    5 non-null int64
Latitude     5 non-null int64
dtypes: int64(3)
memory usage: 200.0 bytes
None
```

This data set is clean; please save it to a data lake:

```
assessName = os.path.join(pathDL,'200-Structured-Zone', '015-04-Assess-Mars-Mines-Location.csv.gz')
locationDF.to_csv(assessName, index=False, encoding='utf-8', compression='gzip')
```

You have completed the Assess Step.

Your data is now ready for the Process Step.

Process Step

The process step is the data engineering that creates the consolidated view on the data you need for the machine learning.

Hub – Time

The hub for time is an extraction from the Mars Times files.

You can load the file from '200-Structured-Zone':

```
assessTimeName = os.path.join(pathDL, '200-Structured-Zone',
'015-04-Assess-Mars-Time.csv.gz')

timeDF=pd.read_csv(assessTimeName, header=0, encoding='utf-8',
compression='gzip')

timeDF.index.name = 'ID'

You can now check the load:

print(timeDF.info())

<class 'pandas.core.frame.DataFrame'>
RangeIndex: 1440000 entries, 0 to 1439999
Data columns (total 5 columns):
MinutesSinceLaunch     1440000 non-null int64
SecondsSinceLaunch     1440000 non-null int64
EarthTime              1440000 non-null float64
MarsTime               1440000 non-null float64
EarthMarsGap           1440000 non-null float64
```

```
dtypes: float64(3), int64(2)
memory usage: 54.9 MB
None
```

You need the 'EarthTime', 'MarsTime' as a Hub for Time Hub in the Data Vault you need to build.

```
timeHub=timeDF[['EarthTime','MarsTime']]
```

```
print(timeHub.shape)
```

```
Result is: (1440000, 2)
```

Next you remove all duplicates:

```
timeFixHub=timeHub.drop_duplicates(subset=None, keep='first',
inplace=False)

print(timeFixHub.shape)
```

Result is: (1440000, 2)

That show you have no duplicates. You can now save the data to the Hub.

```
timeHubName = os.path.join(pathDL,'300-Curated-Zone', 'Hub', 'Time',
'Hub-Time.csv.gz')
timeFixHub.to_csv(timeHubName, index=False, encoding='utf-8',
compression='gzip')
print(timeHubName)
```

Time – Satellite

The complete time data set can be saved as a Time Satellite for future use by the machine learning.

```
timeSatelliteName = os.path.join(pathDL,'300-Curated-Zone', 'Satellite',
'Time', 'Satellite-Time.csv.gz')
timeHub.to_csv(timeSatelliteName, index=False, encoding='utf-8',
compression='gzip')
print(timeSatelliteName)
```

Next, we perform the same tasks on Person.

Hub – Person

You have investigated the person before, so I suggest you save it in the same way you did for Time.

```
assessPersonName = os.path.join(pathDL, '200-Structured-Zone',
'015-04-Assess-Mars-Person.csv.gz')
personDF=pd.read_csv(assessPersonName, header=0, encoding='utf-8',
compression='gzip')
personDF.index.name = 'ID'

print(personDF.head())
personHub=personDF[['Firstname','Lastname']]
print(personHub.head())
```

You can save the Person Hub.

```
personHubName = os.path.join(pathDL,'300-Curated-Zone', 'Hub', 'Person',
'Hub-Person.csv.gz')
personHub.to_csv(personHubName, index=False, encoding='utf-8',
compression='gzip')
print(personHubName)
```

You can save the Person Satellite.

```
personSatelliteName = os.path.join(pathDL,'300-Curated-Zone', 'Satellite',
'Person', 'Satellite-Person.csv.gz')
personDF.to_csv(personSatelliteName, index=False, encoding='utf-8',
compression='gzip')
print(personSatelliteName)
```

Hub – Object

The object Hub holds references to all objects in the data vault.

```
assessObjectName = os.path.join(pathDL, '200-Structured-Zone',
'015-04-Assess-Mars-Mine-Delivery.csv.gz')
objectDF=pd.read_csv(assessObjectName, header=0, encoding='utf-8',
compression='gzip')
objectDF.index.name = 'ID'
```

You need to extract all the objects, and I suggest you use the following process.

```
object1Hub=objectDF[['ObjectHopper']]
object1Hub.columns=['ObjectName']
object1Hub['ObjectType'] = 'MarsHopper'
object1FixHub=object1Hub.drop_duplicates(subset=None, keep='first',
inplace=False)

object2Hub=objectDF[['ObjectOre']]
object2Hub.columns=['ObjectName']
object2Hub['ObjectType'] = 'Ore'
object2FixHub=object2Hub.drop_duplicates(subset=None, keep='first',
inplace=False)

object3Hub=objectDF[['LoadBayID']]
object3Hub.columns=['ObjectName']
object3Hub['ObjectType'] = 'LoadBay'
object3FixHub=object3Hub.drop_duplicates(subset=None, keep='first',
inplace=False)

objectFix0Hub = object1FixHub.append(object2FixHub, sort=False)
objectFixHub = objectFix0Hub.append(object3FixHub, sort=False)
```

You now have a combined set of all the objects. You can save then as an Object Hub.

```
objectHubName = os.path.join(pathDL,'300-Curated-Zone', 'Hub', 'Object',
'Hub-Object.csv.gz')
objectFixHub.to_csv(objectHubName, index=False, encoding='utf-8',
compression='gzip')
print(objectHubName)
```

Next you save the three object Satellites:

```
object1SatelliteName = os.path.join(pathDL,'300-Curated-Zone', 'Satellite',
'Object', 'Satellite-Object-MarsHopper.csv.gz')
object1FixHub.to_csv(object1SatelliteName, index=False, encoding='utf-8',
compression='gzip')
print(object1SatelliteName)

object2SatelliteName = os.path.join(pathDL,'300-Curated-Zone', 'Satellite',
'Object', 'Satellite-Object-Ore.csv.gz')
object2FixHub.to_csv(object2SatelliteName, index=False, encoding='utf-8',
compression='gzip')
print(object2SatelliteName)

object3SatelliteName = os.path.join(pathDL,'300-Curated-Zone', 'Satellite',
'Object', 'Satellite-Object-LoadBay.csv.gz')
object3FixHub.to_csv(object3SatelliteName, index=False, encoding='utf-8',
compression='gzip')
print(object3SatelliteName)
```

Now you need to save the link between the Hopper and the Ore:

```
object1Link=objectDF[['ObjectHopper','ObjectOre','LoadBayID','Tonnage']]

object1Link.reset_index(level=0, inplace=True)
```

The link is saved in the Link for Object to Object types:

```
object1LinkName = os.path.join(pathDL,'300-Curated-Zone', 'Link', 'Object-
Object', 'Link-Object-LoadBay-MarsHopper-Ore.csv.gz')
object1Link.to_csv(object1LinkName, index=False, encoding='utf-8',
compression='gzip')
print(object1LinkName)
```

```
Hub - Location
```

The location hub holds all the information on the physical location of every object or event in the system.

Start by loading the data and then performing the data engineering until you have a hub.

```
assessLocationName = os.path.join(pathDL, '200-Structured-Zone',
'015-04-Assess-Mars-Mines-Location.csv.gz')
locationDF=pd.read_csv(assessLocationName, header=0, encoding='utf-8',
compression='gzip')
locationDF.index.name = 'ID'

LongitudeDF=pd.DataFrame(np.array(range(-180,180)))
LongitudeDF.columns=['Longitude']
LongitudeDF['Key']='1'

LatitudeDF=pd.DataFrame(np.array(range(-90,90)))
LatitudeDF.columns=['Latitude']
LatitudeDF['Key']='1'

locationHub=pd.merge(left=LongitudeDF, right=LatitudeDF, on= "Key",
how= "outer")
locationHub.drop('Key', axis=1, inplace=True)
locationHub.index.name = 'ID'

locationHubName = os.path.join(pathDL,'300-Curated-Zone', 'Hub',
'Location', 'Hub-Location.csv.gz')
locationHub.to_csv(locationHubName, index=False, encoding='utf-8',
compression='gzip')
print(locationHubName)
```

Now save the Location Satellite:

```
locationSatelliteName = os.path.join(pathDL,'300-Curated-Zone',
'Satellite', 'Location', 'Satellite-Location-MarsMine.csv.gz')
locationDF.to_csv(locationSatelliteName, index=False, encoding='utf-8',
compression='gzip')
print(locationSatelliteName)
```

Hub – Event

```
eventHub=objectDF[['Event']]
eventHub.columns=['EventName']
eventFixHub=eventHub.drop_duplicates(subset=None, keep='first',
inplace=False)

print(eventFixHub.shape)

eventHubName = os.path.join(pathDL,'300-Curated-Zone', 'Hub', 'Event',
'Hub-Event.csv.gz')
eventHub.to_csv(eventHubName, index=False, encoding='utf-8',
compression='gzip')
print(eventHubName)
```

The Transform Step, Organize Step, and Report Step will be handled in separate notebooks.

Progress Report

You should at this point look at the great piece of data engineering you completed.

You will now get a progress report on each zone (Figure 15-8) of the data lake.

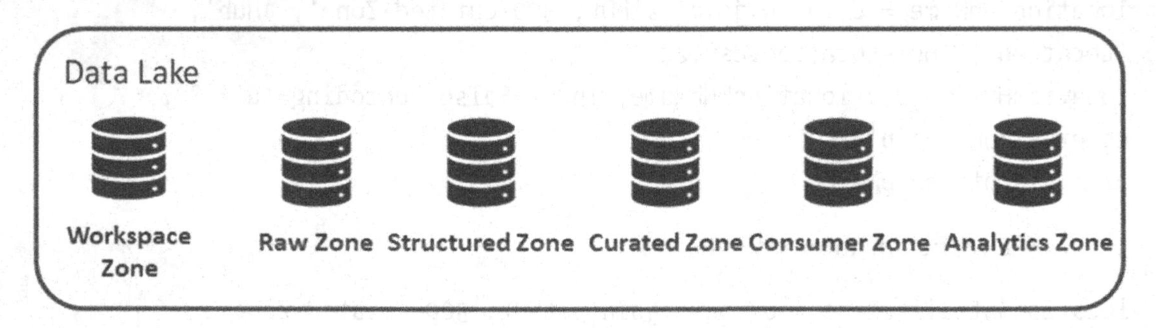

Figure 15-8. *Data Lake Zones*

Raw Zone

```
pathReportname= '../../Results/Chapter 09/999-DL/100-Raw-Zone/'
for root, dirs, files in os.walk(pathReportname, topdown=True):
    for name in files:
        print(os.path.join(root, name))
```

```
../../Results/Chapter 09/999-DL/100-Raw-Zone/015-04-Retrieve-Mars-Hopper.
csv.gz ../../Results/Chapter 09/999-DL/100-Raw-Zone/015-04-Retrieve-Mars-
Mines.csv.gz ../../Results/Chapter 09/999-DL/100-Raw-Zone/015-04-Retrieve-
Mars-Time.csv.gz
```

Structured Zone

```
pathReportname= '../../Results/Chapter 09/999-DL/200-Structured-Zone/'
for root, dirs, files in os.walk(pathReportname, topdown=True):
    for name in files:
        print(os.path.join(root, name))
```

```
../../Results/Chapter 09/999-DL/200-Structured-Zone/015-04-Assess-Mars-
Mine-Delivery.csv.gz ../../Results/Chapter 09/999-DL/200-Structured-
Zone/015-04-Assess-Mars-Mines-Location.csv.gz
../../Results/Chapter 09/999-DL/200-Structured-Zone/015-04-Assess-Mars-
Person.csv.gz ../../Results/Chapter 09/999-DL/200-Structured-Zone/015-04-
Assess-Mars-Robots.csv.gz ../../Results/Chapter 09/999-DL/200-Structured-
Zone/015-04-Assess-Mars-Time.csv.gz
```

Curated Zone

```
pathReportname= '../../Results/Chapter 09/999-DL/300-Curated-Zone/Hub/'
for root, dirs, files in os.walk(pathReportname, topdown=True):
    for name in files:
        print(os.path.join(root, name))
```

```
../../Results/Chapter 09/999-DL/300-Curated-Zone/Hub/Event\Hub-Event.csv.gz
../../Results/Chapter 09/999-DL/300-Curated-Zone/Hub/Location\Hub-Location.
csv.gz
../../Results/Chapter 09/999-DL/300-Curated-Zone/Hub/Object\Hub-Object.csv.gz
../../Results/Chapter 09/999-DL/300-Curated-Zone/Hub/Person\Hub-Person.csv.gz
../../Results/Chapter 09/999-DL/300-Curated-Zone/Hub/Time\Hub-Time.csv.gz
```

That is the end of the notebook, so finish the last steps to see how long it took you and then you are done.

```
nowStop = datetime.datetime.now()
runTime=nowStop-nowStart
print('Start:', nowStart.strftime('%Y-%m-%d %H:%M:%S'))
print('Stop: ', nowStop.strftime('%Y-%m-%d %H:%M:%S'))
print('Time: ', runTime)
```

Well done. You can close your notebook.

Transform Step

You will now start to use your supervised learning skills against the Mars data.

Supervised Learning

You are now ready for the machine learning. The first machine learning we will apply is supervised.

Progress to the next notebook, and you can start seeing the advantage of the data engineering you completed.

Open the Jupyter Notebook in examples under Chapter 15 named: Chapter-015-08-Build-RAPTOR-Mars-02.ipynb

Let's start the clock.

```
import datetime
nowStart = datetime.datetime.now()
```

You need the following libraries to start your unsupervised learning process:

```
from sklearn.svm import LinearSVC
from sklearn.datasets import make_classification
from sklearn.metrics import f1_score, matthews_corrcoef, confusion_matrix
import matplotlib.pyplot as plt
import numpy as np
import os
import pandas as pd
get_ipython().run_line_magic('matplotlib', 'inline')
```

You will work in the Data lake at ***999-DL***.

```
pathDL='../../Results/Chapter 09/999-DL/'
```

You need to load data from the '300-Curated-Zone'.

I suggest you load the Link file between Load Bay, Marshopper, and Ore loaded.

```
objectLinkName = os.path.join(pathDL,'300-Curated-Zone', 'Link', 'Object-
Object', 'Link-Object-LoadBay-MarsHopper-Ore.csv.gz')

object1DF=pd.read_csv(objectLinkName, header=0, encoding='utf-8',
compression='gzip')
object1DF.index.name = 'ID'
```

You can see what data you have by looking at the columns:

```
print(object1DF.columns)
```

You have four columns:

```
Index(['ID', 'ObjectHopper', 'ObjectOre', 'LoadBayID', 'Tonnage'],
dtype='object')
```

Let's describe the data; you have used this before, and this should be easy:

```
print(object1DF.describe())
```

```
                 ID     LoadBayID       Tonnage
count  2.919739e+06  2.919739e+06  2.919739e+06
mean   1.459869e+06  2.501647e+00  5.997769e+01
std    8.428562e+05  1.118402e+00  2.339773e+01
min    0.000000e+00  1.000000e+00  2.000000e+01
25%    7.299345e+05  2.000000e+00  4.000000e+01
50%    1.459869e+06  3.000000e+00  6.000000e+01
75%    2.189804e+06  4.000000e+00  8.000000e+01
max    2.919738e+06  4.000000e+00  1.000000e+02
```

Insight: the Tonnage has a large range of values, you need a scaler.

Now you perform the standard info check:

```
print(object1DF.info())
```

```
<class 'pandas.core.frame.DataFrame'>
RangeIndex: 2919739 entries, 0 to 2919738
Data columns (total 5 columns):
ID               int64
ObjectHopper     object
ObjectOre        object
LoadBayID        int64
Tonnage          int64
dtypes: int64(3), object(2)
memory usage: 111.4+ MB
None
```

You should see no issues here; all looks as expected.

Now you can look at the data itself:

```
print(object1DF.head(10))
```

```
    ID ObjectHopper  ObjectOre  LoadBayID  Tonnage
ID
0    0  Hopper-0001   aluminum          1       78
1    1  Hopper-0001   aluminum          1       78
2    2  Hopper-0002   aluminum          1       64
3    3  Hopper-0002   aluminum          1       64
4    4  Hopper-0002   aluminum          1       64
5    5  Hopper-0001   aluminum          1       78
6    6  Hopper-0003  magnesium          4       55
7    7  Hopper-0003  magnesium          4       55
8    8  Hopper-0004   aluminum          4       54
9    9  Hopper-0004   aluminum          4       54
10  10  Hopper-0005   aluminum          2       45
```

You need a simple scaler on tonnage, so I suggest you use a split at 60 that marks the hopper with a zero or a one if it is loaded correctly.

Tip If you change the allocation facts in the hopper generator in Chapter-015-05-Mars-Mines-01 From tonnage=random.randint(20, 100) to example tonnage=random.randint(20, 150), you will also get a new tonnage rate of bigger than one been overloaded. Or you could change the 60 in scaler to 40 with similar outcomes.

Here is your scaler:

```
def loadrate(t):
    rate = int(t/60)
    return rate
```

You need to perform some data engineering here to apply the scaler:

```
objectSub1DF=object1DF[['ObjectOre','LoadBayID','Tonnage']]
objectSub1DF.drop_duplicates(subset=None, keep='first', inplace=True)
objectSub1DF.columns=['OreName','MineID','Tonnage']
objectSub1DF['TonnageRate'] = objectSub1DF.apply(lambda row:
loadrate(row['Tonnage']), axis=1)
objectSub1DF.drop('Tonnage', axis=1, inplace=True)
objectSub1DF.drop_duplicates(subset=None, keep='first', inplace=True)
```

You can now see the new data set:

```
print(objectSub1DF.head(10))
```

```
      OreName  MineID  TonnageRate
ID
0    aluminum       1            1
6   magncsium       4            0
8    aluminum       4            0
10   aluminum       2            0
13    calcium       4            0
15    silicon       3            0
22  magnesium       2            0
```

```
28       iron        4            1
30       iron        3            1
34    silicon        3            1

objectSatelliteName = os.path.join(pathDL,'300-Curated-Zone', 'Satellite',
'Object', 'Satellite-Object-Ore.csv.gz')
```

Warning The OreName needs a substitution. The LinearSVC algorithm needs numbers to work effectively and efficiently.

You need the ID from the Ore Satellite:

```
object2DF=pd.read_csv(objectSatelliteName, header=0, encoding='utf-8',
compression='gzip')
object2DF.index.name = 'ID'
```

Here is the data.

You now have the missing ID and you can fix the data feature with an ID to column transfer.

Here is the basic information on the data set.

```
print(object2DF.info())

<class 'pandas.core.frame.DataFrame'>
RangeIndex: 6 entries, 0 to 5
Data columns (total 2 columns):
ObjectName    6 non-null object
ObjectType    6 non-null object
dtypes: object(2)
memory usage: 176.0+ bytes
None
```

Here is the sample of the data.

```
print(object2DF.head(10))

ObjectName ObjectType
```

```
ID
0     aluminum         Ore
1   magnesium          Ore
2     calcium          Ore
3     silicon          Ore
4        iron          Ore
5   potassium          Ore
```

Here is the required fix for the feature.

```
objectSub2DF=object2DF[['ObjectName']]
objectSub2DF.drop_duplicates(subset=None, keep='first', inplace=True)
objectSub2DF.reset_index(inplace=True)
objectSub2DF.columns=['OreID','OreName']
objectSub2DF.index.name = 'ID'
```

Good work! You have fixed the ID so that you can use it:

```
print(objectSub2DF.head(20))
```

```
    OreID     OreName
ID
0       0    aluminum
1       1   magnesium
2       2     calcium
3       3     silicon
4       4        iron
5       5   potassium
```

You now simply substitute the Ore ID for the Ore Name.

```
objectDF=pd.merge(left=objectSub1DF, right=objectSub2DF, on='OreName')
objectDF.drop('OreName', axis=1, inplace=True)
objectDF.drop_duplicates(subset=None, keep='first', inplace=True)
```

You have a working base data set for your machine learning.

```
print(objectDF.columns)
```

```
Index(['MineID', 'TonnageRate', 'OreID'], dtype='object')
```

Training Set

You need a training data set for the machine learning to understand the data.

Tip I normally take 70% of the data set to train with as this enables a 30% data set that is unknown to the algorithm, but you can test with it as you know it.

You should take a sample of 70% using this command:

```
objectTrainDF=objectDF.sample(frac=0.7, replace=False)
```

Let's investigate your training set:

```
print(objectTrainDF.shape)
```

```
(34, 3)
```

You have 34 training points.

For the algorithm, you need to split the features ('OreID', 'MineID') from the target ('TonnageRate') for the machine learning to work.

I suggest the following split method using a numpy array.

```
Xtrain=np.array(objectTrainDF[['OreID','MineID']])
ytrain=np.array(objectTrainDF[['TonnageRate'][0]])
```

Test Data Set

Next you need a test set to later test the machine learning's score.

Here is how you get the 30% test set.

```
objectTestDF=objectDF.sample(frac=0.3, replace=True)
#objectTestDF.drop_duplicates(subset=None, keep='first', inplace=True)
```

It's the same as before we split the features and the target.

```
Xtest=np.array(objectTestDF[['OreID','MineID']])
ytest=np.array(objectTestDF[['TonnageRate'][0]])
```

Machine Learning

You can now create the LinearSVC engine.

I suggest max_iter=20000 with dual=True.

```
clf = LinearSVC(random_state=0,
                max_iter=20000,
                dual=True)
```

You can now train the model.

Note All the data engineering comes down to this one command. That is the nature of the IML environment.

```
clf.fit(Xtrain, ytrain)
```

You can now check the success of the machine learning:

```
print('-----------------------------------------------------')
print('Coefficient:', clf.coef_)
print('-----------------------------------------------------')
print('Intercept', clf.intercept_)
print('-----------------------------------------------------')
print('Score', clf.score(Xtrain, ytrain))
print('-----------------------------------------------------')
```

Here are the train results:

```
-----------------------------------------------------
Coefficient: [[ 0.08636684 -0.1258458 ]]
-----------------------------------------------------
Intercept [0.17640288]
-----------------------------------------------------
Score 0.5882352941176471
-----------------------------------------------------
```

You can now test the model against the test data set.

```
IMLresult=np.array(clf.predict(Xtest))
```

Here are the test results:

```
print('--------------------------------------------------')
print('Score - Test Data', clf.score(Xtest, ytest))
print('--------------------------------------------------')
```

Score - Test Data 0.42857142857142855

Now you can test the predictions against the model:

```
print('Test Data Run')
for i in range(IMLresult.shape[0]):
    if ytest[0]==IMLresult[0]:
        outcome='Hit!'
    else:
        outcome='Miss?'
    print('%04d - %04d ore with %02d mine => %0.4f true rate => %0.4f
    predicted - %s' % (i+1, Xtest[i][0],Xtest[i][1],ytest[i],IMLresult[i],
    outcome))
```

Output is:

```
Test Data Run
0001 - 0002 ore with 04 mine => 0.0000 true rate => 0.0000 predicted - Hit!
0002 - 0005 ore with 04 mine => 0.0000 true rate => 1.0000 predicted - Hit!
0003 - 0002 ore with 03 mine => 1.0000 true rate => 0.0000 predicted - Hit!
0004 - 0004 ore with 03 mine => 1.0000 true rate => 1.0000 predicted - Hit!
0005 - 0004 ore with 03 mine => 0.0000 true rate => 1.0000 predicted - Hit!
0006 - 0001 ore with 04 mine => 1.0000 true rate => 0.0000 predicted - Hit!
0007 - 0005 ore with 01 mine => 0.0000 true rate => 1.0000 predicted - Hit!
0008 - 0002 ore with 02 mine => 0.0000 true rate => 1.0000 predicted - Hit!
0009 - 0001 ore with 03 mine => 0.0000 true rate => 0.0000 predicted - Hit!
0010 - 0000 ore with 02 mine => 1.0000 true rate => 0.0000 predicted - Hit!
0011 - 0002 ore with 01 mine => 1.0000 true rate => 1.0000 predicted - Hit!
0012 - 0003 ore with 02 mine => 1.0000 true rate => 1.0000 predicted - Hit!
0013 - 0001 ore with 04 mine => 0.0000 true rate => 0.0000 predicted - Hit!
0014 - 0000 ore with 02 mine => 1.0000 true rate => 0.0000 predicted - Hit!
```

You can now close out the machine learning.

```
nowStop = datetime.datetime.now()
runTime=nowStop-nowStart
print('Start:', nowStart.strftime('%Y-%m-%d %H:%M:%S'))
print('Stop: ', nowStop.strftime('%Y-%m-%d %H:%M:%S'))
print('Time: ', runTime)
```

Unsupervised Learning

The next machine learning model you will be looking at is unsupervised learning.

You can now open the Jupyter Notebook in examples under Chapter 15 named:

Chapter-015-09-Build-RAPTOR-Mars-03.ipynb

```
You can now start the unsupervised learning process.
```

I will guide you through the search for the optimum k for the k-means algorithm.

```
import datetime
nowStart = datetime.datetime.now()
```

You need to activate the graphics engine to run in the foreground.

```
import matplotlib
matplotlib.use('TkAgg')
get_ipython().run_line_magic('matplotlib', 'inline')
```

You then load the libraries to handle the data and the machine learning.

```
import pandas as pd
import os
# The kmeans algorithm is implemented in the scikits-learn library
from sklearn.cluster import KMeans
import numpy as np
from matplotlib import pyplot as plt
```

You will use the data lake at ***999-DL***.

```
pathDL='../../Results/Chapter 09/999-DL/'
```

Alpha Data Set

Load the Hub Location data.

```
locationHubName1 = os.path.join(pathDL,'300-Curated-Zone', 'Hub',
'Location', 'Hub-Location.csv.gz')

location1DF=pd.read_csv(locationHubName1, header=0, encoding='utf-8',
compression='gzip')
location1DF.index.name = 'ID'
location1DF.columns=['Longitude','Latitude']
#location1DF['Mine']=0
```

This is the profile of the data:

```
print(location1DF.shape)
```

Data profile (64800, 2)

Select the alpha data set.

```
A = location1DF
```

You can now start the machine learning on the alpha data set.

I suggest we run the k-means for k=1 to 10 and determine the most optimum k using the Elbow graph.

```
resultA=[]
for k in range (1, 10):
    # Create a kmeans model on our data, using k clusters.  random_state
    helps ensure that the algorithm returns the same results each time.
    kmeans_model = KMeans(n_clusters=k, random_state=1).fit(A.iloc[:, :])
    # These are our fitted labels for clusters -- the first cluster has
    label 0, and the second has label 1.
    labels = kmeans_model.labels_
    # Sum of distances of samples to their closest cluster center
    interia = kmeans_model.inertia_
    print ("k:",k, " cost:", interia)
    resultA.append([k, interia])
```

You can now plot the alpha results:

```
data = np.array(resultA)
x, y = data.T
plt.plot(x, y, linestyle='--', marker='o', linewidth=1.0)

titlestr='Data Set Alpha using the elbow method to determine the optimal
number of clusters for k-means clustering'
plt.title(titlestr)
plt.ylabel('Sum of squared errors')
plt.xlabel('Number of clusters (k)')
plt.grid()
plt.show()
```

Here is the elbow result (Figure 15-9) for the alpha Data set.

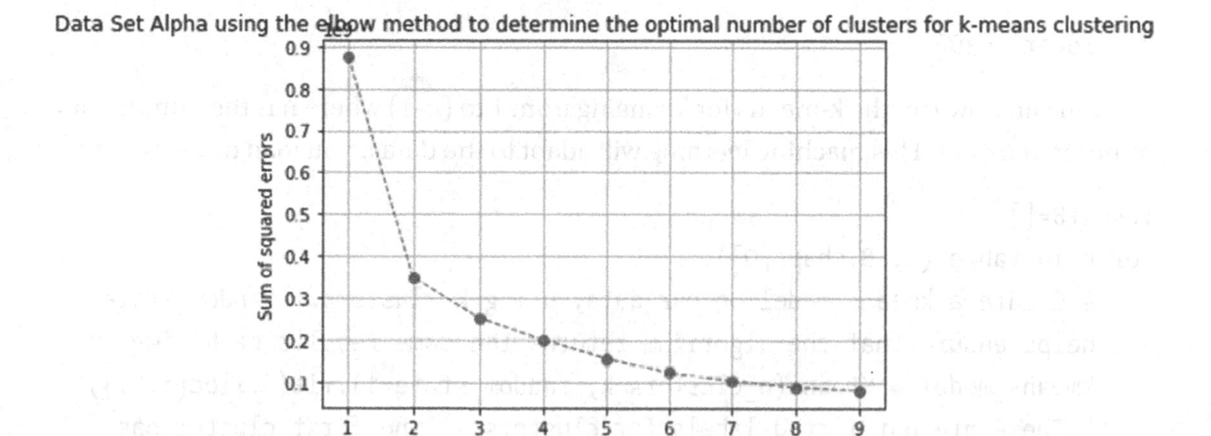

***Figure 15-9.** Unsupervised Alpha Result*

Beta Data Set

The next data set you need is the mine location data, and this will become your beta data set.

You can find the data at this location:

```
locationHubName2 = os.path.join(pathDL,'300-Curated-Zone', 'Satellite',
'Location', 'Satellite-Location-MarsMine.csv.gz')
```

```
location2DF=pd.read_csv(locationHubName2, header=0, encoding='utf-8',
compression='gzip')
location2DF.index.name = 'ID'
location2DF.columns=['MineID','Longitude','Latitude']
#location2DF['Mine']=1
```

Warning There is a mine here that is not a mine. It is your Mars mission control landing pad.

```
is_mine=location2DF['MineID'] > 0
```

```
location3DF=location2DF[is_mine]
```

Load the data as your beta data set.

```
B = location3DF
```

You can now run the k-means for k ranging from 1 to (n-1) where n is the number of mines in the data. This machine learning will adapt to the data if you load more mines.

```
resultB=[]
for k in range (1, B.shape[0]):
    # Create a kmeans model on our data, using k clusters.  random_state
    helps ensure that the algorithm returns the same results each time.
    kmeans_model = KMeans(n_clusters=k, random_state=1).fit(B.iloc[:, :])
    # These are our fitted labels for clusters -- the first cluster has
    label 0, and the second has label 1.
    labels = kmeans_model.labels_
    # Sum of distances of samples to their closest cluster center
    interia = kmeans_model.inertia_
    print ("k:",k, " cost:", interia)
    resultB.append([k, interia])
```

You can now plot the beta results:

```
data = np.array(resultB)
x, y = data.T
plt.plot(x, y, linestyle='--', marker='o', linewidth=1.0)
```

```
titlestr='Data Set Beta using the elbow method to determine the optimal
number of clusters for k-means clustering'
plt.title(titlestr)
plt.ylabel('Sum of squared errors')
plt.xlabel('Number of clusters (k)')
plt.grid()
plt.show()
```

Here are the beta results (Figure 15-10).

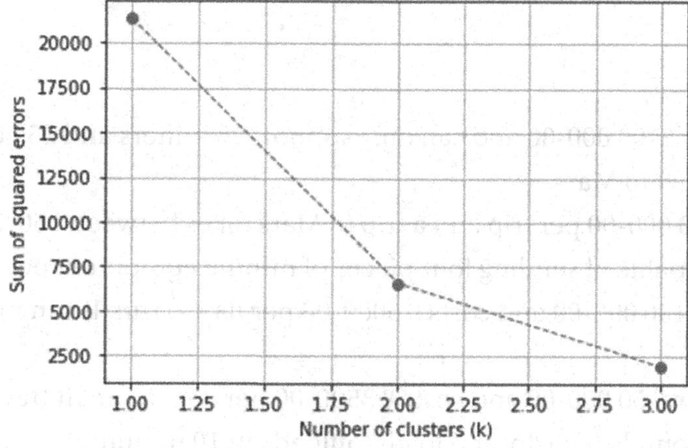

Figure 15-10. *Unsupervised Beta Result*

Insights from this data set:

- The data set is too small as the clusters are limited.
- You need more data points to achieve a good unsupervised learning result.
- That is why there is no clear optimum k value from these results.

Congratulations; you have now completed the unsupervised machine learning. You can now close all the notebooks as the next piece of work is the big final project.

Mars Mission

Now that you have all the knowledge you need, I suggest you look at your requirements again and open notebook: Chapter-015-14-Setup-Mars-Base.ipynb. Remember these?

Requirements

Mars has abundant resources to mine and these can be shipped back to Earth:

Silicon, hydrogen, oxygen, iron, magnesium, manganese, aluminum, calcium, gold, silver, nickel, titanium, platinum, and copper are ready for mining.

Your Costs

A mining spaceport is £100 000-00 and can only support 25 miners and 250 transports. It needs two rockets to get to Mars.

A rocket costs £ 20 000-00 per trip and a trip to Mars takes between 150–300 days. Each rocket is only capable of sending four pieces of mining equipment per rocket.

An ore miner is £ 150 000-00 and costs £ 5000-00 per day to run. It can mine 1000 pounds per day.

An ore transport is £ 50 000-00 and costs £ 3500-00 per day to run. It travels 100 miles per day, carries 250 pounds per trip, and loads/offloads in 10 minutes.

Hydrogen and oxygen extractors are £ 10 000-00 each and cost £ 1500-00 per day to run. An extractor produces 500 pounds of hydrogen or oxygen per day.

Each piece of mining equipment uses 200 pounds of hydrogen and 100 pounds of oxygen per day for power.

A 3D printing plant can use 1000 pounds of silicon to print an extra miner, 750 pounds for an ore transport, 300 pounds for extractor, and 5000 pound per rocket. A new 3D printing plant takes 15 days and 7500 pounds of silicon.

It takes 5 days to make a miner, 3 days for an ore transport, a day for an extractor.

Three-dimensional printing uses 400 pounds of hydrogen and 300 pounds of oxygen per day.

You are only funded to a maximum of £ 1 000 000-00.

During the return flight, each rocket can only carry 500 000 pounds of ore and uses 6000 pounds of hydrogen and 2000 pounds of oxygen per day.

You must return 20% of profits to your shareholders per year.

Your Income

Current Trading prices for 1000 pounds of ore on delivery to Earth:

> Silicon(£1300), iron (£80), magnesium(£5000), manganese (£40), aluminum(£800), calcium(£220), gold(£1200), silver(£15), nickel(£1500), titanium(£30000), platinum(£870), and copper (£2500).

Mars Mission Start

This notebook will build you a complete Mars base with every aspect you need to achieve your project goals.

There are also a few surprises but find them yourself.

Tip Work through the notebook step by step as you complete each of the requirements.

Now I assume you are following the notebook ... Best of luck and enjoy!

Mars Mission Complete

You have just completed the Mars mission simulation.

You can now read the rest of this chapter as I close out a few extra topics you should take note of as you now venture into your new world enabled by IML.

Tip If you go through the examples again, you will spot several parameters and other options. I suggest you use the examples with data of your own or changes in the parameters. This is the way to learn the trade of IML. Practice, practice, and then some more ...

Let's now see what you have learned.

Challenges

You have invested time in completing this book to this point. I suggest you repeat the challenge from Chapter 1.

Look at the environment around you where you spend your daily life and write down three things that you, with your new knowledge, believe are the results of machine learning.

Description	Yes/No

You will now be able to see new opportunities for applying your new skills.

Advice I personally practice with my machine learning up to 12 hours a week, purely on new data sets or problem data sets that I discovered on the Internet.

Just search around, and you will be happy to discover massive amounts of data that is viable for your use.

Question One

Has your insight of what machine learning does around you changed?

I want to predict that the bigger the change in the knowledge of IML, the bigger your insights change, as you would experience at this moment.

Now let's progress to having you write down three things that would be better for you if they were automatically done for you.

Description	Yes/No

Question Two

Could you, with your new knowledge, automate these for yourself?

I would predict you now have the knowledge to apply your own IML methods and techniques to achieve success.

Extra Practice

Try Chapter-015-15-Features-Minerals.ipynb to practice feature importance. What should we mine?

Try Chapter-015-16-Mars-Crime.ipynb to practice features from images and found who stole my cupcake.

Summary

Congratulations; you have now gotten the knowledge to convert your knowledge of machine learning into a real-life project.

This is the end of this book, and you should now be ready to deploy your knowledge against any project you will have to handle.

My last recommendation would be to, at no time, stop your inquisitive mind; and keep on learning new technology, practices, and methodologies.

Thank You

Thank you for spending the time to follow my book and the examples. See you on Mars!

Or any other world of your choice ... The sky is the limit!

May your accomplished knowledge bring you all the good things and great prospects you wish for in your future.

Last and final tip ...

Have fun with Industrialized Machine Learning ... Go and change your world!!!

APPENDIX A

Reference Material

I have compiled additional reference material that provides background information to assist you in understanding the concepts in this book. They are organized by chapter for your convenience.

Chapter 1

Why Python?

Python was developed in 1991 by Guido Van Rossum, and 25 years later, it has evolved into a mainstream language.

The Advantages of Python

Extensive Support Libraries

It offers a huge amount of standard libraries. See: `https://docs.python.org/3/library/`

Integration Feature

Python integrates into the enterprise application layer that makes it easy to develop on all modern operating systems.

Improved Programmer's Productivity

The language has extensive support libraries and clean object-oriented designs that increase a programmer's productivity two- to tenfold .

See: `https://pypi.org/`

A. F. Vermeulen, *Industrial Machine Learning*, https://doi.org/10.1007/978-1-4842-5316-8

Productivity

The strong process integration features, unit testing framework, and enhanced control capabilities underwrite toward the increased speed of the applications and productivity of applications. This an abundant option for building scalable multi-protocol network applications for nearly any ecosystem.

Disadvantages of Python

Python has diverse advantageous features, and programmers prefer this language to other programming languages because it is easy to learn and code.

Python has yet to agree with more strict rules in enterprise development shops.

Difficulty in Using Other Languages

The Python fans become so comfortable with its features and its wide-ranging libraries, so data scientists face complications if they cannot find a 100% compatible library. The lack of understanding of other languages' more rigid requirements then require a steep learning curve to use the other programming languages.

Weak in Mobile Computing

Python has made its presence on numerous desktop and server platforms, but it is perceived as an incompatible language for mobile computing. This is the reason it is uncommon for mobile applications to be built using Python.

Slow in Processing Speed

Python executes with the help of an interpreter and not a compiler, so this causes it to be slower than already assembled and compiled code.

Run-Time Errors

The Python language is dynamically typed so it has various enterprise restrictions because various unexpected errors can show up when the applications are finally run.

Underdeveloped Database Access Layers

When compared with more popular technologies like JDBC and ODBC, the Python's database access layer is established to be slightly immature and basic. Most enterprises require a smoother interaction with complex legacy data. However, the introduction of various libraries that bridge this lack of enterprise-grade data layers is, and will in the future, enrich Python's data capabilities.

I predict that we will see many new developments to enhance the capabilities of the Python language at a native level as worldwide research will deliver new libraries that will support Python as the leading programming language worldwide.

Right now, Python is already a robust programming language and provides an easy usage of the code lines; maintenance will handle the programing needs of the modern Industrialized Machine Learning (IML) as an official programming language.

Why Jupyter Notebook?

Jupyter Notebook was developed in in 2014 as it evolved to support interactive data science and scientific computing across all programming languages.

Project Jupyter exists to develop open source software, open standards, and services for interactive computing across dozens of programming languages. See: `https://jupyter.org/` and `https://jupyter.org/hub`

Why Use Anaconda?

The open source Anaconda Distribution is the easiest way to perform Python data science and machine learning on Linux, Windows, and Mac OS X. With over 11 million users worldwide, it is the industry standard for developing at scale. See: `https://www.anaconda.com/`

To keep your machine learning up to date, use:

```
conda install -c anaconda python  ipython jupyter
```

Note Weekly, I check out the new Python libraries to understand how my fellow machine learning researchers evolve and solve requirements.

Learn from everybody! That is the secret of continuous improvement.

Chapter 2

No extra information for this chapter.

Chapters 3, 4, and 5 – Supervised Learning

Supervised learning is the machine learning (ML) task of deducing a function from labeled training data. The training data consists of a set of training specimens. So, these labeled data sets drive your ML engines.

Bias

I suggest you spend time researching about bias in training data sets as this is a major issue with the current, known training data sets. Bias happens when an algorithm has limited flexibility to learn the true signal from the labeled data set. In the real world of IML, bias will render your solution and your models non-void as high bias results in the prediction being inaccurate.

Investigate the training data sets for low or high bias as one of your first critical steps in you path to well-designed IML.

Variance

Variance denotes an algorithm's sensitivity to specific sets of the training data. The smallest changes in the real-world data in comparison to training data can result in the real world of IML overfitting.

Investigate the training data sets for low or high variance as one of your first critical steps in your path to well-designed IML.

Remember High Bias–Low Variance results in predictions being similar to one another; but on average, they are inaccurate.

Low Bias–High Variance results in overfitting. Different data sets will show different insights given their corresponding data set. Henceforth, the models will predict in a different way.

You will have to handle a trade-off between Bias and Variance.

Chapter 6, 7, and 8 – Unsupervised Learning

The problematic status of unsupervised learning is that of demanding the reliable and repeatable discovery of hidden structure in unlabeled data. Subsequently, the examples given to the learner are unlabeled, so there is no error or reward signal to evaluate a potential solution, and there is a risk that the discovered model is not consistent against true real-life data processing

I suggest you perform an intense Cross-Validation test process.

I suggest you look at:

- Holdout Method
- K-Fold Cross-Validation
- Stratified K-Fold Cross-Validation
- Leave-P-Out Cross-Validation

Chapter 9 – Reinforcement Learning

Reinforcement learning is an area of ML that is evolving at an increasing rate. The reinforcement concept plus the higher computation capabilities of the average computer are creating an ecosystem where the ML can adapt to highly personalized behaviors. ML solutions are now about taking suitable action to maximize reward in a particular situation at speeds that most human activity would not be capable of outperforming. IML can now, for the first time in human existence, adapt to individual preferences of individual humans or systems.

I personally have confidence that reinforcement learning will be one of the crucial development environments in the future of IML.

The most critical subsection of Reinforced Learning (RL) is Multi-Task Learning (MTL), as it targets to solve multiple varied tasks at the same time, by taking advantage of the similarities between different tasks. This can improve the learning efficiency of the RL to levels that are factors better than existing RL achievements.

The MTL will enable IML to adapt on multifaceted solutions with ease.

Chapter 10 – Evolutionary Programming

The original evolutionary programming (EP) techniques were developed by Lawrence Fogel. They were designed at the evolution of artificial intelligence in the sense of developing the ability to predict changes in an environment. Adding this ability to MTL, you have an ecosystem that adapts to the changes in your real-world environment faster than any human can follow.

The combination is the base for the hypothesis that the human race will reach the technological singularity, that is, the point in time at which technological growth becomes uncontrollable and permanently more rapidly developing than humans. The theory is that this will result in an immeasurable rapid set of alterations to human civilization's future. I feel that only the future will tell.

Chapter 11 – Mechatronics

Mechatronics is a fusion of mechanical, electrical, and control engineering. My advice is to compete successfully in a global market; modern manufacturing corporations need you to staff with the ability to integrate electronics, control, software, and mechanical engineering into a range of innovative products and systems. Our future is full of new applications for people with these skills.

In the past, data science and ML performed predictions and humans had the option to enact the predictions. I believe our future IML will use mechatronics to enact the predictions by taking control of physical ecosystems and performing autonomous actions to change the real world.

Many of these systems are already in use, but I forecast that in the next ten years, an exponential aggregate of advances will transfer solutions into everyday life to assist the human race with their daily interactions.

Chapter 12 – Robots

The development in robotics is changing the future labor markets by being introduced into the various business markets.

I suggest you look at the following for extra background.

OpenAI

OpenAI's mission is to ensure that artificial general intelligence benefits all of humanity.

`https://openai.com/`

ROS – Robot Operating System

The Robot Operating System (ROS) is a set of software libraries and tools that help you build robot applications. I have used this operating system to drive many robots I have built.

`http://www.ros.org/` and `https://robots.ros.org/`

I believe within the next five years that we will see major deployments of robots into our everyday lives.

The impact of automation and soft robotics is already on our mobile devices. This will become more a part of our daily routines and lifestyles.

Hyper customized interactions with businesses and each other will become the norm.

The evolution of new manufacturing will support our ever-increasing, personalized world we are developing with our IML solutions.

Chapter 13 – Fourth Industrial Revolution

6C System

The IML is via big data analytics that consists of these 6Cs in the integrated Industry 4.0 and cyber-physical systems environment changing the world we currently know.

The 6C system comprises the following:

- Connection (sensor and networks)

  The evolution of the Internet of Things (IoT) is driving the connection. The introduction of a 5G network will be a major push into an ecosystem of mass communication.

- Cloud (computing and data on demand)

  Combining with cloud and edge computing is allowing IoT into the ecosystem around us.

- Cyber (model and memory)

  Improvements in the secure communication of data will drive the IoT forward.

- Content/context (meaning and correlation)

  This is where IML transforms raw data into business insights and enables the next generation to help the human race to achieve the best world possible.

- Community (sharing and collaboration)

  The amount of sharing and collaboration that is currently trendy has opened as new phases of intuitive partnerships and planet-scale projects to be accomplished with technology.

- Customization (personalization and value)

  The increase in processing power has opened the options to customize IML to fit individual needs.

The Fourth Industrial Evolution with advanced tools (analytics and algorithms) will generate meaningful information at scale.

Sun Models

The sun models is a requirements mapping technique that assists you in recording true requirements at a level so that your nontechnical users can understand the business purpose of your analysis, while you have a stress-free evolution to the comprehensive technical modeling for your data scientist and data engineer to complete their work.

Sun Models are covered in detail in the following books:

- *Practical Hive: A Guide to Hadoop's Data Warehouse System* by Scott Shaw, Andreas François Vermeulen, Ankur Gupta, and David Kjerrumgaard (Apress, 2016).
- *Practical Data Science: A Guide to Building the Technology Stack for Turning Data Lakes into Business Assets* by Andreas François Vermeulen (Apress, 2018).

Chapter 14 – Industrialized Artificial Intelligence

Industrialized AI will start to enforce standards for acceptable and fair AI processing. I suggest you keep track of the regulatory developments to ensure that AI's do not harm people in the short- or even longer-term deployment. I predict we are going to see major changes in our laws to cover some of the new technologies we are deploying.

General Data Protection Regulation

The General Data Protection Regulation 2016/679 is a regulation in European Union law on data protection and privacy for all individual citizens of the European Union and the European Economic Area.

The GDPR sets out seven key principles:

- Lawfulness, fairness, and transparency
- Purpose limitation
- Data minimization
- Accuracy
- Storage limitation

- Integrity and confidentiality (security)
- Accountability
- The GDPR provides the following rights for individuals:
    - The right to be informed.
    - You have to inform a person that you are storing or processing their data.
- The right of access
    - The person is able to access any data you store about them.
- The right to rectification
    - The person has the ability to correct any incorrect information you hold.
- The right to erasure
    - The person can request that you erase all data you hold about them.
- The right to restrict processing
    - The person can opt out of been processed by your machine learning.
- The right to data portability
    - The person has the right to move their data to any other preferred data storage provider or data processor.
- The right to object
    - The person has the right to object against the outcomes of your machine learning. This is a major challenge for your current and future ML work.
- Rights in relation to automated decision-making and profiling

Blue-Green Environment

The ecosystem requires a dual processing environment to ensure your next release is performing as expected with live data.

The blue-green deployment framework uses a proven deployment technique that requires two identical environments:

- The live production environment ("blue") running version n,
- An exact copy of this environment ("green") running version n+1.

You start with a stable Blue Ecosystem with release n: (Figure A-1).

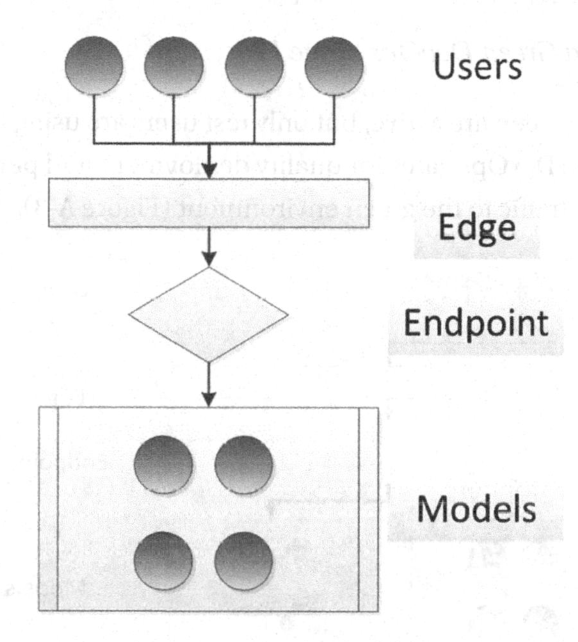

Figure A-1. *Blue Ecosystem*

You need to have a known good and stable ecosystem before you apply a new release.

Now you simply follow the rules and deliver your next great solution.

Perform your DevOps and test on the green environment, monitor technical and business insights' metrics, and check that the whole solution is working correctly (Figure A-2).

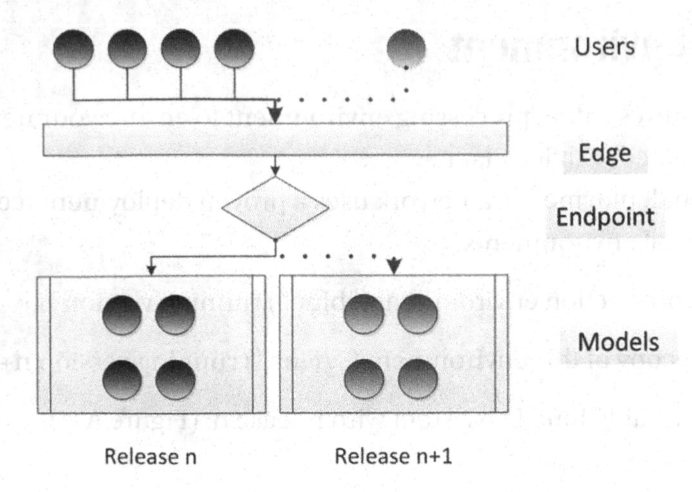

Figure A-2. *Blue and Green DevOps Stage Ecosystem*

Now both blue and green are active, but only test users are using a green ecosystem.

After passing all the DevOps gates for quality deployment and performance, the solution then switches traffic to the green environment (Figure A-3).

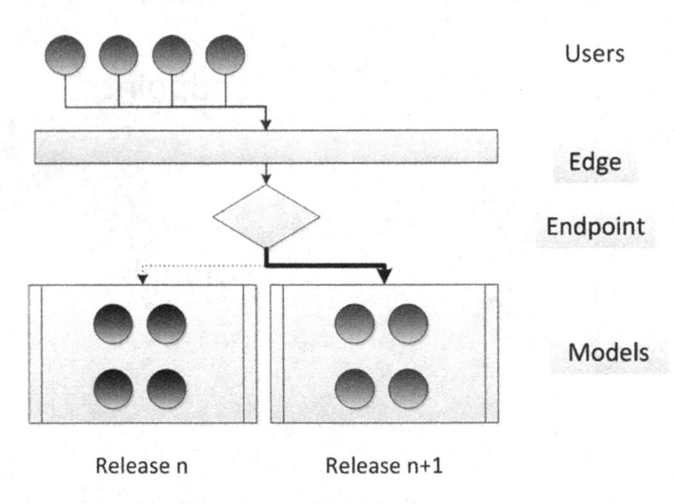

Figure A-3. *Blue offline and Green Online Ecosystem*

If all users are working perfectly, you remove blue, and green is now active.

Advice I normally leave the previous ("blue") solution / release for 72 hours to mitigate the risk of the new release not performing as designed.

Chapter 15 – Industrialized Project

I covered a Mars mission as our industrial final project. It is, however, not as futuristic as many people may believe.

I suggest you read the following to understand how far we are with this concept:

Companies already looking at space mining:

- NASA: `https://mars.nasa.gov/mars2020/`
- ispace: `https://ispace-inc.com/`
- Planetary Resources: `https://www.planetaryresources.com/`
- Deep Space Industries: `http://deepspaceindustries.com/` or `http://bradford-space.com/`
- Kleos Space: `https://kleos.space/`
- OffWorld: `https://www.offworld.ai/`
- Clyde Space: `https://www.clyde.space/`
- Spaceflight Industries: `http://spaceflight.com/`
- Planet Labs: `https://www.planet.com/`
- Blue Origins: `https://www.blueorigin.com/`
- SpaceX: `https://www.spacex.com/`

I picked a space project because I always aimed for the stars. The IML solutions can be deployed in about any part of business or industry. I am currently using my IML for financial services; public safety; health care; and yes, even the space industry.

The stars are only the beginning!!

Enjoy your journey with your new IML knowledge.

Thanks for reading my endeavor to share my knowledge, and please add more of your own insights and pay it forward. Share what you have learned.

Let's change our future together. May all your dreams come true!

Index

A

A. F. Vermeulen, *Industrial Machine Learning*, https://doi.org/10.1007/978-1-4842-5316-8

B

C

D

E

F

G

H

I

J

K

L

M

N

O

P

Q

R

S

Printed in the United States
By Bookmasters

Printed in the United States
By Bookmasters